Advances in Comparative and Environmental Physiology 10

Volumes already published

Vol. 1: NaCl Transport in Epithelia
Edited by *R. Greger* (1988)

Vol. 2: From the Contents: Hibernation – Vapor Absorption
Nutrient Transport – Stable Pollutants (1988)

Vol. 3: Molecular and Cellular Basis of Social Behavior in Vertebrates
Edited by *J. Balthazart* (1989)

Vol. 4: Animal Adaptation to Cold
Edited by *L. C. H. Wang* (1989)

Vol. 5: From the Contents: Osmoregulation – Chemoreception –
Temperature – Spectroscopy – Metalloproteins
Prostaglandins (Eicosanoids) –
Maternal-Fetal Relationships (1989)

Vol. 6: Vertebrate Gas Exchange: From Environment to Cell
Edited by *R. G. Boutilier* (1990)

Vol. 7: From the Contents: Protein Turnover –
Xenobiotic Metabolism Lipoproteins (1991)

Vol. 8: From the Contents: Dormancy in Aquatic Invertebrates –
Root Effect – Red Blood Cell Functions in Fish –
Diving Mammals (1991)

Vol. 9: Volume and Osmolality Control in Animal Cells
Edited by *R. Gilles, E. K. Hoffmann,* and *L. Bolis* (1991)

Vol. 10: Comparative Aspects of Mechanoreceptor Systems
Edited by *F. Ito* (1992)

Vol. 11: Mechanics of Animal Locomotion
Edited by *R. McN. Alexander* (1992)

Volumes in preparation

Vol. 12: Muscle Contraction and Cell Motility:
Molecular and Cellular Aspects
Edited by *H. Sugi* (1992)

Vol. 13: Blood and Tissue Oxygen Carriers
Edited by *Ch. P. Mangum* (1992)

Advances in

Comparative and Environmental Physiology 10

Comparative Aspects of Mechanoreceptor Systems

Guest Editor: F. Ito

With Contributions by
J. Adler • M. Buechner • M.W. Chapleau • J.W. Deitmer
A.H. Delcour • N. Fujitsuka • M.H. Gladden • M.C. Gustin
W. Hamann • F. Ito • C. Kung • B. Martinac • H. Ohmori
B.G. Pickard • J. Ping Ding • H. Römer • B. Rydqvist
F. Sachs • M. Sokabe • J. Tautz • J.G. Widdicombe

With 111 Figures

Springer-Verlag
Berlin Heidelberg New York
London Paris Tokyo
Hong Kong Barcelona
Budapest

Guest Editor:

Professor Fumio Ito
Department of Physiology
School of Medicine
Nagoya University
65 Tsurumai-cho, Showa-ku
Nagoya, Japan

ISBN-13: 978-3-642-76692-3 e-ISBN-13: 978-3-642-76690-9
DOI: 10.1007/978-3-642-76690-9

Library of Congress Cataloging-in-Publication Data
Comparative aspects of mechanoreceptor systems / editor, F. Ito. p. cm. – (Advances in comparative and environmental physiology; 10) Includes bibliographical references and index.
ISBN-13: 978-3-642-76692-3
1. Mechanoreceptors. 2. Physiology, Comparative. 3. Ion channels. I. Ito, Fumio, 1929– . II. Series. QL938.M4C66 1992 591.1'8 – dc20 / 91–32197 CIP

Typesetting: International Typesetters Inc., Manila, Philippines
31/3145-5 4 3 2 1 0 – Printed on acid-free paper

Foreword to the Series

The aim of the series is to provide comprehensive, integrated reviews giving sound, critical and provocative summaries of our present knowledge in environmental and comparative physiology, from the molecular to the organismic level.

Living organisms have evolved a widespread range of basic solutions to cope with the different problems, both organismal and environmental, with which they are faced. A clear understanding of these solutions is of course of fundamental interest for all biologists, zoologically or medically oriented. They can be best comprehended in the framework of the environmental and/or comparative approaches. These approaches demand either wide surveys of animal forms or a knowledge of the specific adaptive features of the species considered. This diversity of requirements, both at the conceptual and technological level, together with the fact that physiology and biochemistry have long been mainly devoted to the service of medicine, can account for the fact that these approaches emerged only slowly amongst the other new, more rapidly growing disciplines of the biological sciences.

The field has now gained the international status it deserves and the organization of a series devoted to it appeared timely to me in view of its actual rapid development and of the interest it arouses for a growing number of biologists, physiologists, and biochemists, independently of their basic, major orientation.

Liège, Belgium *Raymond Gilles*

List of Editors

Those interested in contributing to this series should contact Professor R. Gilles or one of the Series Editors. Review proposals should include a short outline – and a Table of Contents – briefly defining the aims and scope of the review.

Preface

In the past 5 years there has been an enormous increase of evidence that the ion channels activated by mechanical force are common to a wide variety of cell types. Mechanosensitive (MS) ion channels form a small proportion of the total channel population. They are now found in more than 30 cell types from *E. coli*, yeast, to plant, invertebrate, and vertebrate cells, where they occur in virtually all types of cells from bone to smooth muscle, as well as neurons. The majority of MS channels are permeable to monovalent cations and are slightly selective for K^+ over Na^+. However, there are several reports of anion-selective MS channels, MS Ca^{2+} channels, and MS channels with large conductances that do not discriminate markedly between cations and anions.

Recently B. Hille has postulated possible evolutionary relationships between several types of ion channels, with mechanosensitive channels predating even the eukaryotes. Two voltage-gated channel types originate with the stem eukaryotes, as deduced from the presence of voltage-gated K^+ and Ca^{2+} channels in protozoa, algae, or higher plants. Agonist-gated channels as well as voltage-gated Na^+ channels appear with the earliest metazoan animals, as deduced from the presence of Na^+ spikes and fast chemical synapses in cnidaria (coelenterates), ctenophores, and all higher animals.

MS channels have been found in specialized mechanoreceptors such as the crayfish stretch receptor neuron, although most studies have been performed on cell types for which there is no established role for mechanotransduction, such as skeletal muscle fiber. However, there is no evidence for many kinds of mechanoreceptors as to whether the transduction is attributable to the MS channels, because the sensory terminals are too small and too complicated to apply patch-clamp recovering. Thus, the available approach so far is to compare the kinetics of sensory reception in different kinds of mechanosensory organs of various kinds of animal. This approach should allow us to extend our knowledge about the interrelation between MS channels and the properties of mechanoreceptors, the contribution of intracellular second messengers to signal transduction, as well as the metabolism of the receptor channels and their genetic regulation

Nagoya, Japan, January 1992 *Fumio Ito*

Contents

Part I Mechanosensitive Ion Channels

Chapter 3

Mechanosensory Transduction in Ciliates (Protozoa)
J.W. Deitmer

Chapter 4

Towards Molecular Mechanism of Activation
in Mechanosensitive Ion Channels
M. Sokabe and *F. Sachs*

Part II Mechanoreceptor Organs

Chapter 5

Gravity Sensing by Higher Plants
B.G. Pickard and *J. Ping Ding*

Chapter 6

Mechanoreceptors in Respiratory Systems
J.G. Widdicombe and *G. Sant'Ambrogio*

Chapter 7

Cardiovascular Mechanoreceptors
M.W. Chapleau

Chapter 8

Comparative Physiology of Cutaneous Mechanoreceptors
W. Hamann

Chapter 9

Invertebrate Auditory Receptors
H. Römer and *J. Tautz*

Chapter 10

Mechanoelectrical Transduction in Vertebrate Hair Cells
H. Ohmori

Chapter 11

Muscle Mechanoreceptors in Invertebrates
B. Rydqvist

Chapter 12

Muscle Mechanoreceptors in Nonmammalian Vertebrates
F. Ito, M. Sokabe, and *N. Fujitsuka*

Part I

Mechanosensitive Ion Channels

Chapter 1

Mechanosensitive Ion Channels in Bacteria

B. Martinac, A.H. Delcour, M. Buechner, J. Adler, and C. Kung[1]

Contents

1 Introduction

Mechanosensitive ion channels are a class of ion channels which transduce mechanical forces of the cell membrane into an electrical response. These channels are gated by the mechanical force produced by stretching the lipid bilayer and/or the underlying cytoskeleton of cell membranes. They were initially reported by Guharay and Sachs (1984) in chick skeletal muscle and nearly simultaneously by Brehm et al. in embryonic *Xenopus* muscle (Brehm et al. 1984). They are now found in a variety of animal and plant tissues (Ohmori 1984; Brezden et al. 1986; Cooper et al. 1986; Sigurdson et al. 1987; Falke et al. 1988), as well as in fungi (Gustin et al. 1988) and bacteria (Martinac et al. 1987; Zoratti and Petronilli 1988; Berrier et al. 1989). The mechanosensitive channels appear ubiquitous. The only class of animal cells which seem to lack them are undifferentiated cancer cell lines (Sachs 1988). Mechanosensitive ion channels have been suggested to play important physiological roles (touch, hearing, proprioception) in many systems (Sachs 1988). Regulation

[1] Laboratory of Molecular Biology and Departments of Biochemistry and Genetics, University of Wisconsin, Madison, WI 53706, USA

Advances in Comparative and Environmental Physiology, Vol. 10

of cell volume, osmolarity, and growth are also among the processes which are likely to be controlled by these channels.

This review focuses on mechanosensitive channels in bacteria with emphasis on the channels which have been found in *Escherichia coli* (Martinac et al. 1987; Berrier et al. 1989). Most of the information presented comes from our own work.

2 Mechanosensitive Ion Channels in *Escherichia coli*

2.1 Preparations

The mechanosensitive ion channels of *E. coli* have been studied by the patch-clamp recording technique (Hamill et al. 1981) in various giant forms (Martinac et al. 1987; Buechner et al. 1990) and in reconstituted membrane fractions of this gram-negative bacterium (Berrier et al. 1989; Delcour et al. 1989a). Figure 1 summarizes the five preparations of giant forms we used. The upper part of the figure shows three methods which have been used for preparation of giant spheroplasts or giant cells from the behaviorally wild-type strain AW405. Our original preparation (Martinac et al. 1987) used cephalexin, a penicillin analogue, which prevents septation; as a result, *E. coli* cells grow into 50–150 μm long filaments ("snakes"). After they are treated with EDTA and lysozyme to weaken the cell wall, these "snakes" collapse into giant spheroplasts, which are about 5–10 μm in diameter. These spheroplasts are large enough to be used for patch-clamp recording. We found that we can generate giant cells as well as "patchable" bulges in snakes by adding cephalexin and 30–50 mM $MgCl_2$ to the growth medium. No EDTA or lysozyme is necessary. Another way to obtain snakes is by exposing bacteria to UV light. In this way the use of cephalexin can be avoided. The lower part of Fig. 1 shows two methods which have enabled us to obtain giant *E. coli* cells. One preparation makes use of a *lpp*$^-$ *ompA*$^-$ double mutant which lacks the outer membrane proteins, OmpA and lipoprotein, two major links between the cell wall and the outer membrane (Sonntag et al. 1978). In the presence of cephalexin and $MgCl_2$ these cells grow to round giants comparable to giant spheroplasts in size. Finally, we have isolated an osmosensitive mutant which grows to large round cells without any addition of either cephalexin or EDTA-lysozyme. The mechanosensitive ion channels are present in all five preparations (Buechner et al. 1990), and have the characteristic large conductance activated by pressure and by voltage as the channels we originally described (Martinac et al. 1987).

The method we devised for reconstitution of various membrane fractions of the *E. coli* cell envelope is schematically shown in Fig. 2 (Delcour et al. 1989a). After passage of the cells through a French press, the inner and outer membrane fractions were separated by sucrose-gradient centrifugation. Each membrane fraction was mixed at the desired protein-to-lipid ratio with multilamellar azolectin liposomes, obtained by two freeze-thaw cycles in a dry ice/acetone bath. Fusion of the membrane with the azolectin was accomplished by ultra centrifugation. The pellet

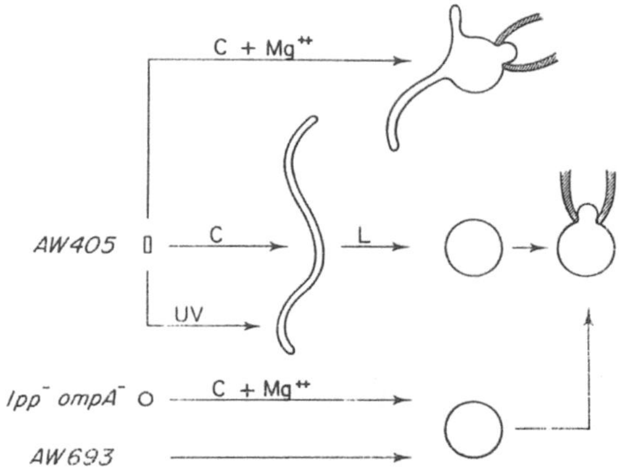

Fig. 1. Five methods of preparing giant forms of *E. coli*. *C* cephalexin; *L* lysozyme; *UV* ultraviolet light; *AW405* wild type; *lpp*⁻ *ompA*⁻ double mutant missing OmpA and lipoprotein; *AW693* osmosensitive mutant. (Buechner et al. 1990)

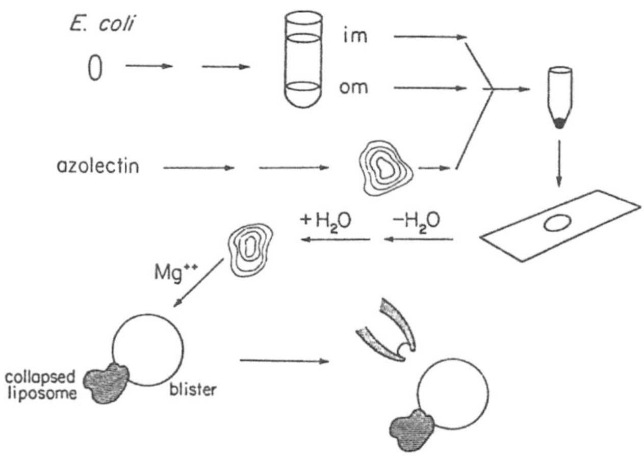

Fig. 2. The scheme of the method used for reconstitution of *E. coli* membrane fractions. *im* inner membrane; *om* outer membrane; -,+ H_2O de-, rehydration. (Delcour et al. 1989a)

was resuspended in MOPS buffer containing 5% ethylene glycol, and 20–40 μl of this suspension were placed on a glass slide. After dehydration in a desiccator, the dried lipid film was rehydrated in a small volume of final recording buffer. The giant liposomes obtained by this dehydration-rehydration cycle (Criado and Keller 1987) were placed in the patch-clamp chamber containing the recording buffer plus 20 mM $MgCl_2$, which causes the liposomes to collapse. Within a few minutes, large, faint, and most likely unilamellar blisters emerged from the sides of the collapsed liposomes. A gentle touch of the electrode onto the blister consistently yielded instantaneous high gigaohm seals. The electrode was usually pulled away from the blister and the patch was excised by brief air exposure. The reconstituted channels have the same basic properties as the channels found in the various giant cell preparations (Martinac et al. 1987; Buechner et al. 1990). A method of reconstitution, similar to the above, was recently reported (Berrier et al. 1989).

2.2 Basic Properties

When suction of a few centimeters Hg is applied to a spheroplast membrane, the activity of one type of mechanosensitive channel is observed. An increase of suction activates more channels, and each tends to stay open for a longer time. Figure 3A shows simultaneous activation of up to five mechanosensitive channels by suction in an on-cell patch of a giant spheroplast prepared in the original way (Martinac et al. 1987). The current trace of a single mechanosensitive ion channel, which was reconstituted in an azolectin liposome, is shown in Fig. 3B. These channels can be activated by positive or negative pressure applied through the pipette onto the membrane patch (Martinac et al. 1987). This symmetric property was also reported for several other mechanosensitive channels in other organisms (Cooper et al. 1986; Gustin et al. 1988). This activation by pressure or suction is reversible (Fig. 3A). The opening probability changes with pressure, following the Boltzmann distribution, in which the free energy available for gating is linearly dependent upon pressure (Fig. 4; Martinac et al. 1987). The mechanosensitive channels are also voltage dependent, and the single channel opening probability can also be fitted to a curve described by the Boltzmann distribution, with free energy dependent on voltage. At positive, depolarizing voltages less pressure is needed to activate the channels,

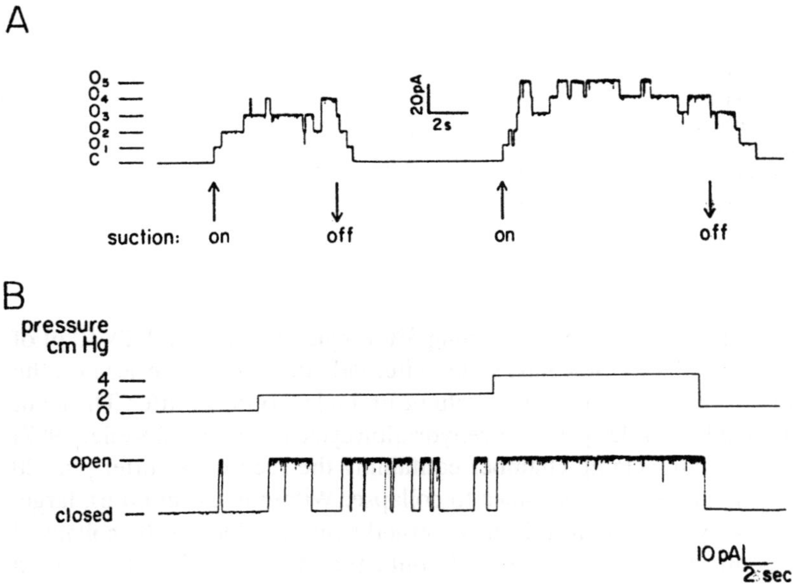

Fig. 3A. Mechanosensitive channels in an on-cell patch of a giant spheroplast were activated by suction through the patch-clamp pipette and they ceased to gate when suction was removed (*arrows*). Pipette voltage was +10 mV. (Martinac et al. 1987). *B* activity of a reconstituted mechanosensitive channel. This channel happened to come from the inner membrane fraction, which was fused with azolectin liposomes at a protein to lipid ratio of 1:100. The pipette voltage was +40 mV. (Delcour et al. 1989a)

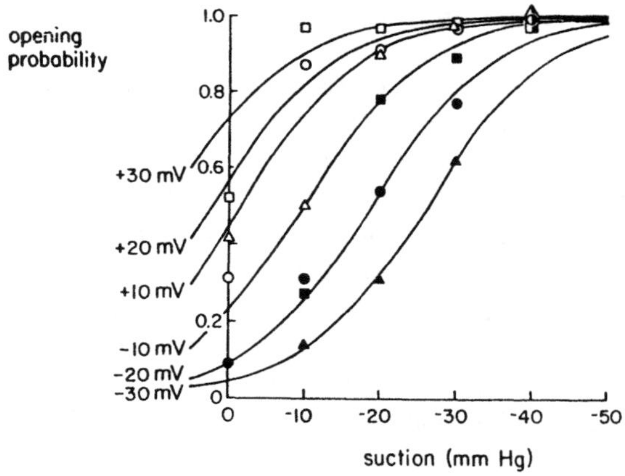

Fig. 4. Opening probability of a mechanosensitive channel plotted versus suction. Data were fitted to a family of Boltzmann curves. Note that less pressure was necessary to activate the channels at positive membrane voltages than at negative voltages. ▲, ●, ■, Δ, O, and ❑ represent –30, –20, –10, +10, +20, and +30 mV, respectively. (Martinac et al. 1987)

which means that at a given suction the channel is more likely to be open upon depolarization. Compared to eukaryotic channels, this channel is slow in its kinetics. Relatively few transitions between the closed and the open state are observed per unit time.

The mechanosensitive channel has a large conductance. In a symmetric solution of about 300 mM salt, the channel conductance at hyperpolarizing voltages is 950 pS; at depolarizing voltages the conductance is 630 pS. The channel has a very low selectivity for ions, although it slightly prefers anions to cations. Organic ions as large as glutamate can carry a current through this channel (Martinac et al. 1987; 1988).

Berrier et al. (1989) found mechanosensitive channels in all three membrane fractions of *E. coli*, i.e. the inner and outer membranes, and contact zones, which connect the two membranes (Bayer 1979). Using the patch-clamp, the authors characterized several types of mechanosensitive channels with large conductances, ranging from 140–1500 pS in 0.1 M KC1. These channels were predominantly found in the inner membrane fraction, although they were also present at lower frequencies in the other two membrane fractions. In the outer membrane fraction the conductance ranged from 500–1500 pS, and in the contact zones it was between 300 and 1500 pS. The channels often associated in clusters of identical conductances, although mechanosensitive channels of different conductances were also observed in the same patch. Since no data were presented on selectivity, pressure dependence of the opening probability, or voltage dependence of the channels, it is at present difficult to assess how these channels are related to the mechanosensitive channel studied in spheroplasts.

2.3 Location of the Channel

Gram-negative bacteria such as *E. coli* have a cell envelope which consists of two membranes, separated by the cell wall (peptidoglycan; Fig. 5). We concluded that the mechanosensitive channel, in cells or in spheroplasts, is located in the outer membrane (Buechner et al. 1990), based on the following evidence and arguments. Giant cells were not treated with EDTA or lysozyme (Fig. 1A, bottom), and therefore they most likely have an intact peptidoglycan layer. Giant spheroplasts and giant cells of AW693 and of the *lpp⁻ ompA⁻* mutant have also two membranes. The integrity of the two membranes was confirmed by thin-section electron microscopy (Ruthe and Adler 1985; Buechner et al. 1990). Since one of the prerequisites for gigaohm-seal formation is a clean patch-clamp pipette tip, it is unlikely that the pipette tip could pass through both the outer membrane and peptidoglycan to form a gigaohm-seal on the inner membrane of giant cells.

We further tested the presence of the mechanosensitive channel in the outer membrane by recording channel activities from inside-out excised patches, which should have their periplasmic side and attached cell wall exposed to the bath solution. These patches were then treated with lysozyme, which weakens the cell wall by breaking glycosidic linkages between N-acetyl-D-muramic acid and N-acetyl-D-glucosamine (Berger and Weiser 1957). If the channel were connected to peptidoglycan, then exposure of the patch to lysozyme should lead to an irreversible change in channel sensitivity to the applied pressure.

Figure 6A shows current traces of a single mechanosensitive ion channel before and after treatment of the patch with lysozyme. In this particular experiment, the membrane patch was exposed after 30 min to a bath solution which contained 5 mg/ml lysozyme. Thereafter, lysozyme was washed out and the channel activity was recorded in the absence of lysozyme. After exposure to lysozyme, activity of the channel *increased* irreversibly at all applied suctions. Interestingly, the channel

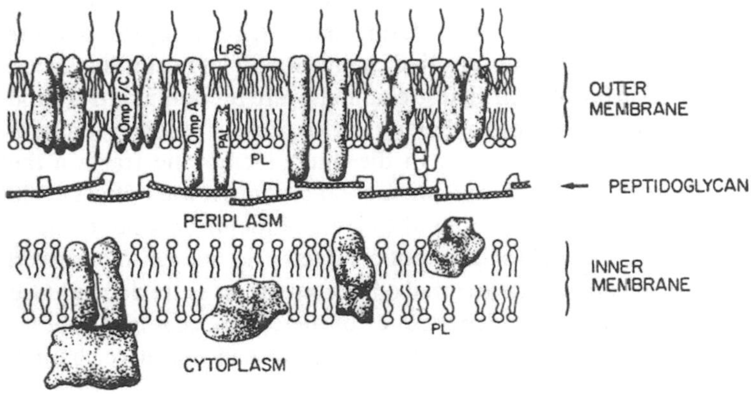

Fig. 5. Diagram of the cell envelope of *E. coli*. The outer and the inner membrane are separated by a cell wall (peptidoglycan). The outer membrane is covalently linked to the peptidoglycan layer. (After Mizushima 1985; Raetz 1986)

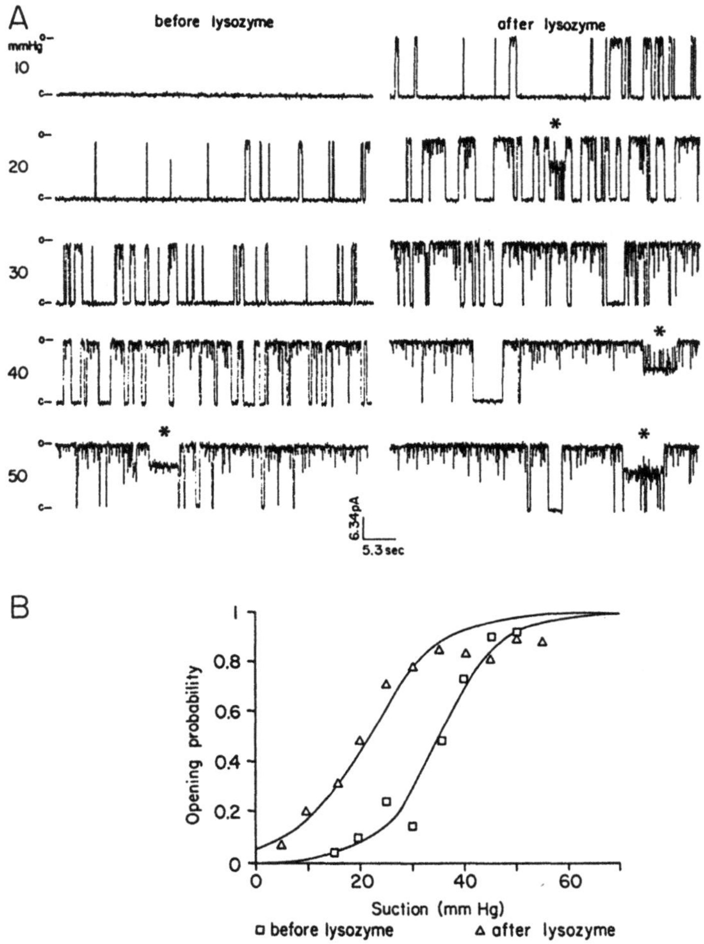

Fig. 6A. Single channel currents recorded in an excised patch of a giant spheroplast before (*left*) and after (*right*) the membrane patch was treated with 5 mg/ml lysozyme (Buechner et al. 1990). Note the frequent occurrence of channel gating at subconducting levels (*asterisks*). *B* Pressure dependence of the opening probability of the single channel shown in *A*. Each data point represents a 2-min-long recording. Data were collected before (*squares*) and after (*triangles*) treatment of the patch with lysozyme, and were fitted to Boltzmann distribution curves

gated more frequently at subconducting levels after exposure to lysozyme than before. It is possible that these subconductances are "pathological" conformational states of the channel protein, and that attachment to the cell wall is important for the proper function of the channel. In Fig. 6B, the opening probability of the same channel was plotted against the applied suction and data were fitted to the Boltzmann exponential functions. In the inside-out excised-patch mode, before lysozyme treatment, the suction necessary for an e-fold change in channel opening probability was 5.7 mm Hg (1 mm Hg = 133 Pa). After the exposure to lysozyme, this curve parameter, indicative of channel response to stretch, did not change significantly, retaining a value of 5.9 mm Hg. The curve shifted in a parallel manner to the left on the pressure axis by 20 mm Hg, however, which indicates a significant lowering of the threshold for the activation of the channel by suction. Lysozyme seems to weaken a restraining force of the membrane without affecting the channel itself. The experiment in Fig. 7 was designed to follow the time course of the effect of lysozyme on the single channel opening probability. Over a period of 1 h the activity of the channel was monitored in the presence of lysozyme in the bath solution and was checked every 5 min at the preset suction. After ca. 10–15 min, the opening probability had almost doubled and thereafter remained approximately constant. The irreversible increase in channel activity occurred in seven different patches which were treated with lysozyme. A control experiment showed that the effect of lysozyme in the presence of N,N′-diacetylchitobiose, a lysozyme inhibitor (Neuberger and Wilson 1967), was less pronounced than when lysozyme was tested alone, which strongly suggested that the channel activity depends on the cell wall structure (Martinac, unpubl. results). The peptidoglycan may play the role of a parallel elastic element in restraining the activity of the mechanosensitive channel. Since the cell wall is covalently linked to the outer membrane, these experiments strongly argue in favor of a mechanosensitive channel location in this membrane.

Berrier et al. (1989) concluded from their work with reconstituted membrane fractions that mechanosensitive channels are located in the inner membrane. They found mechanosensitive channels in the inner membrane fraction 50% of the time. In about 7% of the experiments the channels were present in the outer membrane fraction, and in 15% of the cases in the contact zones. However, such direct comparisons assume the efficiency of incorporation and channel detection to be the same from fraction to fraction, which may not be generally valid. We find it difficult to reconcile their findings with ours, especially because of the different properties of the channels described by the two groups. Except for the various conductance levels, there is no information on the properties of the reconstituted channels in the Berrier et al. report. The 7% occurrence of the mechanosensitive channels in the outer membrane was higher than the occurrence expected from the less than 2% contamination of the outer membrane by the inner membrane fraction. Berrier et

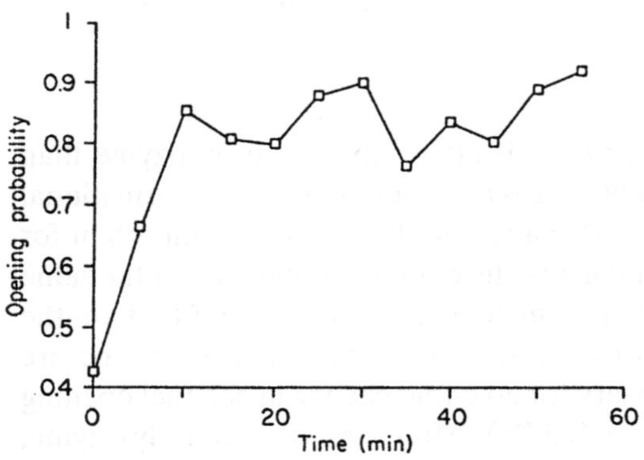

Fig. 7. Time course of the lysozyme effect on the opening probability of a single channel in an inside-out excised patch. Channel activity was monitored in the presence of 5 mg/ml lysozyme in the bath solution. Each data point represents a 2-min-long recording. Suction was 20 mm Hg and pipette voltage was +10 mV

al. (1989) concluded: "Thus it is not excluded that the outer membrane also harbors pressure-sensitive channels." Taking into account that mechanosensitive ion channels were also found in protoplasts of gram-positive bacteria (Zoratti and Petronilli 1988; Delcour and Martinac, unpubl.; Fig. 12B), it is possible that mechanosensitive channels exist in both inner and outer membranes of gram-negative bacteria. They might differ from each other in terms of the conductance and of the pressure- and time-scale on which they operate.

2.4 Mechanosensitive Channels as Reporters of Bilayer Expansion

The two leaflets of the membrane bilayer in living cells are usually asymmetric in lipid composition (Rothman and Lenard 1977). This led Sheetz and Singer (1974) to propose the bilayer couple hypothesis in order to explain the effect of amphipathic compounds (e.g., phenothiazine tranquilizers, local anesthetics) on the shape of human erythrocytes. The hypothesis states that exogenous amphipaths differentially distribute between the two asymmetric leaflets. Positively charged amphipaths (e.g., chlorpromazine, CPZ) should preferentially intercalate into the more negatively charged inner leaflet, and negatively charged amphipaths (e.g., trinitrophenol, TNP) should preferentially insert into the less negatively charged outer leaflet (Fig. 8). In both cases, one of the leaflets should expand relative to the other, causing cup formation or crenation in erythrocytes.

The lipid composition of the outer membrane of *E. coli* is highly asymmetric (Lugtenberg and van Alphen 1983; Nikaido and Vaara 1987), which results in pronounced charge asymmetry (Martinac et al. 1990b). Since the mechanosensitive channels which we study are most likely located in this membrane, the effect of amphipathic drugs on channel activity provides a good test of the bilayer couple hypothesis. Our idea was to see whether the association of exogenous amphipaths with the membrane bilayer should affect the stress/strain relationship in the membrane surrounding the channels and whether the mechanosensitive channels can detect these changes. This is the first test of the hypothesis using the mechanosensitive channel as a molecular reporter of intramembraneous forces, instead of using cell shape as an indicator.

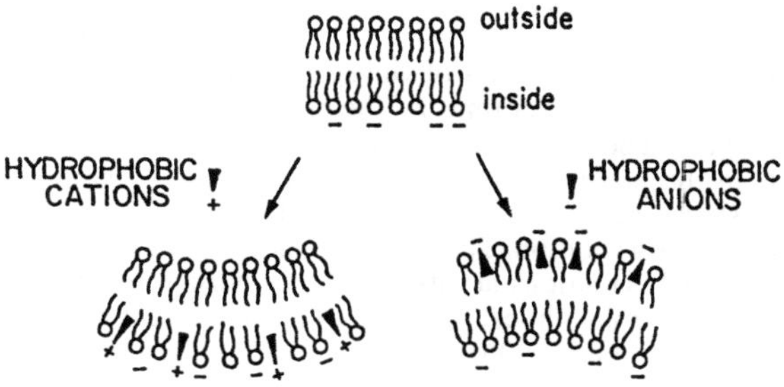

Fig. 8. The bilayer couple model after Sheetz and Singer (1974)

We tested the effect of various amphipathic compounds on mechanosensitive channel activity and found that mechanosensitive channels can indeed serve as reporters of bilayer expansion (Martinac et al. 1990a,b). Figure 9 shows the time-dependent effect of cationic chlorpromazine and anionic trinitrophenol, the two most potent amphipaths tested, on the channel opening probability. Both amphipaths reversibly activated the channel. Recall that the channels can be activated by pressure or suction (Martinac et al. 1987). Since amphipaths with opposite charges should preferentially insert in opposite leaflets of the membrane bilayer, and should expand the membrane bilayer accordingly, there should be a compensatory effect of cationic versus anionic amphipaths. The experiment in Fig. 10 demonstrates such a counteracting effect of cationic chlorpromazine and anionic trinitrophenol on the channel opening probability (Martinac et al. 1990a,b).

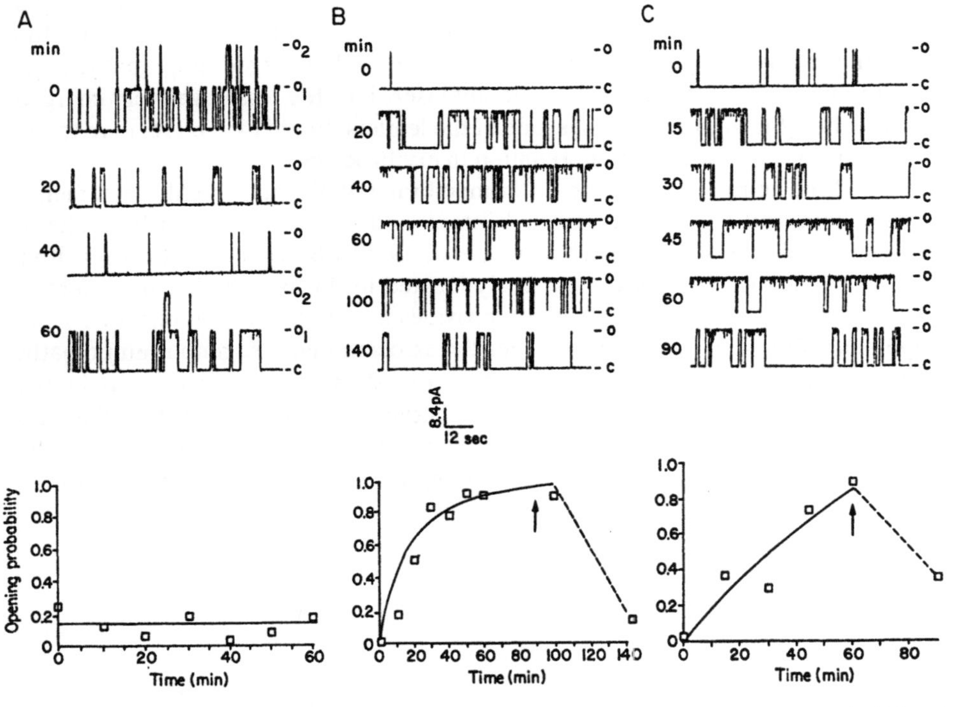

Fig. 9A-C. The effects of chlorpromazine and of trinitrophenol on the opening probability of single channels over time. At the preset suction the channel activity was recorded every 10 min: *A* in the absence of drugs at 15 mmHg; *B* in the presence of 20 μM chlorpromazine at 10 mmHg; and *C* in the presence of 500 μM trinitrophenol at 25 mmHg. Thirty min after trinitrophenol was washed out, the suction was 30 mmHg. *Arrows* show the time when the drugs were perfused out of the bath solution. The *top* portions of the figure show 2-min-long recordings of single-channel current fluctuations at different times during the experiment. The *bottom* portions present the opening probabilities, obtained from the current traces in the top portions, plotted versus time. (Martinac et al. 1990b)

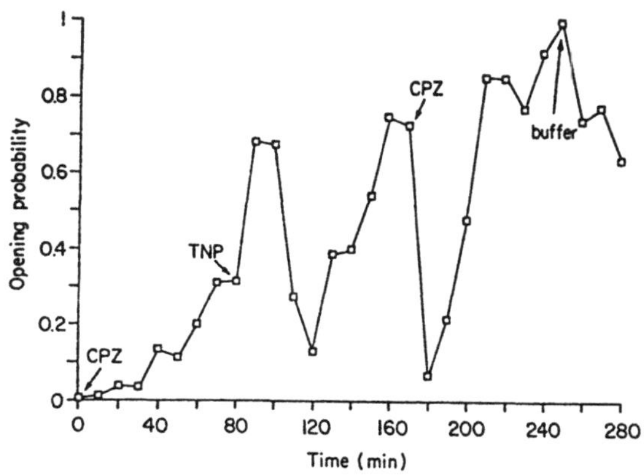

Fig. 10. The compensatory effect of amphipathic compounds with opposite electrical charge on the opening probability of the mechanosensitive channels. The bath was alternately perfused with buffer containing either 20 μM chlorpromazine (*CPZ*) or 500 μM trinitrophenol (*TNP*), or with the buffer without drugs toward the end of the experiment. *Arrows* indicate time points of each solution exchange. Suction was 30 mmHg, and pipette voltage was +10 mV. (Martinac et al. 1990b)

2.5 Physiological Role

The lipid asymmetry of biological membranes might be of special importance for the function of mechanosensitive channels in general, to act as molecular devices capable of detecting the mechanical forces resulting from bilayer expansion. There might be some cell-endogenous amphipaths, with the special task of controlling the bilayer curvature and the level of activity of mechanosensitive channels by associating with and dissociating from one leaflet of the membrane bilayer. The view of mechanosensitive channels as biosensors capable of detecting the intramembraneous forces of the membrane bilayer in living cells reinforces their potentially central role in the regulation of fundamental physiological processes, such as cell growth, exo- and endocytosis, volume and osmoregulation, in which the underlying mechanism includes the expansion of the cell membrane.

Bacteria respond to changes in osmolarity in various ways. Some of the known responses in *E. coli* upon a decrease in osmolarity include the release of small molecules (Britten and McClure 1962); synthesis of membrane-derived oligosaccharides (Kennedy 1982); and an increase in the OmpF to OmpC ratio (Hall and Silhavy 1981). Upon an increase in osmolarity, *E. coli* pumps in K^+ (Epstein and Schultz 1965); synthesizes the osmoprotectants proline and glycine betaine (Le Rudulier et al. 1984); decreases the OmpF to OmpC ratio (Hall and Silhavy 1981); and responds behaviorally by osmotaxis (Li et al. 1988). Because of the possibility that the mechanosensitive channel may be involved in osmoregulation, we tested channel activation by a change in osmolarity. Upon perfusion of the bath with a solution of higher osmolarity than the pipette solution, the activity of the mechanosensitive channels temporarily increased (Fig. 11; Martinac, unpubl.). The osmolarity gradient was apparently large enough to increase membrane stretch in the patch to a sufficient extent for channel activation. The activation was transient

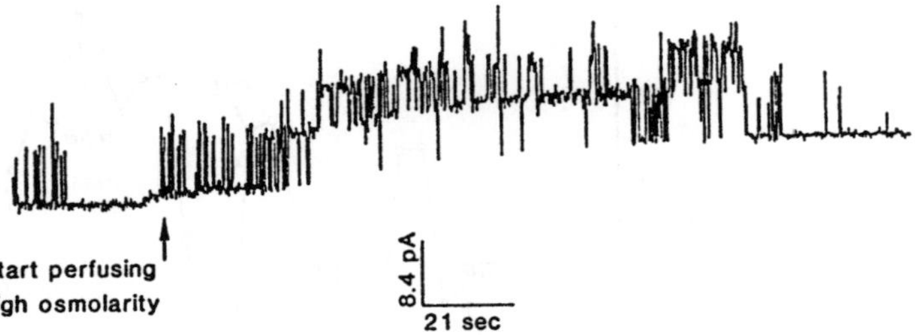

Fig. 11. Transient activation of mechanosensitive channels in *lpp⁻ ompA⁻* giant cells of *E. coli* by an osmolarity gradient of 300 mOsmol salt across an inside-out excised membrane patch. This experiment suggests that the osmolarity difference may stretch the membrane to activate the channels. The applied suction was 25 mmHg, and the pipette voltage was +20 mV. (Martinac, unpubl. results)

in this particular experiment as the perfused bath solution contained only permeant ions. The osmolarity gradient across the membrane consequently dissipated over a few minutes and allowed the membrane patch to relax to the initial level of stretch before bath perfusion. This experiment strongly suggests a possible involvement of the mechanosensitive channel in osmoregulation.

The existence of a mechanosensitive, voltage-dependent osmoregulatory channel in the outer membrane of *E. coli* is at first glance puzzling. The puzzle comes from the traditional picture of the outer membrane as a rigid, static sieve with about 10^5 open water-filled pores (predominantly OmpC and OmpF porins; Hall and Silhavy 1981; Rosenbusch 1974) which allow passage by hydrophilic molecules with a molecular mass less than 600. With that many static pores, the opening and closing of a few mechanosensitive channels seem futile. Also, it is difficult to understand how the outer membrane should sustain an osmotic difference. Yet, the exoskeleton (cell-wall-outer-membrane complex) is not a rigid structure but is able to shrink or swell. Stock et al. (1977) have found that the cytoplasm has an osmolarity equal to or slightly larger than that of the periplasm and the osmotic force is largely experienced by the *outer* membrane. Consequently, the exoskeleton should be the major osmotic barrier. In addition, we have found that several outer membrane channels are regulated by physiological stimuli, such as voltage and/or osmotic pressure (Martinac et al. 1987; Delcour et al. 1989b; Buechner et al. 1990). This dynamic nature of the outer membrane renders the existence of a mechanosensitive channel in it less puzzling.

3 Mechanosensitive Ion Channels in Other Bacteria

Mechanosensitive channels have been found also in gram-positive bacteria. Zoratti and Petronilli (1988) reported the existence of mechanosensitive channels in giant protoplasts of *Streptococcus faecalis* (Fig. 12A). A surprising variety of single-

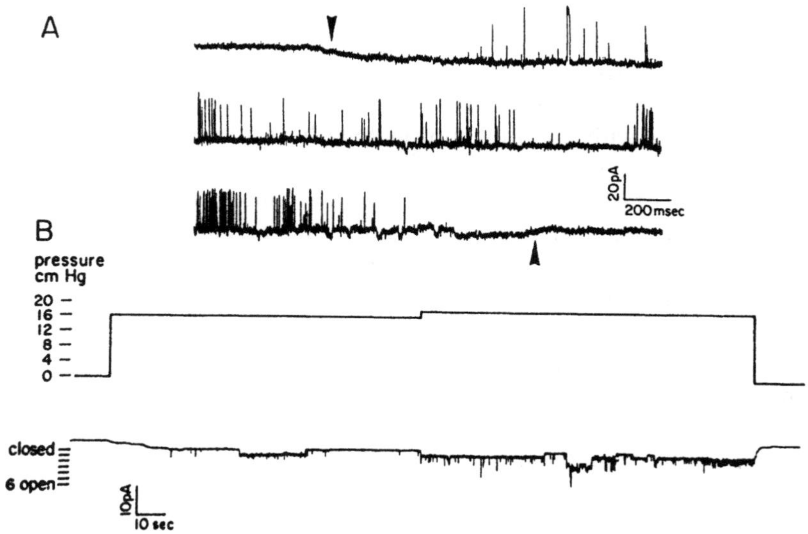

Fig. 12A. Suction-induced channel activity in giant protoplasts of *S. faecalis*. Slight suction was applied (*downward arrowhead*) to the pipette, or released (*upward arrowhead*). Pipette voltage was +20 mV. (Zoratti and Petronilli 1988). *B* Mechanosensitive channels were activated by 16 and 18 cm Hg of suction in giant protoplasts of *B. subtilis*. The protoplasts were made by lysozyme treatment of cells, followed by activation of autolysin for further cell wall digestion. Pipette voltage was –20 mV. (Delcour and Martinac, unpubl. results)

channel conductance sizes, ranging from about 10 pS to several nanosiemens, were observed. The authors favor the model of a high-conductance channel, with a maximal conductance of 1 nS, and possessing a variety of substates. The channel seems to be voltage-dependent and most of its substates conduct both cations and anions. Another characteristic of the channel is a bursting activity with flickering behavior. The authors speculated that, like the *E. coli* channel, the mechanosensitive channel in *S. faecalis* might be involved in osmotic responses, although possible roles in both sensory transduction and transmembrane transport cannot be excluded.

Mechanosensitive channels have been found also in giant protoplasts of *Bacillus subtilis*, another gram-positive bacterium (Fig. 12B; Delcour and Martinac, unpubl.; Zoratti et al. 1990). Our preliminary data suggest that there are some similarities between the *B. subtilis* and *S. faecalis* channels (Fig. 12B). We have found that the mechanosensitive channels in *B. subtilis* have kinetics similar to those in *S. faecalis*. Conductances of different sizes were also present in some of our recordings. The pattern of activity of these channels, however, differs from that of the channels we have found in *E. coli* (Martinac et al. 1987). The existence of mechanosensitive ion channels in both gram-negative and gram-positive bacteria suggests that channels of this type are common in prokaryotes as well as in eukaryotes.

4 Perspective

The physiological functions of mechanosensitive channels in bacteria are still unknown. These channels could play a role in regulation of bacterial morphology, since mutants with aberrant morphology have mechanosensitive channels with changed properties (Li and Martinac, unpubl.), or the channels could function in transduction of osmotic stimuli, since *E. coli* exhibits osmotaxis (Li et al. 1988).

We have made some efforts towards understanding the role and function of mechanosensitive channels in *E. coli*, using genetic methods. Since chlorpromazine affects mechanosensitive channel activity and mediates base-pair substitution and frame-shift mutagenesis in *E. coli* (Obaseiki-Ebor and Akerele 1988), we used it to induce and isolate mutant strains with aberrant properties of mechanosensitive channels. We isolated several chlorpromazine-resistant mutant strains (Martinac et al. 1990a; Li and Martinac, unpubl.). The mechanosensitive channels in one of these mutants differ from those of the wild-type strain in two ways: (1) Chlorpromazine does not affect the opening probability of the channels in the mutant as it does for the wild-type strain (Fig. 9B). (2) The mechanosensitive channels in the mutant have reduced pressure sensitivity as compared with the wild-type channels. The amount of suction necessary for an e-fold change in the channel activity (ratio of probability of being open to probability of being closed) is approximately twice as big for mutant as compared with the wild type. This mutant grows into long filaments in the absence of cephalexin, and probably belongs to the class of *lon* mutations which were first isolated by chlorpromazine resistance by Molnar et al. (1977). We do not know yet what causes the altered properties of the mutant channels. The shape of this mutant and our experiments with lysozyme and amphipaths suggest that the cell wall or outer membrane, or both, are affected by the mutation.

The presence of mechanosensitive channels in prokaryotic membranes and their possible role in such primitive but vital cellular functions as cell-volume control and osmoregulation suggest that these channels might have been among the first to evolve. Bacteria, especially *E. coli*, with its genetic and molecular biological advantages, present a unique opportunity not only for studying the biophysics and the molecular biology of mechanosensitive channels, but also for answering questions on the evolution of ion channels in general.

Acknowledgments. We thank Pat Hanson, Christine Hirsher and Leanne Olds for expert technical assistance. We would like to thank also Mario Zoratti for sharing original illustrations and for helpful discussions. This work was supported by NIH grant DK39121 and by a grant from the Lucille P. Markey Trust.

References

Bayer ME (1979) The fusion sites between outer membrane and cytoplasmic membrane of bacteria: their role in membrane assembly and virus infection. In: Inouye M (ed) Bacterial outer membranes: biogenesis and function. Wiley, New York, pp 167–202

Berger LR, Weiser RS (1957) The β-glucosaminidase activity of egg-white lysozyme. Biochim Biophys Acta 26:517–521

Berrier C, Coulombe A, Houssin C, Ghazi A (1989) A patch-clamp study of inner and outer membranes and of contact zones of *E. coli*, fused into giant liposomes. Pressure-activated channels are localized in the inner membrane. FEBS Lett 259:27–32

Brehm P, Kullberg K, Moody-Corbett F (1984) Properties of nonjunctional acetylcholine receptor channels on innervated muscle of *Xenopus laevis*. J Physiol (Lond) 350:631–648

Brezden BL, Gardner DR, Morris CE (1986) A potassium selective channel in isolated *Lymnea stagnalis* heart muscle cells. J Exp Biol 123:175–189

Britten RJ, McClure FT (1962) The amino acid pool of *Escherichia coli*. Bact Rev 26:292–335

Buechner M, Delcour AH, Martinac B, Adler J, Kung C (1990) Ion channel activities in the *Escherichia coli* outer membrane. Biochim Biophys Acta 1024:111–121

Cooper KE, Tang JM, Rae JL, Eisenerg RS (1986) A cation channel in frog lens epithelia responsive to pressure and calcium. J Membr Biol 93:259–269

Criado M, Keller B (1987) A membrane fusion strategy for single-channel recordings of membranes usually non-accessible to patch-clamp pipette electrodes. FEBS Lett 224:172– 176

Delcour AH, Martinac B, Adler J, Kung C (1989a) Modified reconstitution method used in patch-clamp studies of *Escherichia coli* ion channels. Biophys J 56:631–636

Delcour AH, Martinac B, Adler J, Kung C (1989b) Voltage-sensitive ion channel of *Escherichia coli*. J. Membr Biol 112:267–275

Epstein W, Shultz SG (1965) Cation transport in *Escherichia coli*. V. Regulation of cation content. J Gen Physiol 49:221–234

Falke LC, Edwards KL, Pickard BG, Misler S (1988) A stretch-activated anion channel in tobacco protoplasts. FEBS Lett 237:141–144

Guharay G, Sachs F (1984) Stretch-activated single ion channel currents in tissue-cultured embryonic chick skeletal muscle. J Physiol 352:685–701

Gustin MC, Zhou X-L, Martinac B, Kung C (1988) A mechanosensitive ion channel in the yeast plasma membrane. Science 242:762–765

Hall MN, Silhavy TJ (1981) Genetic analysis of the major outer membrane proteins of *Escherichia coli*. Ann Rev Genet 15:91–142

Hamill OP, Marty A, Neher E, Sakmann B, Sigworth FJ (1981) Improved patch-clamp techniques for high-resolution current recordings from cells and cell-free membrane patches. Pfluegers Arch Eur J Physiol 391:85–100

Kennedy EP (1982) Osmotic regulation and the biosynthesis of membrane-derived oligosaccharides in *Escherichia coli*. Proc Natl Acad Sci USA 79:1092–1095

Le Rudulier D, Strom AR, Dandekar AM, Smith LT, Valentine RC (1984) Molecular biology of osmoregulation. Science 224:1064–1068

Li C, Boileau AJ, Kung C, Adler J (1988) Osmotaxis in *Escherichia coli*. Proc Natl Acad Sci USA 85:9451–9455

Lugtenberg B, van Alphen L (1983) Molecular architecture and functioning of the outer membrane of *Escherichia coli* and other gram-negative bacteria. Biochim Biophys Acta 737:51–115

Martinac B, Buechner M, Delcour AH, Adler J, Kung C (1987) Pressure-sensitive ion channel in *Escherichia coli*. Proc Natl Acad Sci USA 84:2297–2301

Martinac B, Delcour AH, Buechner M, Adler J, Kung C (1988) Pressure-induced membrane tension activates a mechanosensitive ion channel in the outer membrane of the bacterium *Escherichia coli*. Biophys J 55:494a

Martinac B, Tengowski MW, Adler J, Kung C (1990a) Effect of amphipaths on pressure sensitivity of the mechanosensitive ion channel in *Escherichia coli*. Biophys J 57:323a

Martinac B, Adler J, Kung C (1990b) Mechanosensitive ion channels of *E. coli* activated by amphipaths. Nature 348:261–263

Mizushima S (1985) Structure, assembly, and biogenesis of the outer membrane. In: Nanninga N (ed) Molecular cytology of *Escherichia coli*. Academic Press, London, pp 39–75

Molnar J, Holland IB, Mandi Y (1977) Selection of *lon* mutants in *Escherichia coli* by treatment with phenothiazines. Genet Res Camb 30:13–20

Neuberger A, Wilson BM (1967) Inhibition of lysozyme by N-acyl-D-glucosamine derivatives. Nature (Lond) 215:524–525

Nikaido H, Vaara M (1987) Outer membrane. In: Neidhardt FC, Ingraham JL, Magasanik B, Brooks Low K, Schaechter M, Edwin Umberger H (eds) *Escherichia coli* and *Salmonella typhimurium*. Cellular and molecular biology, Vol. I. Am Soc Microbiol, Washington, DC, pp 7–15

Obaseiki-Ebor EE, Akerele JO (1988) The mutagenic activity of chlorpromazine. Mutat Res 208:33–38

Ohmori H (1984) Studies of ionic currents in the isolated vestibular hair cell of the chick. J Physiol (Lond) 350:561–581

Raetz C (1986) Molecular genetics of membrane phospholipid synthesis. Ann Rev Genet 20:253–295

Rothman JE, Lenard J (1977) Membrane asymmetry. The nature of membrane asymmetry provides clues to the puzzle of how membranes are assembled. Science 195:743–753

Rosenbusch JP (1974) Characterization of the major envelope protein of *Escherichia coli*. Regular arrangement of the peptidoglycan and unusual dodecyl-sulphate binding. J Biol Chem 249:8019–8029

Ruthe H-J, Adler J (1985) Fusion of bacterial spheroplasts by electric fields. Biochim Biophys Acta 819:105–113

Sachs F (1988) Mechanical transduction in biological systems. CRC Crit Rev Biomed Eng 16:141–169

Sheetz MP, Singer SJ (1974) Biological membranes as bilayer couples. A molecular mechanism of drug-erythrocyte interactions. Proc Natl Acad Sci USA 71:4457–4461

Sigurdson WJ, Morris CE, Brezden BL, Gardner DR (1987) Stretch activation of a K^+ channel in molluscan heart cells. J Exp Biol 127:191–209

Sonntag I, Schwarz H, Hirota Y, Henning U (1978) Cell envelope and shape of *Escherichia coli*: multiple mutants missing the outer membrane lipoprotein and other major outer membrane proteins. J Bacteriol 136:280–285

Stock JB, Rauch B, Roseman S (1977) Periplasmic space in *Salmonella typhimurium* and *Escherichia coli*. J Biol Chem 252:7850–7861

Zoratti M, Petronilli V (1988) Ion-conducting channels in a gram-positive bacterium. FEBS Lett 240:105–109

Zoratti M, Petronilli V, Szabo I (1990) Stretch-activated composite ion channels in *Bacillus subtilis*. Biochem Biophys Res Commun 168:443–450

Chapter 2

Mechanosensitive Ion Channels in Yeast. Mechanisms of Activation and Adaptation

M.C. Gustin[1]

Contents

1 Introduction

The ability to sense and respond to mechanical stresses in the environment is almost certainly a universal feature of all cells. Although there are specialized cells (e.g., hair cells in the cochlea) for sensing mechanical stimuli, many if not all cells appear capable of detecting changes in pressure on their cell membrane. The mechanosensors for most mechanical stimuli appear to reside in the cell surface membrane. Physiological experiments have demonstrated that the activity of certain ion channels (mechanosensitive channels; Howard et al. 1988; Sachs 1989; Morris 1990), solute transporters (Wood 1988), and enzymes (Watson 1990) is regulated by mechanical stress on the cell membrane. The molecular mechanism of how mechanical stress activates such membrane proteins remains a complete mystery.

The yeast *Saccharomyces cerevisiae* has several advantages as a system for studying the molecular mechanism of mechanosensitive (MS) ion channel activation. Patch-clamp electrophysiological studies of the plasma membrane in yeast spheroplasts reveal MS channels that are activated by, and adapt to, stretching of

[1]Department of Biochemistry and Cell Biology, Rice University, Houston, TX 77251, USA

Advances in Comparative and Environmental Physiology, Vol. 10

the membrane (Gustin et al. 1988). MS channels can be studied using several different recording modes, including whole-cell mode. The power of yeast molecular biology allows one to genetically manipulate the composition of the plasma membrane lipid bilayer (Carman and Henry 1989) as well as the cytoskeleton (Drubin 1990), two cellular structures implicated in MS channel regulation. Yeast have only three channels in their plasma membrane: (1) a depolarization-activated K^+-specific channel (Gustin et al. 1986; Gomez-Lagunas et al. 1989); (2) the nonselective MS channel; and (3) a multiple-conductance state, cation-specific channel (Gustin, unpubl. data). The latter channel is seen only rarely. Therefore, if K^+ is removed from recording solutions, the MS channel can be studied by itself in electrophysiological recordings, even in whole-cell mode. A disadvantage of yeast spheroplasts (and plant cells, Schroeder 1988) for patch clamp studies is that formation of a gigaohm seal is more difficult than for animal cells. High gigaohm seals ($> 10\ G\Omega$) will form readily, but take longer and require more suction than in animal cells (Saimi et al. 1990).

Using the knowledge and insight gained from patch-clamp studies of yeast MS channels, the following topics concerning MS channels will be covered in this review: (1) patch-clamp methodology for measuring MS channel activation; (2) the relationship between pressure and tension in MS channel activation; (3) molecular hypotheses concerning mechanosensitive membrane proteins; (4) adaptation in mechanosensitive cells; and (5) the possible function of MS channels in eukaryotic organisms with cell walls such as fungi and plants. Both stretch-activated and stretch-inactivated MS channels have been seen in animal cells (Morris 1990). This chapter will be confined to a discussion of the stretch-activated MS channels in yeast and related channels in other organisms.

2 Mechanisms of Activation of Mechanosensitive Channels

2.1 Patch Versus Whole-Cell Recordings

In the patch-clamp technique, two types of recording configuration are generally employed for studies of ion channels: the patch configuration (either on-cell or excised) and the whole-cell configuration (Hamill et al. 1981). Both of these configurations have been used to good effect in the study of yeast MS channels (Gustin et al. 1988). However, this is an exception. Most studies of MS channels use exclusively the patch configuration. There are two reasons for concentrating on the patch configuration. First, many patches contain a single channel. This allows the gating and conductance properties of the channel to be studied in detail, using well-established methods (Colquhoun and Hawkes 1983; Hille 1984). A second reason is that observation of current fluctuations with a unit conductance in membrane patches is a reassuring indication that a MS channel, and not some nonspecific hole of fluctuating conductance, is being opened by pressure. This problem has been studied in greater detail in ascidian oocytes where pressure-induced quantal current fluctuations in patch recordings have been showed to be due to changes in the conductance of the patch membrane, and not in the parallel conductance of the gigaohm seal (Moody and Bosma 1989).

Despite these advantages, there are two arguments for using the whole-cell configuration as well as the patch configuration for studying MS channels. First, the geometry of the membrane in the whole cell configuration is more easily observable and, for a spherical cell, has a single radius of curvature, simplifying the analysis of mechanical stress applied to the cell membrane. In contrast, the membrane in a patch can have multiple radii of curvature which can change upon application of pressure (Sachs 1989). The more intact whole-cell configuration may also be a better system than the excised patch for studying cellular regulation of MS channels (Sect. 3).

The mechanical stimulus used to activate MS channels in the patch configuration is usually hydrostatic pressure (inward or outward) applied through the recording pipette with a mechanical device. Studies in yeast spheroplasts indicate that the same procedure can be used to stimulate MS channels in the whole-cell recording configuration (Gustin et al. 1988). Only outward pressure is effective in the latter configuration, indicating that cell swelling is the stimulus. In whole-cell recordings of yeast spheroplasts, reduction of bath osmolarity to increase internal osmotic pressure is only rarely capable of activating MS channels. In the rare cases where external hypotonic solutions do stimulate yeast MS channel currents in whole-cell recordings, the effect is only transient.

Different results have been obtained for peripheral blood cells and epithelial cells where there are ion channels activated by cell swelling. Such ion channels (volume-sensitive channels) vary in selectivity, being either K^+-selective (Hudson and Schultz 1988; Sackin 1989), Cl^--selective (Cahalan and Lewis 1988; McCann et al. 1989; Worrell et al. 1989) or cation-nonselective (Ubl et al. 1988; Falke and Misler 1989; Bear 1990). In membrane patches, these channels, like yeast MS channels, are activated by applied suction. However, two observations suggest that the mechanism of channel activation for volume-sensitive channels and for yeast MS channels may be different. First, cell swelling (resulting from a reduction in external osmolarity or uptake of an osmotically active solute) activates volume-sensitive channels in on-cell patches (Hudson and Schultz 1988; Sackin 1989; Ubl et al. 1988; Falke and Misler 1989; Bear 1990). In contrast, on-cell patch recordings of yeast MS channels show no effect of reducing extracellular osmolarity. Second, in T lymphocytes (Cahalan and Lewis 1988), airway epithelial cells (McCann et al. 1989), and colonic epithelial cells (Worrell et al. 1989), reducing external osmolarity activates volume-sensitive channels (Cl^--selective) in whole-cell recordings. As previously noted, similar changes in external osmolarity rarely have any effect on whole-cell MS channel currents in yeast spheroplasts. It is unclear whether these differences are the result of a difference in activation mechanisms, mechanical sensitivity, or the membrane-attached cytoskeleton determining the tension on the bilayer. Lymphocytes and yeast are the same size, thus ruling out different membrane tensions for the same pressure (Sect. 2.2). Since an increase in internal osmotic pressure can theoretically be dissipated by water movement up the patch pipette in whole-cell mode, the membrane tension cannot be calculated simply from the predicted osmotic pressure and cell diameter.

2.2 Pressure versus Tension

When pressure is applied to the cell membrane, the membrane is deformed or stretched, creating tension. The relationship between applied pressure normal to the membrane and the tension developed in the plane of the membrane is given by the empirical relation known as Laplace's law. For a spherical cell at equilibrium,

$$T_m = \frac{Pd}{4}, \qquad (1)$$

where T_m = membrane tension, P = applied pressure and d = cell diameter. Is it the force acting normal to the membrane (pressure) or that in the plane of the membrane (tension; Fig. 1) that regulates the opening of MS channels? The following observations have been used to support the notion that MS channels are regulated by T_m and not P.

1. MS channels in membrane patches are opened by applying either positive or negative pressure. This is seen for yeast MS channels (Fig. 2) and for virtually all other stretch-activated MS channels (Morris 1990) that have been tested. If a channel is acted upon by unequal pressure across the membrane (such as being "pushed" through the membrane), positive and negative pressure ought to have the opposing, not the same effect on the opening of MS channels. Thus, the observation that pressure in either direction has the same effect on opening of MS channels argues against a direct effect of pressure.
2. Measurements of the pressure-dependence of MS channel P_{open} have been shown in several cases to be well fit to an exponential (Boltzman) function of P^2 and not P (Guharay and Sachs 1984; Cooper et al. 1986). This matched a prediction of a model proposed by Sachs (Sect. 2.3) for regulation of MS channels by membrane tension and not pressure per se (Sachs 1989). This test of whether a channel is regulated by tension has been questioned recently (Morris 1990). Analysis of data on MS channel gating in the outer membrane

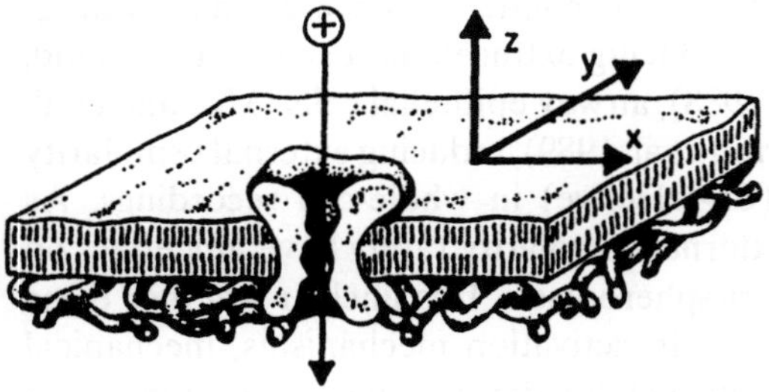

Fig. 1. Forces critical to activation of MS channels. The diagram shows a hypothetical MS channel inserted into a lipid bilayer. When pressure is applied to the membrane, two forces act upon MS channels: pressure (force along the *Z* axis, normal to the membrane) and membrane tension (the restorative force in the *XY* plane of the stretched membrane). In many cells, the tension in the phospholipid bilayer depends on the relative stiffness of the bilayer and the subtending, membrane-attached, filamentous structures of the cytoskeleton

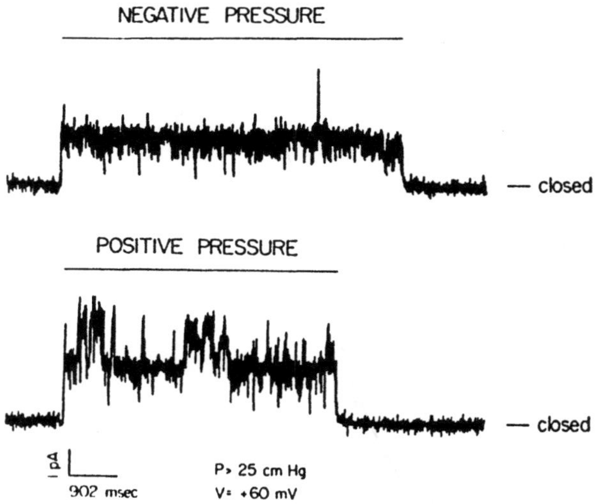

Fig. 2. MS channels currents in a membrane patch excised from a yeast spheroplast. Both positive and negative pressure cause MS channel activation. Voltage shown is the pipette voltage. Solutions were (in mM): pipette and bath – 120 CsCl, 5 $MgCl_2$, 0.1 EGTA, 5 CsHEPES pH 7.2. Temp. = 22 °C

of *E. coli* indicate a better fit to a Boltzmann function of P, not P^2 (Martinac et al. 1987). In yeast, where there is independent evidence of regulation by membrane tension, the pressure dependence of MS channel P_{open} can be well fit to an exponential function of either P or P^2 (Fig. 3). P_{open} data for animal cell MS channels can be similarly well fit, with either P or P^2 Boltzman functions for the same channel (Erxleben 1989; Morris 1990).

3. Laplace's law predicts that for a channel regulated by membrane tension, P_{open} will be a function of not only the pressure but also the geometry of the cell or patch. Observations on MS channel currents in whole-cell recordings of yeast spheroplasts (Gustin et al. 1988) bear out this prediction. MS channel currents (normalized to cell area) are smaller in smaller cells for the same applied pressure. This observation is consistent with MS channel regulation by membrane tension. This analysis assumes that the membranes of small and large cells have nearly identical mechanical properties. Supporting this assumption is the observation that the pressure required to lyse a spheroplast is inversely proportional to cell size (Gustin et al. 1988).

When the internal pressure of a cell is too high, the surface membrane is disrupted or lysed and cell death ensues. The internal pressure necessary to produce a lytic tension in the membrane depends on cell geometry (Laplace's law). The common finding of MS channels in large-diameter egg cells (Methfessel et al. 1986; Moody and Bosma 1989) suggests that such channels function to regulate internal osmolarity and thus keep internal pressure low in these fragile cells. Such osmoregulation may be a particularly important selective advantage for creatures that lay their eggs in freshwater.

A final point about the physical basis of MS channel gating concerns membrane thinning during the application of a pressure gradient across the cell membrane. During the membrane stretching, there may also be an accompanying thinning of

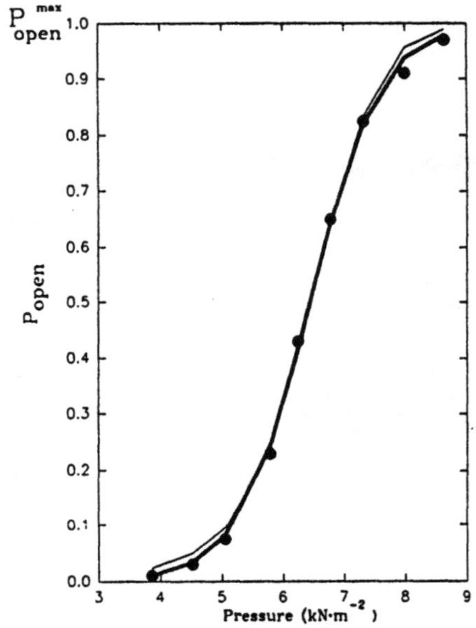

Fig. 3. Pressure dependence of MS channel currents in a whole-cell recording from a yeast spheroplast. Whole-cell MS channel currents were measured at different pressures and divided by the maximal, saturating MS channel current to yield the MS channel open probability, P_{open}. The maximal P_{open} was arbitrarily set to a value of 1. The data was fitted using a Marquardt-Levenberg algorithm (Sigmaplot 4.0 program) to one of the following two modified Boltzmann equations: $P_{open} = 1/1 + \exp(-m_1P = b_1)$ (*heavy line*) or $P_{open} = 1/1 + K_e\exp(-m_2P^2 + b_2)$ (*thin line*), where m_1, m_2 (proportionality constants) = 1.741, 0.138, respectively; b_1, b_2 (pressure-independent constants) = 11.21, 5.78, respectively; and P = pressure. 1 cmHg = 1.33 KN m^{-2}. Solutions were (in mM): pipette – 100 CsGlutamate, 5 $MgCl_2$, 1 EGTA, 3 ATP, 210 mannitol, 10 HEPES pH 7.2; bath – 10 CsGlutamate, 390 mannitol, 2 $MgCl_2$, 1 $CaCl_2$, 10 MES pH 5.5. Temp. = 22 °C

the bilayer, depending on whether the volume of the phospholipid bilayer remains constant (Fettiplace et al. 1971; White 1974). The amount of membrane thinning would therefore be proportional to the increase in area during stretching. Based on the effect of pressure gradients on the dielectric breakdown voltage of giant algal cells, Zimmerman (1978) has proposed that membrane thinning occurs during membrane stretching and may be the physical basis of the turgor sensing in these cells. The amount of membrane thinning expected for a physiological pressure of 1 bar in these walled cells is approximately 0.2–0.4 nm. This change, although small in comparison to the total thickness of the bilayer (approx. 6.0 nm), is comparable in percentage to the expected change in membrane area. Perhaps MS channels are actually regulated by the thickness of the membrane and not by membrane stretching per se.

2.3 Molecular Hypotheses of MS Channel Activation

Although there is no biochemical data concerning the molecular composition of MS channels and how their activity is regulated, several biophysical models of the gating mechanism have been proposed. The predictions of these models help to guide electrophysiological experiments on the gating of these interesting channels. Three different molecular models have been proposed to account for the regulation of MS channel activity by mechanical force on the cell membrane. Model #1, first proposed by Sachs (Guharay and Sachs 1984), predicts a Boltzmann function for MS channel gating, wherein the ratio of P_{open} and P_{closed} is an exponential function of the square of the membrane tension. This prediction is a consequence of the following reasoning: "membrane tension does work on a Hooke's law elastic element, and this free energy, ΔG, is used to open the channel" (Sachs 1989). The following three equations of model 1 yield, respectively:

1. The tension, T_m, from stretching the elastic element,

$$T_m = K_A \frac{\Delta A}{A}, \tag{2}$$

where K_A = area elastic constant of the channel, ΔA = change in area of the channel in the plane of the membrane for a given T_m, A = area of the channel in the plane of the membrane.

2. The free energy, ΔG, available from stretching the membrane,

$$\Delta G = A\Delta G_A = A \int K_A \frac{\Delta A}{A} = \frac{1}{2} AK_A \left(\frac{\Delta A}{A} \right)^2 = \frac{AT_m^2}{2K_A}, \tag{3}$$

where ΔG_A = free energy per unit area, all other terms defined above.

3. The open probability, P_{open}, of the MS channel,

$$P_{open} = P_{max}/1 + K_e e^{\frac{fAT_m^2}{2K_A kT}}, \tag{4}$$

where P_{max} = maximum open probability of the MS channel, K_e = equilibrium constant controlling the tension-independent probability, f = an efficiency factor describing the fraction of work available for channel gating, k = Boltzmann constant, T = absolute temperature, all other terms defined above.

The underlying assumption of model 1 is that an MS channel behaves analogously to an elastic spring, i.e., there is a linear relation between tension and stretching of the channel. In model 2, proposed by Morris (1990), only one state (open or closed) of the MS channel is compliant (able to be stretched), otherwise these two models are very similar. However, the underlying assumption of both models that MS channels behave as Hookean elastic elements appears to be at odds with the available data. Kinetic analysis of single channel recordings of MS channels reveal only a limited number of closed and open states with very rapid transitions between states (Guharay and Sachs 1984; Cooper et al. 1986; Sigurdson et al. 1987). The lack of evidence for a continuum of channel states invalidates the analogy of a

MS channel and a linear Hookean spring and, most importantly, the derivation of free energy density (ΔG_A) from the integration of a linear equation [as in Eq. (3)] describing channel strain ($\Delta A/A$).

Model 3, proposed by Howard et al. (1988), predicts that the MS channel open probability is an exponential function of the membrane tension. According to this model, each state of the channel has a distinct cross-sectional area, A, in the plane of the membrane. Thus, the free energy differences, ΔG, between the open and closed states of the channel (for a simplified two-state channel) is given as

$$\Delta G = \Delta G_0 + T_m \Delta A \, , \tag{5}$$

where ΔG_0 = free energy difference between the closed and open states of the MS channel in the absence of membrane tension, ΔA = difference in cross-sectional area between the closed and open states of the MS channel. The open probability of the MS channel, P_{open}, is given by the Boltzmann relation:

$$P_{open} = P_{max}/1 = K_e e^{T_m \Delta A/kT} \, . \tag{6}$$

Two features of this model are noteworth: (1) in contrast to the former models, this model of MS channels conforms to the observation of a limited number of closed and open states with very rapid transitions between states. (2) A powerful prediction of this model for the structure of an MS channel is that ΔA can be directly calculated from the slope of the Boltzman relation [Eq. (6)]. In practice, however, there appear to be several obstacles to application of this model to analysis of MS channel gating.

First, the actual tension on the channel and thus, the calculated value of ΔA, is difficult to determine. Membrane tension is very dependent on geometry of a membrane patch, which can change during recordings of MS channel currents (Sachs 1989). A more important problem for analyzing the membrane tension dependence of MS channel function is that plasma membranes often have an attached cytoskeleton (Chasis and Shohet 1987). The mechanical properties of a membrane are thus a function not only of the phospholipid bilayer wherein the MS channels reside, but also of the attached cytoskeleton (Evans et al. 1976). Elastic proteins of a membrane-parallel cytoskeleton will reduce tension in the bilayer, thereby resulting in the underestimation of ΔA. If the critical force for MS channel gating is tension in the lipid bilayer, then this tension will be reduced according to the following relation:

$$T_{bilayer} = T_m - T_{cytoskeleton} \, . \tag{7}$$

Another uncertainty in quantitative measurements of MS channel regulation by mechanical force is whether membranes behave elastically over the range of applied pressure. Is there a linear relationship between stress (pressure) and strain (change in membrane area – ΔA)? Viscous properties of a membrane-attached cytoskeleton may affect the development of tension in the bilayer (Hochmuth and Waugh 1987). Direct measurements of membrane stretching during activation of MS channels, although difficult, may help to reveal more details of this critical stress-strain relationship. Two methods, measurements of membrane capacitance changes with a phase-lock amplifier (Lindau and Neher 1988; Sachs 1987) and video-enhanced microscopic measurements of positional changes in surface-attached gold particles (Sheetz et al. 1989), appear to have the resolution necessary to measure the small area changes (less than 1–2% of total) expected.

There is another complication. Probably the most pressing question regarding molecular mechanisms of MS channel function is whether MS channels "sense" the stretching of the bilayer or the cytoskeleton. Certainly MS channels respond to stretching of the membrane, but what part of the membrane? If MS channels are sensing the stretching of the lipid bilayer, then it ought to be possible to reconstitute active MS channels in lipid vesicles, removed from cytoskeletal proteins. However, nothing has yet been reported on MS channel reconstitution experiments.

The role played by the cytoskeleton in stretch activation of MS channels is still unclear. In addition to reducing tension development in the bilayer, a membrane-attached cytoskeleton has been suggested to act more directly in the activation of MS channels, particularly for "volume-activated" MS channels (Hudson and Schultz 1988). Small changes in cell volume that activate channels in membrane patches would be unlikely to significantly stretch a mechanically isolated patch membrane (Hudson and Schultz 1988; although see Sackin 1989). The activation of channels by cell swelling in on-cell patches has led to the suggestion (Hudson and Schultz 1988) that membrane stretching outside the patch may be coupled to channel activation in the patch by a cytoskeletal linkage. If MS channels sense the stretching of the cytoskeleton, then cytoskeleton-disrupting chemical or genetic manipulations ought to inhibit the activation of MS channels. The only relevant experiment on this question is one study using an actin-disrupting drug. Actin-based cytoskeletal lattices are known to be disrupted by the drug cytochalasin B. Treatment of cultured cells with cytochalasin B does not inhibit the activation of MS channel in membrane patches, but rather decreased the pressure required for activation (Guharay and Sachs 1984). These results are consistent with a critical role for the phospholipid bilayer in stretch activation of MS channels, but need to be buttressed with additional experiments on this question.

3 Mechanisms of Adaptation of Mechanosensitive Channels

Pressure activation of MS channels will, in certain cases, depend on the duration of the pressure pulse. For example, application of a step change in pressure to the yeast plasma membrane causes channel open probability (P_{open}) to rapidly increase and then more slowly decrease (Fig. 4; Gustin et al. 1988).

The decrease in P_{open} is observed even as the stimulating pressure is held constant. For yeast MS channels, this phenomena (a decline in MS channel activity in response to a sustained and constant mechanical stress) is termed adaptation. This phenomena is not limited to yeast, but has also been observed in fibroblasts

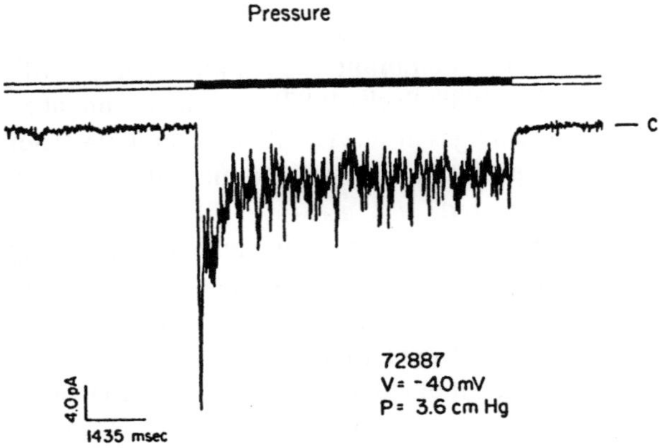

Fig. 4. Adaptation of whole cell MS channel currents in a yeast spheroplast. Constant outward, cell-inflating pressure was applied during the time period indicated by the *blackened line*. Solutions and temperature as in Fig. 3

(Stockbridge and French 1988), oocyges (Moody and Bosma 1989), and *E. coli* (Martinac, unpubl. results). Adaptation of MS channels in yeast is enhanced by negative interior membrane potentials, large stimulating pressures, and the presence of calcium in the external media (Gustin et al. 1988). None of these conditions are absolutely required. For instance, adaptation can be observed at positive interior membrane potentials if calcium is present outside the cell and large stimulating pressures are used. A yet-unexplained feature of MS channel adaptation in yeast is that the phenomena is readily apparent in whole-cell recordings, but rarely so in excised patch recordings.

The decrease in MS channel P_{open} with large sustained pressures appears to be due to a change in the pressure sensitivity of the channels (Fig. 5). For the whole-cell recording shown, the pressure was increased step-wise (pressure off, pressure on, pressure off, higher pressure on, pressure off, etc.) until a nearly maximal level of MS channel current was observed. During this experiment, the membrane potential was held at +40 mV to reduce the extent of adaptation. At the highest pressures, some adaptation was observed. When the pressure was stepped upward a second

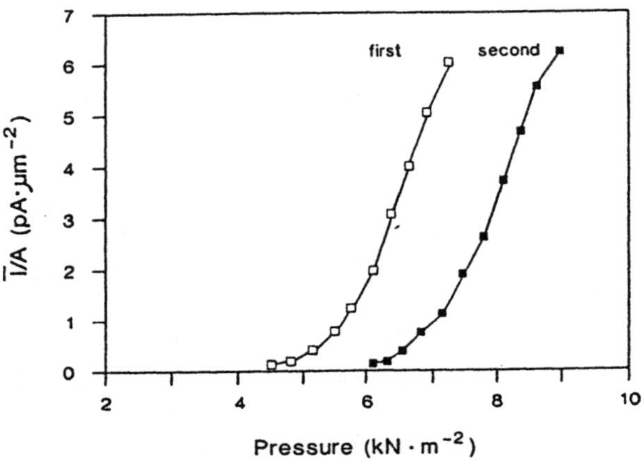

Fig. 5. Pressure dependence of MS channel currents in a yeast spheroplast before (*first*) and after (*second*) adaptation. See text for experimental details. Whole-cell outward currents were measured and normalized to cell area using cell capacitance (1 pF = 100 μm^2). Solutions and temperature as in Fig. 3. V = +40 mV

time, the entire pressure-current curve shifted rightward without an apparent decrease in maximum current. This result suggests that adaptation results not from a decrease in MS channels capable of responding to pressure, but from a change in sensitivity of the ensemble of channels. These features of MS channel adaptation in yeast, namely, calcium and voltage dependence, nonlinear dependence on mechanical stimulus, and sensitivity change, resemble those seen for adaptation of displacement-sensitive MS channels in hair cells (Eatock et al. 1987; Assad et al. 1989; Crawford et al. 1989). It is therefore possible that similar mechanisms may underly both phenomena. Adaptation of MS channels in yeast could be the result of alterations in (1) the response of the channels themselves; (2) the mechanical properties of the membrane; or (3) geometric factors. Adaptation of MS channels in yeast is correlated with an increase in lytic tension of the cell; it takes a higher pressure to "pop" the cell after adaptation has occurred. This observation suggests that changes in membrane elasticity or geometry accompany adaptation or are perhaps causal.

4 Physiological Function of Mechanosensitive Channels in Walled Organisms

The discovery of MS channels in yeast and related organisms (fungi and plants) (Falke et al. 1988; Gustin et al. 1988; Cosgrove and Hedrich 1990; Moran 1990) raises the important question of the physiological function of these interesting channels. By analogy with animal cells (Howard et al. 1988; Erxleben 1989), MS channels might play a central role in mechanosensation. Plants and fungi are capable of sensing if they have been pushed or prodded (Edwards and Pickard 1987). Cells from these organisms sense and respond to changes in turgor pressure (Gutknecht et al. 1978). Cellular responses to such stimuli are important for cell growth and access to important nutrients and energy sources. One special complication for mechanosensation in fungi and plants is the mechanical properties of the cell wall. Mechanical stress on the plasma membrane in which the MS channels reside is limited by the stiff cell wall. Since the patch-clamp technique cannot be applied if the cell wall is intact, three questions concerning MS channel function remain unanswered. Are physiologically relevant mechanical forces acting on cells with an intact wall capable of activating MS channels? Are plants and fungi mechanosensitive? Are MS channels necessary for the cellular responses of fungi and plants to physiologically relevant mechanical stimuli? The purpose of the following discussion is to briefly survey current knowledge and thinking, and to discuss what experimental approaches have the best chance at yielding answers to these difficult questions.

4.1 Membrane Stretching in Plants and Fungi

Stretching of the cell membrane appears to be the important stimulus for activation of MS channels (Sect. 2.2). With this in mind, the previous question can be restated: can the plasma membrane stretch underneath the cell wall of a fungal or plant cell? Two lines of evidence suggest that membrane stretching can occur despite the cell wall: (1) there is ample evidence that plant and fungal cells can expand; (2) electrical measurements of membrane thickness indicate that pressure gradients across the cell membrane cause membrane thinning, as expected for a membrane being stretched.

Expansion of surface area in plant and fungal cells occur during normal cell growth (Taiz 1988) and in response to sudden increases in turgor pressure (Gutknecht et al. 1978). The major determinant of expansion is the cell wall. Both reversible (elastic) and irreversible (plastic) expansion of the wall can take place. Cell growth is a function of plastic changes in the cell wall. This same wall can elastically expand to accommodate increases in cell turgor and volume while preventing cell lysis. During the slow process of cell growth, the volume of walled cells increases in a turgor pressure- dependent process (Green et al. 1971; Ortega et al. 1989). If the turgor pressure exceeds the yield threshold of the cell wall, plastic expansion of the wall occurs and cell growth is observed. The increased cell area of the growing cell is most likely due to the insertion of new membrane by an exocytotic mechanism (Sievers and Schnepf 1981), rather than by membrane stretching. Nonetheless, the cell membrane may have a small amount of constant tension because of the steady turgor pressure and less stiff cell wall of the growing cell (Bartnicki-Garcia and Lipman 1972).

The plant or fungal cell is also capable of rapid but limited elastic expansion when turgor pressure is suddenly increased. It is under these circumstances that stretching of the plasma membrane is more likely to occur. To study this phenomena, the development of experimental systems such as large algal cells (Gutknecht et al. 1978), and techniques such as the pressure probe (Zimmermann et al. 1969) for measuring and manipulating turgor pressure have been invaluable. Decreasing the osmolarity of the extracellular media (Gutknecht et al. 1978) or experimental injection of oil into the cell causes a sudden increase in turgor pressure and rapid swelling of the cell (Zimmermann and Steudle 1974). Generally, this swelling does not exceed a few percent of total cell volume. The relationship between pressure changes and volume changes is determined by the elastic properties of the cell wall as in the following equation (Philip 1958):

$$\Delta P = \varepsilon \frac{\Delta V}{V}, \quad (8)$$

where ΔP = change in turgor pressure, ε = volumetric elastic modulus (a measure of the mechanical properties of the cell wall), ΔV = change in cell volume, and V = total cell volume.

In large algal cells, the value of ΔP can range up to 8 bar and that of ε up to 300–600 bar (Zimmermann 1978). ε itself is not a constant but is pressure and volume dependent, becoming larger at higher pressure and volume (Zimmermann and Steudle 1974). The important reason for why cell expansion occurs is simple: the cell walls are stiff, but the forces are large.

Is the increase in cell area during turgor pressure-induced cell swelling large enough for significant stretching of the cell membrane to occur? Area changes during swelling of giant algal cells are calculated to be up to 2–3% of total area (Zimmerman 1978). The limits to cell expansion are similar in magnitude to those allowed for expansion of a lipid bilayer. Measurements of membrane stretching by micropipette aspiration in plant protoplasts (Wolfe and Steponkus 1981) indicate that the cell surface membrane cannot be stretched beyond 2–3% of total surface area without cell lysis occurring. Thus, the small scale of membrane area changes observed in response to increased turgor pressure matches well that allowed for stretching of biological membranes.

There is no direct evidence of plasma membrane stretching in a plant or fungal cell with an intact wall. However, increasing turgor pressure does appear to cause membrane thinning in giant algae at the same time cell expansion is occurring (Coster et al. 1977). The correlation of membrane area expansion and membrane thinning is consistent with a constant volume membrane that is being stretched. The method for assessing the thickness of a membrane assumes that biological membranes act as parallel-plate capacitors (Cole 1962). If the voltage across the membrane/capacitor, applied as short pulses, is steadily increased, the voltage at which dielectric breakdown (current flow across the capacitor) will occur depends on the thickness of the capacitor. Using this method, it was found that the dielectric breakdown voltage is inversely proportional to the turgor pressure, i.e., turgor pressure causes membrane thinning (Coster et al. 1977). Based on this evidence of turgor pressure-dependent membrane thinning, Zimmermann (1978) has proposed that the turgor sensor "senses" membrane thinning. According to this model, the thinning of the membrane can be caused either by membrane stretching when a pressure gradient exists across the membrane, or by membrane compression at high absolute pressure (for an opposing view see Gutknecht et al. 1978).

4.2 Mechanosensitivity of Plants and Fungi

Plants and fungi appear to sense and respond to a variety of mechanical stimuli. There are many examples of specialized mechanosensation in these organisms: touch receptors, gravity-sensing receptors, and the ability to sense surface topography. Both rapid and slow responses to touch have been observed. In the case of prey capture by the Venus flytrap (Burdon-Sanderson 1882) and certain nematode-snaring fungi (Commandon and de Fonbrune 1939) the stimulus appears to be purely mechanical, in that the response can be elicited by a metal needle instead of the prey. Touching a hair on the inside of the modified leaf of the Venus flytrap elicits a rapid transient depolarization (an action potential) of the receptor cell membrane potential (Benolken and Jacobsen 1970). Two hairs need to be touched in rapid succession to trigger trap closure. The nematode-snaring fungi (Barron 1977) form, from a branch of a hyphae, rings of three cells each. When a worm enters the ring there is a short delay of 1–2 s and then the cells swell in 0.1 s to constrict around the worm. Ring closure can only be elicited by touching the inner, not outer, surface of the ring. Thus, rapid mechanosensory responses triggered by spatially localized receptors are found not only in animals but also in plants and fungi. On a slower time scale, mechanical stimulation of plants causes a decrease in growth in the direction of the stimulus (Jaffe 1973). The same touch stimulus that induces this response, called thigmomorphogenesis, up-regulates the expression of several genes, including that which codes for calmodulin (Braam and Davis 1990). How fast the gene expression response to touch is, is not known, except that large changes in gene expression can already be observed by the first time point taken, i.e., 10 min. A more subtle mechanosensory response is that to gravity observed in certain plants and fungi (Dennison 1961; Edwards and Pickard 1987). Downward growth of plant roots depends on gravity. A specific section of the root, the cap, is necessary for this response (Jackson and Barlow 1981). Correlated with graviception are spatially localized changes in the electric field around the root (Iwabuchi et al. 1989), suggesting changes in membrane ion transport. Finally, micrometer-scale features etched onto a silicon wafer are able to direct the direction of growth and eventual differentiation of hyphae extending from germinating spores of the fungus *Uromyces* (Hoch et al. 1987).

In addition to these specialized examples of mechanosensory responses, fungal and plant cells sense and respond to changes in turgor pressure (Green et al. 1971; Gutknecht et al. 1978; Caldwell et al. 1986; Ortega et al. 1989; although see Nakagawa et al. 1974; Higgins et al. 1987). The importance of turgor pressure for normal cell growth in walled organisms has been revealed by elegant studies on tip growth in sporangiophores of *Phycomyces* (Ortega et al. 1989). The cells of *Phycomyces* are large enough to allow a pressure probe to be inserted into the cell without perturbing normal growth (Cosgrove et al. 1987). This probe was used to change intracellular pressure (on top of the normal turgor pressure) while continuously observing the effect on tip growth rate. The results indicate that growth rate and internal pressure are tightly coupled. If turgor pressure is lowered, growth rate rapidly slows. If turgor pressure is stepped up slightly, growth rapidly increases. A large step-up in turgor pressure has a biphasic effect on growth rate: an initial stimulation followed by inhibition of growth rate to a lower level than before the step. Thus, too low or too high a turgor pressure slows growth. Since the irreversible extension of the cell wall is a function of the turgor pressure (Dale and Sutcliffe 1986) above the "yield threshold" of the cell wall, the response to lower pressure is easily understood. However, the reason for slowing of growth at higher turgor pressure is not yet clear, but may be the result of "strain hardening" of the cell wall (Green et al. 1977).

Studies on turgor sensing and response have been carried out mainly in large-diameter cells where direct measurement and manipulation of turgor pressure is possible (Gutknecht et al. 1978). Turgor pressure is the outwardly directed excess hydrostatic pressure which is equal and opposite to the inwardly directed pressure of the cell wall. The source of this excess pressure is an internal osmolarity exceeding that outside the cell. Turgor pressure is essential for cell growth and thus is regulated by homeostatic mechanisms. When extracellular osmolarity changes, turgor pressure changes. Changes in turgor pressure elicit alterations in membrane permeability and metabolism of major cytoplasmic solutes, thus restoring the osmotic gradient and the turgor pressure. The specific solute whose cytoplasmic concentration is regulated by changes in extracellular osmolarity varies from species to species (Chamberlin and Strange 1989). Cells are capable of both up-regulating and down-regulating turgor pressure. Both small- and large-diameter cells use similar mechanisms to regulate internal osmolarity in response to external osmotic changes (Hellebrust 1976). Whether small-diameter cells, like the better-studied large-diameter cells, sense turgor pressure per se, rather than some related parameter such as cell volume or the concentration of a specific solute, remains to be determined. Finally, how cells sense turgor is not known. Specifically, the molecules that transduce changes in pressure gradients across the cell membrane into changes in solute permeability or gene expression have not yet been identified.

4.3 MS Channels as Mechanotransducers

Indirect evidence indicates that plant and fungal plasma membranes can be stretched, possibly during mechanostimulation. However, the molecular mechanism of the mechanosensor involved in mechanosensation in plants and fungi is unknown. Changes in turgor pressure elicit changes in membrane permeability. Touch can elicit action potentials in certain plants. Stretch-activated MS channels have the correct functional attributes to participate as mechanotransducers in these physiological responses, but there is little evidence to support this hypothesis. The only experimental evidence linking MS channels to mechanosensory phenomena is the finding that Gd^{3+} and not the related ion, La^{3+}, is an inhibitor of both MS channels and root gravitropism (Millet and Pickard 1988). To test whether MS channels act as turgor sensors or underlie other mechanosensory phenomena in plants and fungi, new tools will be needed. In particular, specific high-affinity inhibitors of MS channels or mutants defective in MS channel function are needed.

Only Gd^{3+} ions appear to be relatively effective in inhibiting MS channels in yeast (Gustin et al. 1988), as well as in other organisms (Yang and Sachs 1989). The use of externally applied Gd^{3+} ions as a specific probe of the physiological role of MS channels in yeast and other organisms is limited by at least three factors: (1) In yeast, Gd^{3+} is an effective inhibitor from the inside and not the outside of the cell (Zhou and Gustin, unpubl. results; however, see Yang and Sachs 1989). (2) Gd^{3+} has complex, poorly understood effects on MS channels and cell membranes. For instance, Gd^{3+} appears to increase the lytic tension of yeast spheroplasts. (3) Gd^{3+} and phosphate have a low solubility product. This last property can lead to the underestimation of Gd^{3+} activity in phosphate-containing growth media or possibly to $[Gd^{3+}]$-dependent limitation of phosphate required for growth.

An alternative approach to probing the role of MS channels in mechanosensation in plants and fungi is to use genetics. If MS channels play an important role in graviception in roots, for example, a mutational analysis should turn up the MS channel gene as one of those required for proper graviception. Conversely, an

MS channel inhibitor such as Gd^{3+} could be used to isolate mutants defective in MS channel function (see limitations of Gd^{3+} above). If MS channel function is required for graviception, then such mutants ought to be defective in graviception.

In order to forge a genetic link between MS channels and mechanosensation in plants or fungi, the choice of experimental system is important. Three requirements of such a system are:

1. Methods for the isolation and analysis of mutants using genetic screens or selections based on a well-described physiological response to a mechanical stimulus;
2. Availability of well-developed classical and molecular genetic approaches; and
3. Methods for assaying MS channel function, in particular, patch-clamp analysis of single channel activity and whole-cell currents.

Two experimental systems, graviception (Moore and Evans 1986) in *Arabidopsis thaliana* and osmoregulation in the yeast *Saccharomyces cerevisiae* (Brown 1978), come closest to fulfilling each of these requirements.

Mutants defective in root gravitropism have been isolated in *Arabidopsis* (Bullen et al. 1990). The genetics and molecular biology of *Arabidopsis* are relatively well-developed (compared to other plants; Goldberg 1988). There is, therefore, a realistic chance of using mutants to isolate graviception genes and analyzing the products of these genes. Moreover, *Arabidopsis* leaf protoplasts have been patch-clamped (Lew 1990). Still to be developed in this system is a method for isolating and patch-clamping the specific root cells involved in graviception.

Yeast respond to changes in external osmolarity by changing the cytoplasmic concentration of glycerol (Reed et al. 1987). An upshift in external osmolarity causes an equivalent upshift in internal glycerol (Maiorella et al. 1984; Meikle et al. 1988). It has been proposed that this is a response to altered turgor pressure (Meikle et al. 1988). Mutants defective in this response have been isolated and are being used to clone putative turgor-response genes by complementation (Brewster and Dwyer, unpubl. results). Once these genes are isolated, powerful methods of yeast molecular biology can be used to assess the function of the respective gene products (Botstein and Fink 1988). This, plus the ability to do patch-clamp analysis of MS channels (Gustin et al. 1988), should allow the role of such channels in yeast turgor regulation to be determined. What might this role be? One mechanism by which cytoplasmic glycerol concentration is regulated is by rapid changes in membrane permeability to glycerol (Blomberg and Adler 1989). When external osmolarity is down-shifted (presumably increasing turgor pressure), glycerol permeability increases dramatically. MS channel are more likely to open at high turgor pressure. Thus, rapidly reacting MS channels may regulate glycerol permeability directly, by acting as the glycerol "permease", or indirectly, by allowing ions or other solutes that regulate glycerol permeability to cross the membrane.

5 Summary

This chapter has discussed MS channels and their possible physiological function in walled eukaryotic cells such as yeast. A more fundamental question is, how do any and all cells sense mechanical forces in their environment? MS channels may be just one of many mechanisms cells employ in mechanosensation. Is the same physical parameter "sensed" by each mechanosensor? For certain molecular mechanisms, thinning rather than stretching may be the important parameter. Or for others, tension in the membrane-attached cytoskeleton rather than that in the bilayer may be regulatory. Only when we are able to examine the structure of different mechanosensors will these issues become clearer. With mutants, molecular genetics, and assays of mechanosensor function, that time may not be far away.

Acknowledgments. I wish to thank Ching Kung, in whose lab much of this work was done, Boris Martinac for helpful discussions, and Jay Brewster for help with illustrations. This work supported by a grant from the NSF.

References

Assad JA, Hacohen N, Corey DP (1989) Voltage dependence of adaptation and active bundle movement in bullfrog saccular hair cells. Proc Natl Acad Sci USA 86:2918–2922

Barron GL (1977) The nematode-destroying fungi. Can Biol Publ, Guelph, Ontario

Bartnicki-Garcia S, Lipman E (1972) The bursting tendency of hyphal tips of fungi: presumptive evidence for a delicate balance between wall synthesis and wall lysis in apical growth. J Gen Microbiol 73:487–500

Bear CE (1990) A nonselective cation channel in rat liver cells is activated by membrane stretch. Am J Physiol 258:C421-C428

Benolken RM, Jacobsen SL (1970) Response properties of a sensory hair excised from Venus's flytrap. J Gen Physiol 56:64–82

Blomberg A, Adler L (1989) Roles of glycerol and glycerol-3-phosphate dehydrogenase (NAD^+) in acquired osmotolerance of *Saccharomyces cerevisiae*. J Bacteroil 171:1087–1092

Botstein D, Fink GR (1988) Yeast: an experimental organism for modern biology. Science 240:1439–1443

Braam J, Davis RW (1990) Rain-, wind-, and touch-induced expression of calmodulin and calmodulin-related genes in *Arabidopsis*. Cell 60:357–364

Brown AD (1978) Compatible solutes and extreme water stress in eukaryotic microorganisms. Adv Microbiol Physiol 7:181–242

Bullen BL, Best TR, Gregg MM, Barsel SE, Poff KL (1990) A direct screening procedure for gravitropism mutants in *Arabidopsis thaliana* (L.) Heynh. Plant Physiol 93:525–531

Burdon-Sanderson J (1882) On the electromotive properties of the leaf of *Dionaea* in the excited and unexcited states. Phil Trans R Soc Lond B 173:1–53

Cahalan MD, Lewis RS (1988) Role of potassium and chloride channels in volume regulation by T lymphocytes. In: Gunn RB, Parker JC (eds) Cell physiology of blood. Rockefeller, New York, pp 281–301

Caldwell JH, Van Brunt J, Harold FM (1986) Calcium-dependent anion channel in the water mold, *Blastocladiella emersonii*. J Membr Biol 89:85–97

Carman GM, Henry SA (1989) Phospholipid biosynthesis in yeast. Annu Rev Biochem 58:635–669

Chamberlin ME, Strange K (1989) Anisosmotic cell volume regulation: a comparative view. Am J Physiol 257:C159-C173

Chasis JA, Shohet SB (1987) Red cell biochemical anatomy and membrane properties. Annu Rev Physiol 49:237–248
Cole KS (1962) The advance of electrical models for cells and axons. Biophys J 2:101–119
Colquhoun D, Hawkes AG (1983) The principles of the stochastic interpretation of ion-channel mechanism. In: Sakmann B, Neher E (eds) Single-channel recording. Plenum Press, New York, p 135–175
Commandon J, de Fonbrune P (1939) De la formation et du fonctionnement des pieges des champignons predateurs des nematodes. Recherches effectuees a l'aide de la micromanipulation et de la cinematographie. CR Acad Sci Paris 207:304–305
Cooper KE, Tang JM, Rae JL, Eisenberg RS (1986) A cation channel in frog lens epithelia responsive to pressure and calcium. J Membr Biol 93:259–269
Cosgrove DJ, Hedrich R (1990) Stretch-activated cation, anion, and calcium channels coexist in guard cells. Plant Physiol 93:17
Cosgrove DJ, Ortega JKE, Shropshire Jr W (1987) Pressure probe study of the water relations of *Phycomyces blakesleeanus* sporangiophores. Biophys J 51:413–423
Coster HGL, Steudle E, Zimmermann U (1977) Turgor pressure sensing in plant cell membranes. Plant Physiol 58:636–643
Crawford AC, Evans MG, Fettiplace R (1989) Activation and adaptation of transducer currents in turtle hair cells. J Physiol 419:405–434
Dale JE, Sutcliffe JF (1986) Water relations of plant cells. In: Plant physiology, Vol. IX: Water and solutes in plants. Academic Press, New York, pp 1–48
Dennison DS (1961) Tropic responses of *Phycomyces* sporangiophores to gravitational and centrifugal stimuli. J Gen Physiol 45:23
Drubin DG (1990) Actin actin-binding proteins in yeast. Cell Motil Cytoskel 15:7–11
Eatock RA, Corey DP, Hudspeth AJ (1987) Adaptation of mechanoelectrical transduction in hair cells of the bullfrog sacculus. J Neurosci 7:2821–2836
Edwards KL, Pickard BG (1987) Detection and transduction of physical stimuli in plants. In: Wagner E, Greppin H, Millet B (eds) The cell surface in signal transduction. Springer, Berlin Heidelberg New York Tokyo, pp 41–66
Erxleben C (1989) Stretch-activated current through single ion channels in the abdominal stretch receptor organ of the crayfish. J Gen Physiol 94:1071–1083
Evans E, Waugh R, Melnick L (1976) Elastic area compressibility modulus of red cell membrane. Biophys J 16:585–595
Falke LC, Misler S (1989) Activity of ion channels during volume regulation by clonal N1E115 neuroblastoma cells. Proc Natl Acad Sci USA 86:3919–3923
Falke LC, Edwards KL, Pickard BG, Misler S (1988) A stretch-activated anion channel in tobacco protoplasts. FEBS Lett 237:141–144
Fettiplace R, Andrews DM, Haydon DA (1971) The thickness, composition, and structure of some lipid bilayers and natural membranes. J Membr Biol 5:277–96
Goldberg RG (1988) Plants: novel developmental processes. Science 240:1460–1467
Gomez-Lagunas F, Pena A, Lievano A, Darszon A (1989) Incorporation of ionic channels from yeast plasma membranes into black lipid membranes. Biophys J 56:115–119
Green PB, Bauer K, Cummins WR (1977) Biophysical model for plant cell growth: auxin effects. In: Jungreis AM et al. (eds) Water relations in membrane transport in plants and animals. Academic Press, New York, pp 30–45
Green PB, Erickson RO, Buggy J (1971) Metabolic and physical control of cell elongation rate: in vivo studies in *Nitella*. Plant Physiol 47:423–430
Guharay F, Sachs F (1984) Stretch-activated single ion channel currents in tissue-cultured embryonic chick skeletal muscle. J Physiol 352:685–701
Gustin MC, Martinac B, Saimi Y, Culbertson MR, Kung C (1986) Ion channels in yeast. Science 233:1195–1197
Gustin MC, Zhou XL, Martinac B, Kung C (1988) A mechanosensitive ion channel in the yeast plasma membrane. Science 242:762–765
Gutknecht J, Hastings DF, Bisson MA (1978) Ion transport and turgor pressure regulation in giant algal cells. In: Giebisch G, Tosteson DC, Ussing HH (eds) Membrane transport in biology, vol III. Springer, Berlin Heidelberg New York Tokyo, pp 125–174
Hamill OP, Marty A, Neher E, Sakmann B, Sigworth F (1981) Improved patch-clamp techniques for high resolution current recordings from cells and cell-free patches. Pflügers Arch Eur J Physiol 391:85–100

Hellebrust JA (1976) Osmoregulation. Annu Rev Plant Physiol 27:485–505
Higgins CF, Cairney J, Stirling DA, Sutherland L, Booth IR (1987) Osmotic regulation of gene expression: ionic strength as an intracellular signal. Trends Biochem Sci 12:339–344
Hille B (1984) Ionic channels of excitable membranes. Sinauer Assoc Sunderland, Massachusetts
Hoch HC, Staples RC, Whitehead B, Comeau J, Wolf ED (1987) Signalling for growth orientation and cell differentiation by surface topography in *Uromyces*. Science 235:1659–1662
Hochmuth RM, Waugh RE (1987) Erythrocyte membrane elasticity and viscosity. Annu Rev Physiol 49:209–219
Howard J, Roberts WM, Hudspeth AJ (1988) Mechanoelectrical transduction by hair cells. Annu Rev Biophys Biophys Chem 17:99–125
Hudson RL, Schultz SG (1988) Sodium-coupled glycine uptake by Ehrlich ascites tumor cells results in an increase in cell volume and plasma membrane channel activities. Proc Natl Acad Sci USA 85:279–283
Iwabuchi A, Yano M, Shimizu H (1989) Development of extracellular electric pattern around *Lepidium* roots: its possible role in root growth and gravitropism. Protoplasma 148:94–100
Jackson MB, Barlow PW (1981) Root geotropism and the role of growth regulators from the cap: a reexamination. Plant Cell Environ 4:107–123
Jaffe MJ (1973) Thigmomorphogenesis: the response of plant growth and development to mechanical stimulation. Planta 114:143–157
Lew RR (1990) Ion channels of *Arabidopsis*. Plant Physiol 93:49
Lindau M, Neher E (1988) Patch-clamp techniques for time resolved capacitance measurements in single cells. Pflügers Arch Eur J Physiol 411:137–146
Maiorella BL, Blanch HW, Wilke CR (1984) Feed component inhibition in ethanolic fermentation by *Saccharomyces cerevisiae*. Biotechnol Bioeng 26:1155–1166
Martinac B, Buechner M, Delcour AH, Adler J, Kung C (1987) Pressure-sensitive ion channel in *E. coli*. Proc Natl Acad Sci USA 84:2297–2301
McCann JD, Li M, Welsh MJ (1989) Identification and regulation of whole-cell chloride currents in airway epithelium. J Gen Physiol 94:1015–1036
Meikle AJ, Reed RH, Gadd GM (1988) Osmotic adjustment and the accumulation of organic solutes in whole cells and protoplasts of *Saccharomyces cerevisiae*. J Gen Microbiol 134:3049–3060
Methfessel C, Witzmann V, Takahashi T, Mishina M, Numa S, Sakmann B (1986) Patch clamp measurements on *Xenopus laevis* oocytes: currents through endogenous channels and implanted acetylcholine receptor and sodium channels. Pflügers Arch Eur J Physiol 407:577–588
Millet B, Pickard BG (1988) Gadolinium ion is an inhibitor suitable for testing the putative role of stretch-activated ion channels in geotropism and thigmotropism. Biophys J 53:155a
Moody WJ, Bosma MM (1989) A nonselective cation channel activated by membrane deformation in oocytes of the ascidian *Boltenia villosa*. J Membr Biol 107:179–188
Moore R, Evans ML (1986) How roots perceive and respond to gravity. Am J Bot 73:574–587
Moran N (1990) Stretch-activated channels in plasmalemma of pulvinar motor cells. Plant Physiol 93:17
Morris CE (1990) Mechanosensitive Ion Channels. J Membr Biol 113:93–107
Nakagawa S, Kataoka H, Tazawa M (1974) Osmotic and ionic regulation in *Nitella*. Plant Cell Physiol 15:457–468
Ortega JKE, Zehr EG, Keanini RG (1989) In vivo creep and stress relaxation experiments to determine the wall extensibility and yield threshold for the sporangiophores of *Phycomyces* Biophys J 56:465–475
Philip JR (1958) The osmotic cell, solute diffusability, and the plant water economy. Plant Physiol 33:264–271
Reed RH, Cudek JA, Foster R, Gadd GM (1987) Osmotic significance of glycerol accumulation in exponentially growing yeasts. Appl Environ Microbiol 52:2119–2123
Sachs F (1987) Baroreceptor mechanisms at the cellular level. Fed Proc, Fed Am Soc Exp Biol 46:12–16
Sachs F (1989) Ion channels as mechanical transducers. In: Stein WD, Bronner F (eds) Cell shape: determinants, regulation and regulatory role. Academic Press, New York, pp 63–92
Sackin H (1989) A stretch-activated K^+ channel sensitive to cell volume. Proc Natl Acad Sci USA 86:1731–1735
Saimi Y et al. (1990) Recording of ion channels of microorganisms. Methods Enzymol (in press)

Schroeder JI (1988) K^+ transport properties of K^+ channels in the plasma membrane of *Vicia faba* guard cells. J Gen Physiol 92:667–683
Sheetz MP, Turney S, Qian H, Elson EL (1989) Nanometre-level analysis demonstrates that lipid flow does not drive membrane glycoprotein movements. Nature (Lond) 340:284–288
Sievers A, Schnepf E (1981) Morphogenesis and polarity of tubular cells with tip growth. In: Kiermayer O (ed) Cytomorphogenesis in plants. Springer, Berlin Heidelberg New York, pp 263–299
Sigurdson WJ, Morris CE, Brezden BL, Gardner DR (1987) Stretch activation of a K^+ channel in molluscan heart cells. J Exp Biol 127:191–209
Stockbridge LL, French AS (1988) Stretch-activated cation channels in human fibroblasts. Biophys J 54:187–190
Taiz L (1984) Plant cell expansion: regulation of cell wall mechanical properties. Annu Rev Plant Physiol 35:585–657
Ubl J, Murer H, Kolb H-A (1988) Ion channels activated by osmotic and mechanical stress in membranes of opossum kidney cells. J Membr Biol 104:223–232
Watson PA (1990) Direct stimulation of adenylate cyclase by mechanical forces in S49 mouse lymphoma cells during hypoosmotic swelling. J Membr Biol 265:6569–6575
White SK (1974) Comments on electrical breakdown of bimolecular lipid membranes as an electro-mechanical instability. Biophys J 14:155–158
Wolfe J, Steponkus PL (1981) The stress-strain relation of the plasma membrane of isolated plant protoplasts. Biochim Biophys Acta 643:663–668
Wood JM (1988) Proline porters effect the utilization of proline as nutrient or osmoprotectant for bacteria. J Membr Biol 106:183–202
Worrell RT, Butt AG, Cliff WH, Frizzell RA (1989) A volume-sensitive chloride conductance in human colonic cell line T84. Am J Physiol 256:C1111-C1119
Yang XC, Sachs F (1989) Block of stretch-activated ion channels in *Xenopus* oocytes by gadolinium and calcium ions. Science 243:1068–1971
Zimmermann U (1978) Physics of turgor- and osmoregulation. Ann Rev Plant Physiol 29:121–148
Zimmermann U, Steudle E (1974) Hydraulic conductivity and volumetric elastic modulus in giant algal cells: pressure and volume dependence. In: Zimmermann U, Dainty J (eds) Membrane transport in plants. Springer, Berlin Heidelberg New York Tokyo, pp 64–71
Zimmermann U, Hade H, Steudle E (1969) Kontinuierliche Druckmessung in Pflanzenzellen. Naturwissenschaften 56:634

Chapter 3

Mechanosensory Transduction in Ciliates (Protozoa)

J. W. Deitmer[1]

Contents

1 Mechanoreception and Cellular Behaviour

Most living cells are mechanosensitive in that mechanical strain exerted on their plasma membrane results in a transient change in electrical conductance of the membrane. In cells unspecialized for mechanoreception, a mechanically induced "injury" would result in a membrane potential change due to traumatic leak currents. In specialized mechanoreceptor tissue, however, the mechanical energy input results in the opening and/or closing of specific sensory ion channels in the membrane; this may provide a very high sensitivity to a mechanical stimulus. The gating of these mechanosensitive channels results in a defined change in ion conductance, and consequently, in a shift of the membrane potential. The steps of mechanoelectrical coupling are summarized by the following scheme: mechanical energy input – deformation of the sensitive structure – gating of mechanosensory ion channels – change in membrane ion conductance – receptor current flow – receptor potential – voltage-sensitive modification and integration – motor output.

[1]Abteilung für Allgemeine Zoologie, FB Biologie, Universität Kaiserslautern, Postfach 3049, 6750 Kaiserslautern, FRG

Advances in Comparative and Environmental Physiology, Vol. 10

Unicellular organisms can react to a variety of environmental stimuli, although specialized sense organs are missing. Sensory functions such as responses to touch, light, chemicals and temperature are integrated by the single cell. The cell membrane plays a crucial role in integrating and processing sensory stimuli, which lead to a specific behavioural response.

Ciliated protozoa have been found particularly suitable objects to study mechanoreception at a cellular level. Most ciliates are free-swimming cells driven by ciliary organelles. The cell surface of *Paramecium*, for example, is covered by some 3000–6000 single cilia. Although each cilium acts as an autonomous oscillator, all cilia beat in a largely coordinated way. The hypotrichous ciliate *Stylonychia* has two types of compound ciliary organelles, membranelles and cirri, each of which consists of a cluster of 20–80 cilia. Single cilia are found only in restricted areas. Locomotion is produced mainly by the ventral and marginal cirri and by a row of membranelles at the anterior end of the cell, which also provide a "feeding" water current towards the oral groove.

When a ciliate cell is mechanically stimulated, either by running into an obstacle during forward or backward locomotion or by being touched by a moving probe, it will commonly react in the manner of an escape. A dual behavioural response of *Paramecium* to stimulation was described many decades ago (Jennings 1906). When a cell meets an obstacle while swimming forwards, it reverses the direction of its ciliary power stroke and swims backwards for a while. When, on the other hand, it is touched at the posterior end, for instance by another moving object, forward swimming is initiated or accelerates. Thus, in either case, the cell tends to "avoid" the source of mechanical stimulation. Differing behavioural responses to mechanical stimuli applied to the anterior or posterior end of the cell indicate that the cell is able to discriminate between different sites of stimulus reception.

2 Physiological Responses to Mechanical Stimulation

The mechanisms of the differential behaviour following mechanical stimulation were analyzed with electrophysiological techniques. The recording of membrane potential and membrane currents using intracellular microelectrodes, as first performed by Kamada (1934), gave the answers to mechanosensory transduction processes.

Naitoh and Eckert (1969), in correlating electrical membrane responses with anterior and posterior mechanical pulses, rediscovered the bipolar mechanosensitivity in *Paramecium*. When the tip of a fine glass capillary was moved onto the anterior surface of the cell membrane, a depolarization of the membrane was recorded; a membrane hyperpolarization occurred upon stimulation of the posterior part of the cell (Fig. 1). The bipolarity of the mechanoreceptor responses is represented by two types of mechanically sensitive ion channels localized in the anterior part ("anterior channel") and in the posterior part ("posterior channel") of the cell. Moving the site of stimulus application from anterior to posterior (or vice versa) produced receptor potentials which gradually changed amplitude and polarity (Fig. 1). The greatest sensitivity, indicated by the largest response amplitude, is at the anterior and at the posterior cell end, while the de- and hyperpolarizing responses decrease as the site of stimulation approaches the middle part of the cell body. This is due to overlapping gradients of opposite ion conductances (see Sect. 5).

While the hyperpolarization is a genuine receptor potential, the stimulus-induced depolarization following stimulation at the anterior cell end consists of a depolarizing receptor potential, equivalent to a generator potential, and an action potential.

Voltage-clamp experiments, using the two-intracellular-microelectrode technique, have provided further evidence for the bipolar nature of the mechanoelectrical response. An inward current is recorded upon anterior stimulation, and an outward current upon posterior stimulation. These currents are pure receptor responses, since no voltage-gated ion channels are activated while the membrane is clamped at its resting potential. Receptor potentials of either polarity, and, under voltage-clamp, receptor inward and outward currents, remained unimpaired after deciliation of the cell. In fact, the cilia were found to be devoid of mechanoreceptor channels. Transmission of mechanical distortion from the cilia to the somatic membrane is a passive process involving substantial delay (Machemer-Röhnisch and

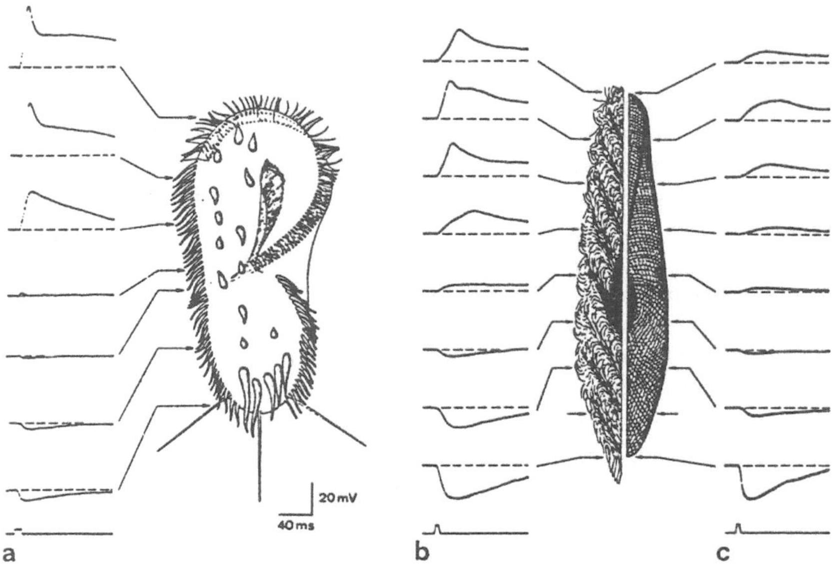

Fig. 1a-c. Localization of mechanosensitivity in *Stylonychia* (*a*) and *Paramecium* (*b, c*) as defined by the membrane potential response to mechanical stimulation at different sites (*arrows*, anterior *upward*, posterior *downward*). After deciliation of *Paramecium* (*c*) the mechanoreceptor responses are unaffected, although the voltage-dependent responses have disappeared. (After Machemer and Deitmer 1985)

Machemer 1984). It is the cell soma which includes the mechanoreceptor channels in its membrane and, thereby, constitutes the substrate of sensory transduction.

Differential mechanosensitivity, in providing the basis for bidirectional escape responses, appears to be characteristic for many free-swimming ciliates, including *Paramecium* (Naitoh and Eckert 1969a); *Dileptus* (Doroszewski 1970); *Euplotes* (Epstein and Eckert 1973); *Stylonychia* (De Peyer and Machemer 1978a); and *Tetrahymena* (Onimaru et al. 1979). A bipolarity in mechanoreceptor responses is absent from *Stentor* (Wood 1982, 1985) and *Didinium* (Hara and Naitoh 1980; Hara and Asai 1980). Comparative aspects of the mechanosensitivity of ciliates have previously been reviewed (Naitoh 1984; Machemer and Deitmer 1985).

3 Properties of Mechanosensitive Ion Channels

3.1 Anterior Ion Channels

The receptor response following mechanical stimulation of the anterior cell end is a membrane depolarization or, during voltage-clamp, a transient inward current. The reversal potential of the receptor potential and receptor current was found between zero and +15 mV (Naitoh and Eckert 1969a; De Peyer and Machemer 1978a; Ogura and Machemer 1980; De Peyer and Deitmer 1980). Ion substitution experiments revealed that the current through the anterior channel may be carried by several divalent cations such as Ca^{2+}, Mg^{2+}, Sr^{2+} and Ba^{2+} (Fig. 2A). The slope of the reversal potential of the anterior receptor current was similar for varying Ca^{2+} and Mg^{2+} concentrations, suggesting that the channel is equally permeable to Ca^{2+} and Mg^{2+} (Fig. 2B). This is important, since freshwater may contain Ca^{2+} and Mg^{2+} ions in similar concentrations. Manganese, cadmium and cobalt, on the other hand, inhibited the current through the anterior channel in *Stylonychia* (Deitmer 1982). Manganese was reported to permeate this channel in *Paramecium* (Satow et al. 1983).

The conductance increase upon anterior mechanical stimulation suggested, in addition, the activation of a K^+ current. The change in reversal potential of the receptor current with the external Ca^{2+} and K^+ concentrations indicated a conductance ratio, Ca^{2+} conductance over K^+ conductance, of approximately 3:1 (Ogura and Machemer 1980). It is still unknown whether or not the receptor conductance found for both Ca^{2+} and K^+ reflects the selectivity of a single channel, being somewhat more permeable to Ca^{2+} than to K^+, or that of two populations of ion channels, one being selective for Ca^{2+} (Sr^{2+}, Ba^{2+}, Mg^{2+}) and the other for K^+. In the latter case, contributions of the two ion conductances to the anterior response could then be accounted for by different densities and/or different conductivities of the two types of channels.

In summary, the anterior mechanoreceptor response is largely due to an increase in the membrane conductance to divalent cations, with little discrimination between Ca^{2+} and Mg^{2+}. A minor, but significant contribution to this response derives from an increased K^+ conductance.

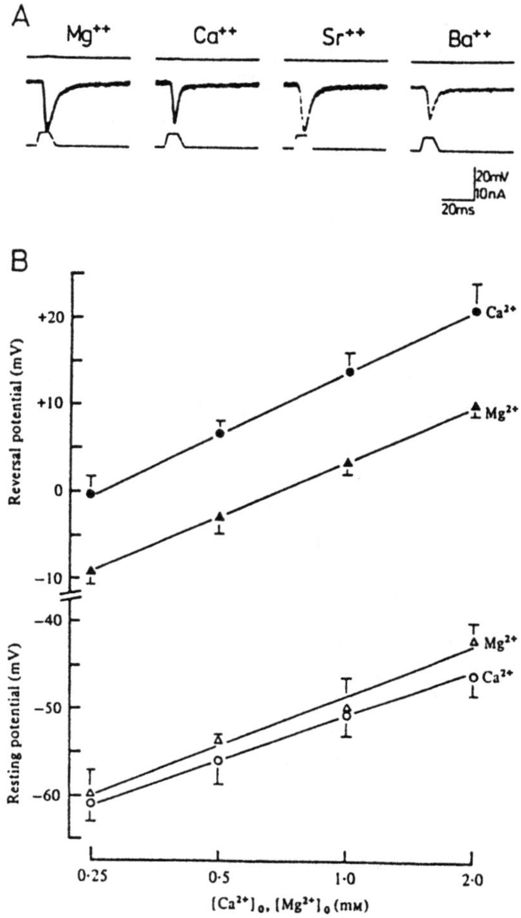

Fig. 2A. Receptor currents following mechanical stimulation of the cell anterior in *Stylonychia* using different divalent cations as charge carriers. *B* Reversal potential of anterior mechanoreceptor response in *Stylonychia* in the presence of Ca^{2+} or Mg^{2+} as charge carriers. The *slope* indicates similar permeability of the anterior channel to Ca^{2+} and Mg^{2+}. (*A* after Deitmer 1983; *B* after De Peyer and Deitmer 1980)

3.2 Posterior Ion Channels

3.2.1 Selectivity and Blockade

The receptor response following mechanical stimulation of the posterior membrane ("posterior mechanoreceptor response") is a hyperpolarization, or, during voltage clamp, an outward current. The ion involved in this response was found to be K^+. Naitoh and Eckert (1969a, 1973) reported a slope of about 50 mV of the reversal potential of the posterior receptor potential in *Paramecium* for a tenfold change in the external K^+ concentration. They concluded that K^+ ions are the main charge carriers during the posterior receptor response. Later studies in *Stylonychia* and *Paramecium* reported a slope of 56–58 mV of reversal potential of either receptor potentials (De Peyer and Machemer 1978a) or receptor current (Ogura and Machemer 1980; Deitmer 1982). The dependence of the reversal potential of the posterior mechanoreceptor current from the external K^+ concentration in *Stylonychia* is shown in Fig. 3. Hence, K^+ ions are the only charge carriers through posterior mechanoreceptor channels.

The reversal potential, V_{rev}, of the posterior receptor response was between −85 and −90 mV in *Paramecium* and in *Stylonychia* in the presence of 1 mM K^+ (De Peyer and Machemer 1978a; Ogura and Machemer 1980; Deitmer 1982). Due to the almost ideal Nernstian slope, V_{rev} is virtually identical with E_k. This allows the calculation of the intracellular K^+ concentration according to the Nernst equation:

$$[K^+]_i = [K^+]_0 \times \exp(V_{rev} \times F/RT),$$

where $[K^+]_i$ and $[K^+]_0$ are the intracellular and extracellular concentrations of K^+, and R, T and F have their usual thermodynamic meanings. The intracellular K^+ concentrations of these cells in artificial freshwater were calculated to be between 32 and 37 mM. Due to the Nernstian slope of the receptor current at various external K^+ concentrations, it can be assumed that the *intracellular* K^+ concentration remained constant during these experiments.

The posterior mechanoreceptor response is inhibited by some well-known blockers of K^+-selective membrane channels. Naitoh and Eckert (1973) reported that tetraethylammonium (TEA^+, 0.5–4 mM) reduced the receptor potential amplitude, presumably by obstructing current flow through the mechanoreceptor channel. This was confirmed by more detailed voltage-clamp studies. In the presence of 1 mM TEA^+ and 2 mM K^+, mechanically induced conductance was reduced to 20% without a change in the reversal potential (Fig. 4). Lowering the external K^+ concentration (e.g., to 1 mM) increased the inhibitory effect of TEA^+, suggesting a competitive interaction of K^+ and TEA^+ with the receptor channel (Deitmer 1982).

4-aminopyridine (4-AP, 0.5 mM), but not external Cs^+ (1 mM), procaine (1 mM) or tetramethylammonium (1 mM), was found to inhibit the posterior mechanoreceptor current (Deitmer 1982). The reported permeation of TEA^+ and Ca^{2+} in the presence of high external TEA^+ concentrations (4–10 mM) in *Para-*

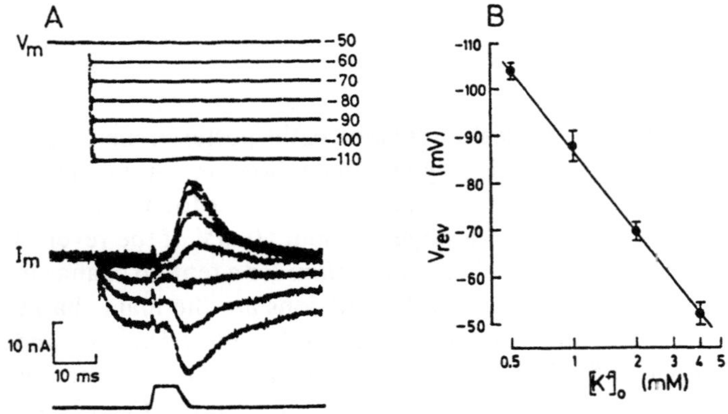

Fig. 3A. The receptor current following mechanical stimulation at the cell posterior in *Stylonychia* at different membrane potential. The current reverses near −90 mV. *B* Dependence of the reversal potential of the receptor current, V_{rev}, on the external K^+ concentration (slope 58 mV). (After Deitmer 1982)

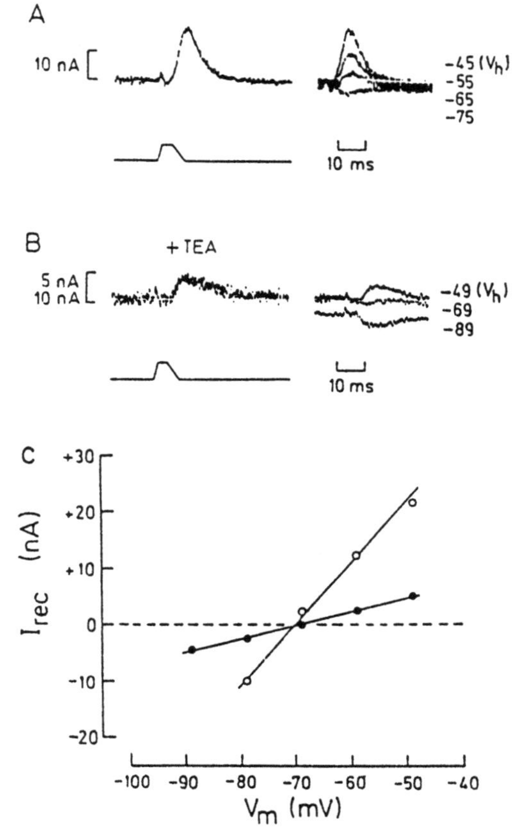

Fig. 4A-C. Inhibition of the posterior mechanoreceptor current in *Stylonychia* by tetraethylammonium (TEA, 1 mM). *A* Receptor currents and their reversal; *B* the same mechanical stimulation in the presence of TEA; *C* plot of the receptor current amplitude versus the membrane potential in the absence (*open circles*) and presence (*filled circles*) of 1 mM TEA. The reversal potential remains constant, but the *slope*, indicating the receptor conductance decreases greatly in the presence of TEA. (After Deitmer 1982)

mecium (Naitoh and Eckert 1973) could not be confirmed in *Stylonychia* (Deitmer 1982) and does not correspond to the Nernstian slope established for the posterior receptor response in *Paramecium.*

There is also no indication that Li^+, Na^+ or Cs^+ can carry charges through the posterior mechanoreceptor channel. When these ions were added or exchanged for K^+ in the external solution, reversal potentials only changed according to altered external K^+ concentrations (Deitmer 1982). Preliminary experiments suggested, however, that Rb^+ does permeate the posterior mechanosensitive channel (Deitmer, unpubl.). Thus, this channel is highly selective for K^+ (and Rb^+) ions, comparably to many voltage-gated K^+ channels (Hille 1978), but in contrast to the nonselective monovalent cation channel described for the crayfish stretch receptor (Edwards et al. 1981), for hair cells of the vertebrate inner ear (Corey and Hudspeth 1979) and for Pacinian corpuscles (Akoev 1982).

Varying the external divalent cation concentration or replacing Ca^{2+} with Mg^{2+}, Sr^{2+}, Ba^{2+} or Mn^{2+} has no significant effect on the posterior receptor response (Deitmer 1982). Experiments employing intracellular microelectrodes filled with ethylene glycol tetra-acetic acid (EGTA), a potent Ca^{2+} chelator, indicated that intracellular EGTA does not affect the receptor response (Deitmer, unpubl.). It seems unlikely, therefore, that this K^+-dependent response is affected or even regulated by intracellular Ca^{2+}, as shown for other types of K^+-selective channel (Meech 1978; Deitmer and Eckert 1985).

3.3 Time Characteristics

The rise time of the posterior receptor current is weakly voltage-dependent; it decreases linearly with membrane depolarization. An *e*-fold decrease was extrapolated to occur with a 410-mV depolarization (Deitmer 1981). Interestingly, a discontinuity between the time to peak and voltage was observed beyond the reversal potential: during subsequent inward current, the time to peak decreased by about 50% before rising again with further hyperpolarization.

The decay time constant of receptor currents allows an estimate of the mean open lifetime of the activated ion channels. In ciliates this time constant has been found to be quite different for the anterior and the posterior receptor current and among various genera. The decay time constant of the anterior receptor current is $\geq$ 25 ms in *Paramecium* (Ogura and Machemer 1980). In *Stylonychia*, two time constants of 1.8 ms and 7.2 ms were determined when Ca^{2+} carried the current, and a single time constant of 7.6 ms was determined when the current was carried by Mg^{2+}. A similar single exponential time course was also evident when Ba^{2+} or Sr^{2+} acted as charge carriers. This suggests that the fast decay time constant of the Ca^{2+} receptor current is specifically due to Ca^{2+} (De Peyer and Deitmer 1980). It is as yet unclear whether this phenomenon can be explained in terms of a Ca^{2+}-dependent inactivation, as suggested for the voltage-gated Ca^{2+} channels in *Paramecium* (Brehm and Eckert 1978), for one of the two voltage-gated Ca^{2+} channels in *Stylonychia* (Deitmer 1984, 1986) and for Ca^{2+} channels of other tissue (Eckert and Chad 1984).

The decay time constant of the posterior receptor current is 7–8 ms for both *Paramecium* (Ogura and Machemer 1980; Machemer-Röhnisch and Machemer 1984) and *Stylonychia* (Deitmer 1981, 1982). It was shown to be voltage-dependent; it increased *e*-fold with a membrane hyperpolarization of 110 mV. A voltage dependence was also found for the response of the crustacean stretch receptor (Swerup et al. 1983). In this preparation the time constant was smallest at about –50 mV, and increased with both de- and hyperpolarization of the membrane.

It is interesting that a voltage dependence similar to that of the ciliate mechanoreceptor is exhibited by the acetylcholine-induced endplate current at the frog neuromuscular junction (Gage 1976). Voltage sensitivity of decay time constants or of mean open lifetime of channels may therefore represent an intrinsic property of certain ion channels and may be unrelated to the mode of stimulation.

4 Electrical Excitation: Voltage-Dependent Ion Channels

Sensory reception and behavioural action are mediated by the electrical membrane response. Following anterior mechanical stimulation, this includes the action potential elicited by the depolarizing receptor potential. In order to understand the sensory-motor coupling leading to the appropriate behaviour in unicellular organisms, some fundamentals of the voltage-dependent membrane processes involved need to be discussed.

Comparable to nerve and muscle, electrical excitation in ciliates is represented by the action potential. Action potentials have been investigated in detail in two different types of ciliates, the holotrich *Paramecium* and the hypotrich *Stylonychia*. The ionic channels underlying the action potential in ciliates are similar in properties and function to those expressed in a neuron of an invertebrate or vertebrate organism. Therefore, free-swimming ciliates have been referred to as being "swimming sensory cells" (Machemer and De Peyer 1977) or "swimming neurons" (Kung and Saimi 1985).

The action potential, elicited by depolarization of the membrane, such as the depolarizing mechanoreceptor response, is graded to the stimulus strength with variable rates of rise and fall (Fig. 5A), and reaches peak amplitudes between 20 mV and 50 mV (Naitoh and Eckert 1968; De Peyer and Machemer 1977; Deitmer and Machemer 1982; Pape and Machemer 1986). The duration of the action potential ranges from 30–150 ms. Ion substitution experiments have shown that the depolarizing phase of the action potential is Ca^{2+}-dependent; the repolarizing phase is mainly due to K^+ conductances.

Depolarization of the cell membrane from the resting (= holding) potential induces a transient Ca^{2+} inward current under voltage-clamp. In *Stylonychia*, this inward current can be separated into two components, Ca^{2+} current I and Ca^{2+} current II (Fig. 5B). Properties of these two types of Ca^{2+} channels, which have now been described in a great variety of excitable cells (Deitmer 1986, 1988; McClesky et al. 1986) are listed in Table 1.

Deciliation experiments on the holotrichous ciliate *Paramecium* had shown that the voltage-dependent Ca^{2+} channels reside in the membrane of the cilia, but not in the cell body membrane (Ogura and Takahashi 1976; Dunlap 1977; Machemer and Ogura 1979). From the different types of compound cilia forming functionally different motor organelles in *Stylonychia*, the possibility of an inhomogenous distribution of Ca^{2+} channels in the membrane was tested. For this, cells were cut transversely, or parts of the cells were removed to obtain cell fragments. The remaining, resealed anterior or posterior cell parts were used for intracellular recording (Deitmer 1984). In cells without the anterior part, i.e. without their membranellar band, the second action potential component and inward current I are absent. When the posterior half or third of a cell is removed to record from an anterior cell fragment with its membranellar band intact, the action potential still has two peaks, and both inward current I and current II can be observed.

In a cell which had released the anterior part including its membranellar band, the second action potential component disappeared, and inward current I could not be recorded. The inward current-voltage relationship revealed only one maximum at −18 mV, which was typical for that of inward current II (Deitmer 1984).

This strongly suggests that the two voltage-dependent Ca^{2+} channels reside in different parts of the cell membrane, one in the membrane of the membranelles,

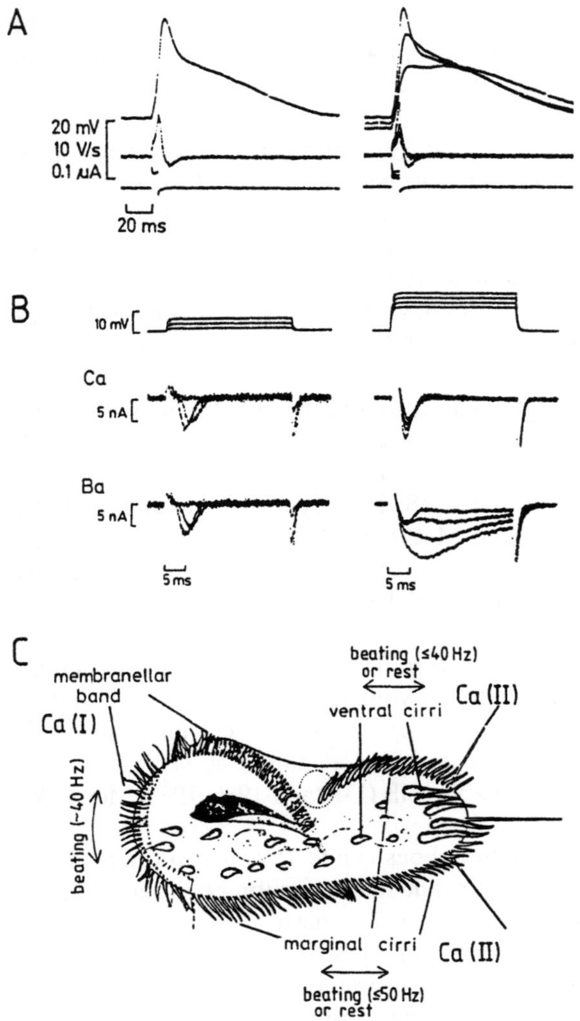

Fig. 5A-C. Voltage-gated membrane responses and ciliary motor responses in *Stylonychia*. *A* Action potentials (*upper traces*) evoked by intracelullar current pulses (*lower traces*) showing graded and all-or-none components (first derivative: *middle traces*). *B* Ca^{2+} and Ba^{2+} currents through voltage-dependent channel I (*left columns*) and channel II (*right columns*), showing different kinetics and inactivation properties when carried by these different divalent cations. *C* Ventral view of *Stylonychia* indicating the ciliary organelles, their maximum beating frequency, and the localization of Ca^{2+} channel I and II

and the other in the membrane of the cirri (Fig. 5C). Recording from cells which were in the process of encystment, during which period the cells resorb all their ciliary organelles, confirmed this interpretation. In these rounded cells with almost all membraneless and cirri resorbed, current II was reduced to less than 10%, while current I was completely absent (Deitmer 1987). In addition, young cells of less than 30 min after cell division have a relatively larger amount of current I than current II, suggesting that the inward currents are coupled to the morphogenetic development of the ciliary organelles (Deitmer et al. 1986).

It is therefore tempting to speculate that inward current I, restricted to the membranelles, is involved in controlling membranellar activity, while inward current II controls the beating activity of ventral and marginal cirri. The two Ca^{2+}

Table 1. Properties of the two voltage-dependent calcium channels

Property	Ca^{2+} channel I	Ca^{2+} channel II
Selectivity	Ca^{2+}, Sr^{2+}, Ba^{2+}	Ca, Sr^{2+}, Ba^{2+}
Activation	Fast, transient	Fast, (longer component)
kinetics	–47 mV (V_h = –50 mV)	-33 mV
$V_{0.5}$	(varies with V_h)	
Inactivation		Ca^{2+}-dependent
mechanism	Voltage-dependent	44 ms, 218 ms
kinetics	20–60 ms	
Channel blockers	Co^{2+}, Mn^{2+}	Co^{2+}, Mn^{2+}
	Cd^{2+} ($K_{0.5}$–10^{-4} M)	Cd^{2+} ($K_{0.5}$–10^{-5} M)
	Con A (5 nM)	
Peak current, nA (% total)	5 – 10 nA (20%)	25–40 nA (80%)
Stimulates I_k(Ca)	No	Yes
Localization	Membranelles	Cirri

channels might thus provide the means for separate gears for independent motor control of membranelles and cirri.

The two voltage-dependent Ca^{2+} channels bring about a depolarization which "amplifies" the depolarization caused by the mechanosensitive divalent cation channel in the anterior cell membrane. The concurrent Ca^{2+} influx through the ciliary (voltage-dependent), but not the somatic (mechanosensitive) channels leads to the ciliary motor apparatus, and hence to behavioural response (Fig. 6). The localization of the different types of ion channels is therefore of great functional importance for the integration of sensory stimuli. The consequences of inhomogeneous ion channel distribution for sensory-motor coupling in unicellular organisms provide a unique example for functional cell compartmentalization.

5 The Significance of Ion Channel Localization

The distribution in the surface membrane of two types of mechanosensory channels in the manner of two opposing gradients is meaningful for locomotion and the feeding responses of ciliates. Collision of the anterior cell end with an obstacle leads to the opening of Ca^{2+} (Mg^{2+}) receptor channels, so that the membrane depolarizes. Electrotonic spread of the depolarizing receptor potential into the cilia evokes an action potential. The Ca^{2+} receptor current entering the somatic space does not activate even the neighbouring cilia (Deitmer 1983; De Peyer and Machemer 1978b). It is the depolarization and the following voltage-dependent Ca^{2+} influx into the ciliary space which triggers the activation of cilia to beat in reverse and to initiate the retreat of the cell from the hindering object (Fig. 6).

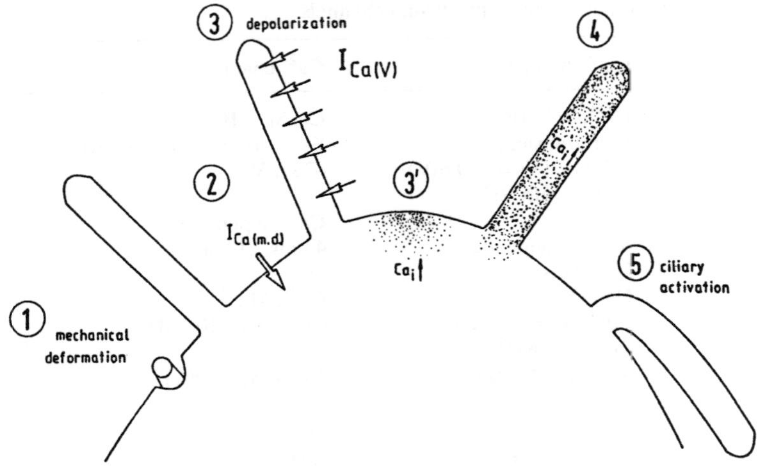

Fig. 6. Localization of anterior mechanoreceptor Ca^{2+} current flow [Ica(m.d.)] and voltage-gated Ca^{2+} current [Ica(v)] and the sequence of events during senso-electro-mechanical coupling. Mechanical stimulation of the cell anterior *(1)* produces a local Ca^{2+} receptor current through the cell body membrane (2), which leads to a membrane depolarization, and hence activation of voltage-gated Ca^{2+} current through the ciliary membrane (*3*), and a local Ca^{2+} transient near the somatic membrane (*3'*). The Ca^{2+} influx into the cilia results in a large increase of intraciliary Ca^{2+} (*4*), which activates the ciliary axoneme (*5*)

The opening of the posterior mechanoreceptor channels serves to generate a membrane hyperpolarization, which spreads electrotonically to the cilia. The nature of the charge carrier has no signalling function. It is interesting to note that, in the chain of events, there occur virtually no membrane currents across the ciliary membranes. These membranes are thought to be largely devoid of resting conductances for ions and, therefore, implicate only minor cable losses. The appropriate passive properties of the narrow ciliary cable guarantee transmission of a voltage signal throughout a 10–50 μm cilium (Deitmer 1984; Machemer 1986).

The nature of the coupling between hyperpolarization of the ciliary membrane and activation of microtubular sliding of the axoneme remains yet to be resolved. Claims of evidence for the involvement of cyclic nucleotides, such as cyclic AMP, as a mediator in the process (Gustin et al. 1983) have not been confirmed (Hennessey et al. 1985; Nakaoka and Machemer 1990).

Although the entire somatic cell surface is sensitive to touch, the receptor responses are minimal or even missing near the latitudes of the cytostome. Here, the inward and outward receptor currents are shortcircuiting (Ogura and Machemer 1980). A mechanosensitivity of the cytopharynx has not been directly tested, but may be inferred from experiments with suspension feeders like *Stylonychia* and *Paramecium* which demonstrate food selection by means of morphological and dynamic criteria (Karpenko et al. 1977; Fenchel 1980). A reduced responsiveness to mechanical disturbances of the peristomal surface area may suppress "escape responses" of the cell following a bombardment by food particles.

The action potential of a ciliate serves to supply the ciliary space with a brief "Ca^{2+} pulse". Voltage amplification has no major function in the relatively compact ciliate cell body, nor is the Ca^{2+} action potential code for the transmission of frequency signals comparable to nervous tissue. Ca^{2+}-dependent reorientation of the ciliary cycle and peak rates of beating of the cilia can start within a few milliseconds after onset of the depolarization (Machemer 1974; De Peyer and Machemer 1982a), which is characteristic of the avoiding reaction in ciliates. Behavioural responses following depolarization can, hence, be directly related to the Ca^{2+} and K^+ channels modulating the action potential.

In *Paramecium,* where the voltage-dependent Ca^{2+} channels are localized on the ciliary membrane only (Ogura and Takahashi 1976; Dunlap 1977), the current flows through the ciliary membrane to produce a Ca^{2+} increase within the intraciliary space which activates the axoneme (Fig. 6). The ciliary beating activity (towards the cell anterior), and hence backward swimming of the cell, is directly related to the Ca^{2+} current during the action potential.

In *Stylonychia*, where two types of Ca^{2+} channel are distributed unevenly in the membranes of membranelles and cirri, a differential regulation of the Ca^{2+} influx through these channels appears to control the membranellar and cirral motility. The functionally different ciliary organelles employ two distinct types of Ca^{2+} channels, presumably to achieve different Ca^{2+} transients in the two cell compartments, and hence independent motor control of these organelles. In fact, each group of these ciliary organelles displays a characteristic beating activity. The cirral frequency and directional response is a function of the membrane potential (Machemer 1974; De Peyer and Machemer 1982a,b) much comparable to electromotor coupling of cilia in *Paramecium* (Machemer 1974, 1977). The membranelles, on the other hand, which generate feeding currents and filter for appropriate food size, do not vary their frequency with the membrane potential (Deitmer et al. 1984).

The low-threshold Ca^{2+} channel I may, in addition, play a role in modulating spontaneous electrical activity in *Stylonychia* (Deitmer 1984), comparable to neurons (Llinas and Yarom 1981; Carbone and Lux 1984). Stochastic fluctuations of the membrane potential by a few millivolts can trigger action potentials. Spontaneous action potentials, arising from the resting potential at intervals of about 1-3 seconds, determine the behaviour of an undisturbed *Stylonychia* cell, which varies between stationary feeding and rapid "back-and-forth" jerks (Machemer and Deitmer 1987). The rate of generation of spontaneous action potentials was markedly reduced after the inhibition of Ca^{2+} channel I by concanavalin A (Ivens and Deitmer 1986). Thus, a loss of the Ca^{2+} channel I function results in a change in cellular behaviour.

In summary, both mechanoreceptor and voltage-sensitive ion channels are distributed in a highly organized pattern. This uncouples the ion flux during the mechanoreceptor response from having a direct effect on ciliary motor control. The signal is the *membrane potential change* which ultimately results in the change of ciliary motor output. Although the mechanisms leading from the hyperpolarizing receptor potential to an augmentation of the ciliary beating frequency is still unknown, the *reversal* of ciliary beating due to an intraciliary Ca^{2+} transient is well established.

6 Concluding Remarks

Mechanoreception is an important sensory function of ciliated protozoa, helping them to deal with their physical environment. Unlike metazoan organs for mechanoreception, which are often anatomically complex and difficult to investigate at the cellular level, ciliates, as "swimming mechanoreceptors", are comparatively simple structures. They permit experiments employing direct stimulation and electrical recording. The ciliary motor output, in addition, is a most sensitive and reliable indicator of the state of excitation of the cell. As unicellular organisms, ciliates are surrounded by the experimental solution, which appears to pose no problems for the diffusion of ions to and from the cell membrane. Furthermore, ciliates are easily cloned, mass-cultured and genetically manipulated (Kung 1979).

Mechanoelectrical transduction processes can thus be studied at the single-cell level. The ion channels involved in mechanoreception have been identified and their distribution in the somatic membrane has been grossly assayed. The channels have been characterized, although still incompletely, by their voltage sensitivity, time course and ion selectivity.

At present, we begin to understand how the sensory-electro-motor coupling may work including regional membrane specializations. In the future, even more attention will be paid to structural and functional diversities of the eucaryotic cell. The biophysical analysis of individual ion channels, both receptor and voltage-sensitive channels, will further elucidate their molecular properties. The cooperation of all these channels result in the membrane responses leading to the observed cellular behaviour. The membrane is the integrating structure of all electrical processes occurring. Biochemical studies using enriched ciliary membrane fractions may eventually be successful in characterizing the molecular structure and function of the channel proteins. The use of behavioural mutants with altered channel function effectively complements these biophysical and biochemical approaches.

References

Akoev GN (1982) The effect of Mg and Ca on the excitability of Pacinian corpuscles. Brain Res 239:39–49

Brehm P, Eckert R (1978) Calcium entry leads to inactivation of calcium channel in *Paramecium*. Science 202:1203–1206

Carbone E, Lux HD (1984) A low voltage-activated calcium conductance in embryonic chick sensory neurons. Biophys J 46:413–418

Corey DP, Hudspeth AJ (1979) Ionic basis of the receptor potential in a vertebrate hair cell. Nature (Lond) 281:675–677

Deitmer JW (1981) Voltage and time characteristics of the potassium mechanoreceptor current in the ciliate *Stylonychia*. J Comp Physiol 141:173–182

Deitmer JW (1982) The effects of tetraethylammonium and other agents on the potassium mechanoreceptor current in the ciliate *Stylonychia*. J Exp Biol 96:239–249

Deitmer JW (1983) Ca channels in the membrane of the hypotrich ciliate *Stylonychia*. In: Grinnell A, Moody WJ (eds) The physiology of excitable cells. Liss, New York, pp 51–63

Deitmer JW (1984) Evidence for two voltage-dependent calcium currents in the membrane of the ciliate *Stylonychia*. J Physiol 355:137–159
Deitmer JW (1986) Voltage dependence of two inward currents carried by calcium and barium in the ciliate *Stylonychia mytilus*. J Physiol 380:551–574
Deitmer JW (1987) Loss of electrical excitability during encystment of the hypotrichous ciliate *Stylonychia mytilus*. Naturwissenschaften 74:140–142
Deitmer JW (1988) Multiple types of calcium channels: Is their function related to their localization? In: Grinnell AD, Armstrong D, Jackson MB (eds) Calcium and ion channel modulation. Plenum Press, New York, pp 19–32
Deitmer JW, Eckert R (1985) Two components of Ca-dependent potassium current in identified neurones of *Aplysia californica*. Pflügers Arch 403:353–359
Deitmer JW, Machemer H (1982) Osmotic tolerance of Ca-dependent excitability in the marine ciliate *Paramecium calkinsi*. J Exp Biol 97:311–324
Deitmer JW, Machemer H, Martinac B (1984) Motor control of three types of ciliary organelles in the ciliate *Stylonychia*. J Comp Physiol A 154:113–120
Deitmer JW, Ivens I, Pernberg J (1986) Changes in voltage-dependent calcium currents during the cell cycle of the ciliate *Stylonychia*. Exp Cell Res 162:549–554
De Peyer JE, Deitmer JW (1980) Divalent cations as charge carriers during two functionally different membrane currents in the ciliate *Stylonychia*. J Exp Biol 88:73–89
De Peyer JE, Machemer H (1977) Membrane excitability in *Stylonychia*: properties of the two-peak regenerative Ca-response. J Comp Physiol 121:15–32
De Peyer JE, Machemer H (1978a) Hyperpolarizing and depolarizing mechanoreceptor potentials in *Stylonychia*. J Comp Physiol 127:255–266
De Peyer JE, Machemer H (1978b) Are receptor-activated ciliary motor responses mediated through voltage or current? Nature (Lond) 276:285–287
De Peyer JE, Machemer H (1982a) Electromechanical coupling of cilia. I. Effects of depolarizing voltage steps. Cell Motil 2:483–496
De Peyer JE, Machemer H (1982b) Electromechanical coupling of cilia. II. Effects of hyperpolarizing voltage steps. Cell Motil 2:497–508
Doroszewski M (1970) Responses of the ciliate *Dileptus* to mechanical stimuli. Acta Protozool 7:353–362
Dunlap K (1977) Localization of calcium channels in *Paramecium caudatum*. J Physiol 171:119
Eckert R, Chad JE (1984) Inactivation of Ca channels. Prog Biophys Mol Biol 44:215–267
Edwards C, Ottoson D, Rydqvist B, Swerup B (1981) The permeability of the transducer membrane of the crayfish stretch receptor to calcium and to other divalent cations. Neuroscience 6:1455–1460
Epstein M, Eckert R (1973) Membrane control of ciliary activity in the protozoan *Euplotes*. J Exp Biol 58:437–462
Fenchel T (1980) Suspension feeding in ciliate protozoa: functional response and particle size selection. Microbiol Ecol 6:1–11
Gage PW (1976) Generation of end-plate potentials. Physiol Rev 56:177–247
Gustin MC, Bonini MN, Nelson DL (1983) Membrane potential regulation of cAMP: control mechanism for the swimming behavior in *Paramecium*. Soc Neurosci Abstr 9:167
Hara R, Asai H (1980) Electrophysiological responses of *Didinium nasutum* to *Paramecium* capture and mechanical stimulation. Nature (Lond) 283:869–870
Hara R, Naitoh Y (1980) Electrophysiological responses of *Didinium nasutum* to mechanical and electrical stimulation. Zool Mag Tokyo 89:450
Hennessey TM, Machemer H, Nelson DL (1985) Injected cyclic AMP increases ciliary beat frequency in conjunction with membrane hyperpolarization. Eur J Cell Biol 36:153–156
Hille B (1978) Ionic channels in excitable membranes. Current problems and biophysical approaches. Biophys J 22:283–294
Ivens I, Deitmer JW (1986) Inhibition of a voltage-dependent Ca current by concanavalin A. Pflügers Arch 406:212–217
Jennings HS (1906) Behavior of the lower organisms. Columbia Univ Press, New York
Kamada T (1934) Some observations on potential differences across the ectoplasm membrane of *Paramecium*. J Exp Biol 11:94–102
Karpenko AA, Railkin AI, Seravin LN (1977) Feeding behaviour of unicellular animals. II. The role of prey mobility in the feeding behaviour of protozoa. Acta Protozool 16:333–344

Kung C (1979) Biology and genetic of *Paramecium* behaviour. In: Brakefield XO (ed) Topics in neurogenetics. Elsevier, New York, pp 1–26
Kung C, Saimi Y (1985) Ca^{2+} channels of *Paramecium*: a multidisciplinary study. Membr Transport 23:45–66
Llinas R, Yarom Y (1981) Electrophysiology of mammalian inferior olivary neurones in vitro. Different types of voltage-dependent ionic conductances. J Physiol 315:549–567
Machemer H (1974) Frequency and directional responses of cilia to membrane potential changes in *Paramecium*. J Comp Physiol 92:293–316
Machemer H (1977) Motor activity and bioelectric control of cilia. Fortschr Zool 24:195–210
Machemer H (1986) Electromotor coupling in cilia. In: Lüttgau HCh (ed) Membrane control of cellular activity. Progr Zool 33, Fischer, Stuttgart, pp 205–250
Machemer H, Deitmer JW (1985) Mechanoreception in ciliates. In: Ottoson D (ed) Progress in sensory physiology, vol 5. Springer, Berlin Heidelberg New York Tokyo, pp 81–118
Machemer H, Deitmer JW (1987) From structure to behaviour: *Stylonychia* as a model system for cellular physiology. In: Corliss JD, Patterson DJ (eds) Progress in protistology, vol 2. Biopress, Bristol, pp 213–330
Machemer H, De Peyer JE (1977) Swimming sensory cells: electrical membrane parameters, receptor properties and motor control in ciliated Protozoa. Verh Dtsch Zool Ges Erlangen 1977:86–110
Machemer H, Ogura A (1979) Ionic conductances of membranes in ciliated and deciliated *Paramecium*. J Physiol 296:49
Machemer-Röhnisch S, Machemer H (1984) Receptor current following controlled stimulation of immobile tail cilia in *Paramecium caudatum*. J Comp Physiol A 154:263–271
McClesky EM, Fox AP, Feldman D, Tsien RW (1986) Different types of calcium channels. J Exp Biol 124:177
Meech RW (1978) Calcium-dependent potassium activation in nervous tissues. Annu Rev Biophys Bioeng 7:1–18
Naitoh Y (1984) Mechanosensory transduction in protozoa. In: Colombetti G, Lenci F (eds) Membranes and sensory transduction. Plenum Press, New York, pp 113–135
Naitoh Y, Eckert R (1968) Electrical properties of *Paramecium*: modification by bound and free cations. Z Vergl Physiol 61:427–452
Naitoh Y, Eckert R (1969a) Ionic mechanisms controlling behavioral responses in *Paramecium* to mechanical stimulation. Science 164:963–965
Naitoh Y, Eckert R (1969b) Ciliary orientation: controlled by cell membrane or by intracellular fibrils? Science 166:1633–1635
Naitoh Y, Eckert R (1973) Sensory mechanisms in *Paramecium*. II. Ionic basis of the hyperpolarizing mechanoreceptor potential. J Exp Biol 54:53–65
Nakaoka Y, Machemer H (1990) Effects of cyclic nucleotides and intracellular Ca on voltage-activated ciliary beating in *Paramecium*. J Comp Physiol A 166:401–406
Ogura A, Machemer H (1980) Distribution of mechanoreceptor channels in the *Paramecium* surface membrane. J Comp Physiol 135:233–242
Ogura A, Takahashi K (1976) Artificial decilication causes loss of calcium-dependent responses in *Paramecium*. Nature (Lond) 264:170
Onimaru H, Naitoh Y, Ohki K, Nozawa Y (1979) Electrophysiological studies on the membrane of *Tetrahymena*. Dobutsugaku Zasshi (Zool Mag Tokyo) 88:529
Pape C, Machemer H (1986) Electrical properties and membrane currents in the ciliate *Didinium*. J Comp Physiol A 158:11–124
Satow Y, Murphy AD, Kung C (1983) The ionic basis of the depolarizing mechanoreceptor potential of *Paramecium tetraurelia*. J Exp Biol 103:253–264
Swerup C, Rydqvist B, Ottoson D (1983) Time characteristics and potential dependence of early and late adaptation in the crustacean stretch receptor. Acta Physiol Scand 119:91–99
Wood DC (1982) Membrane permeabilities determining resting, action and mechanoreceptor potentials in *Stentor coeruleus*. J Comp Physiol 146:537–550
Wood DC (1985) The mechanism of tubocurarine action on mechanoreceptor channels in the protozoan *Stentor coeruleus*. J Exp Biol 117:215–235

Chapter 4

Towards Molecular Mechanism of Activation in Mechanosensitive Ion Channels

M. Sokabe[1] and *F. Sachs*[2]

Contents

1 Introduction

1.1 Mechanosensors

Mechanosensing is ubiquitous in plant and animal kingdoms. Higher animals are equipped with specialized mechanosensors such as hair cells in the inner ear, stretch receptors, or touch sensors that contribute to the conscious sensations of external mechanical stimuli. There is another class of mechanosensors, including muscle spindles and tendon organs, that permits smooth and coordinated motion through

[1] Department of Physiology, Nagoya University School of Medicine, Nagoya 466, Japan
[2] Department of Biophysics, SUNY at Buffalo, Buffalo NY 14214, USA

Advances in Comparative and Environmental Physiology, Vol. 10

unconscious feedback of internal mechanical variables to the central nervous system. The most vital mechanical sensors, however, are those of the hollow viscera. These sensors measure such variables as blood pressure and filling of the bladder, the intestines, the lungs, and the heart. Some visceral sensors are connected to the nervous system and others are not. Smooth muscles and endothelial cells in blood vessels respond directly to changes in tension, thus forming one component of local regulation of blood flow (Johnson 1979; Lansman et al. 1987; Palmer et al. 1987). This last example shows that mechanosensing is not an exclusive property of the nervous system. Individual cells need to sense mechanical stimuli. Until the first report on stretch-activated (SA) channels from a nonsensory cell (skeletal muscle cells; Guharay and Sachs 1984), the role of mechanical sensing in nonspecialized cells had been overlooked. Since then, many papers have been published on mechanosensitive ion channels in nonsensory cells ranging from prokaryotes (Martinac et al. 1987), to plant cells (Falke et al. 1988), to the cells of higher animals. It is now recognized that most cells possess mechanosensitive ion channels. Though we know little about the physiological functions of these ion channels in nonsensory cells, their ubiquitous distribution suggests that they have a fundamental role in cells. They may be involved in osmoregulation (Christensen 1987; Ubl et al. 1988; Sackin 1989), control of cell-growth (Sachs 1989), and control of cell division (Medina and Bregestovski 1988).

1.2 Mechanosensitive Ion Channels Form a New Major Family

Mechanosensitive channels show a variety of conductances (20–1000 pS) and ion selectivities (generic cation, anion, or potassium selective; see recent reviews of Sachs 1988, 1989; Morris 1990). At least two classes of gating have been discovered: stretch-activated channels (SAC; Guharay and Sachs 1984) and stretch-inactivated channels (SIC; Morris and Sigurdson 1989). The activation and inactivation in most mechanosensitive ion channels is an even function of pipette pressure, i.e., these channels cannot discriminate between convex and concave membranes. However, it was very recently discovered that a new class mechanosensitive ion channels in cultured glia are activated only by positive pipette pressure corresponding to bending toward the cytoplasm (Ding et al. 1989). Six types of gating modes could be imagined; the SAC and SIC classes could have three subclasses: (1) a class sensitive to both negative and positive pressures; (2) a class sensitive only to positive pressure; and (3) a class sensitive only to negative pressure (Fig. 1). As mentioned above, three gating modes have been reported. The variety of properties (conductance, ion selectivity, gating mode) of mechanosensitive ion channels suggests the existence of a family of ionic channels, comparable to the other families of voltage-dependent and ligand-operated channels. This seems to be reasonable, since cells are exposed not only to electrical and chemical stimuli, but also to mechanical stimuli, including osmotic, gravitational, and inertial forces. Cells need these three major families of ion channels to detect external and internal signals. In the course of evolution, specialized sensory cells developed that emphasized the sensory properties of different groups of channels and accessory structures tuned to

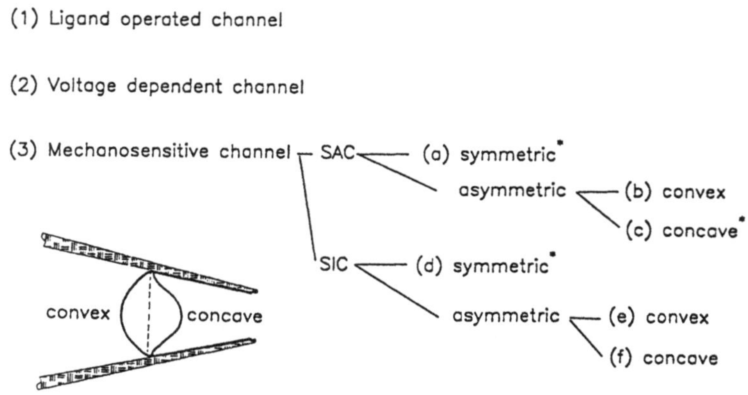

Fig. 1. Three major families of ion channels and putative six sub classes of mechanosensitive ion channels (*a-f*). * indicates the subclasses already found

efficiently collect external signals. Accessory structures include the stereociliary tufts of hair cells and the lengthy cilia of olfactory epithelium. The channels in specialized cells may, however, be the same as those in nonspecialized cells.

Since the recent progress on mechanosensitive ion channels has already been extensively reviewed (Sachs 1988, 1989; Morris 1990), we will concentrate here on the gating mechanism of these channels. We will discuss: (1) methodological difficulties in studying mechanosensitive ion channels; (2) characteristics of SA channel gating; (3) thermodynamic models for SA channel gating; and (4) measurements of tension in patch membranes.

2 Difficulties in Studying Mechanosensitive Ion Channels

2.1 Patch-Clamp Study of Mechanosensitive Ion Channels

The major biophysical interest for mechanosensitive ion channels is the mechanism by which the mechanical stimulus couples to the channel gating. Before discussing this problem, a short look at the experimental methods for studying mechanosensitive ion channels is necessary. There are basically three ways to measure channel currents: (1) classical voltage clamping (Deitmer 1981); (2) whole-cell patch-clamping (Gustin et al. 1988); and (3) single-channel patch-clamping (cf, Guharay and Sachs 1984). Each method has its own merits and demerits (Morris 1990). The first method preserves a relatively intact cell, but quantifying the mechanical stimulus and isolation of mechanosensitive currents from other currents is not simple. The second method permits us some control of the mechanical stimulus by inflating cells of known curvature (although only positive pressure can be used), but again the isolation of mechanosensitive currents from other currents is difficult. The last method, single-channel patch-clamping (cell-attached patch and excised patch), is superior to the other two methods because it can resolve single channel currents which, by their discrete nature, give us confidence that the responses arise from the mechanosensitive ion channels rather than from some diffuse leakage process. Moreover, we can control the mechanical stimulus (positive or negative pressure in the pipette) in a quantitative way. The method is poor, however, in determining the physiological role of the ion channels in cellular functions, and there is a possibility that the process of forming the patch may modify cytoskeletal or other structures attached to the plasma membrane (see Sect. 3.1).

2.2 Pressure or Tension? Theoretical Models

Most studies on the mechanosensitive ion channels have used the patch clamp, using pressure as a stimulus. Then do these channels actually sense the pressure difference across membranes? Probably not. Several lines of evidence suggest that mechanosensitive ion channels sense membrane tension rather than pressure. First, the channel activity is generally an even function of applied pressure, i.e., the most mechanosensitive ion channels can be activated both positive and negative pressures (Martinac et al. 1987; Morris 1990). Positive and negative pressures both produce positive tension. Second, the sensitivity of the channel activity to pressure is variable from patch to patch. Even for the same channel in the same patch, the sensitivity varies with time (e.g., Cooper et al. 1986). If the pressure is a direct stimulus, such variability is not expected.

According to Laplace's law (applicable to thin uniform membranes), the membrane tension is a function of transmembrane pressure and the membrane curvature. If we assume that patch membranes form a section of sphere, the tension, T, can be calculated as

$$T = PD/4, \qquad (1)$$

where P is the transmural pressure and D is the diameter of the sphere (Fig. 2). D is not constant but varies with pressure (Sokabe and Sachs 1988, 1990) and perhaps with the diameter of the pipette at the location of the patch. Moreover, as we apply suction repeatedly to stimulate the channels, the location of the patch may change (Sokabe and Sachs 1990). By measuring tension using images of the patch, we have obtained strong evidence that the direct stimulus for the channel is tension but not pressure (Sokabe et al. 1991). However, membrane tension may not be the only candidate for a direct stimulus for mechanosensitive ion channels. Tension acts tangential to membranes. It is possible that a force normal to the membrane can also modulate channel activity. The above-mentioned channel in glia that is sensitive only to positive pressure (Ding et al. 1989) suggests this possibility. The asymmetric sensitivity to pressure observed in this channel can be explained by assuming a component of the force normal to the membranes, probably via cytoskeleton. Since this type of channel is uncommon, however, we will confine our discussion to the standard stretch-activated ion channels (SAC) with symmetric sensitivity to pressure.

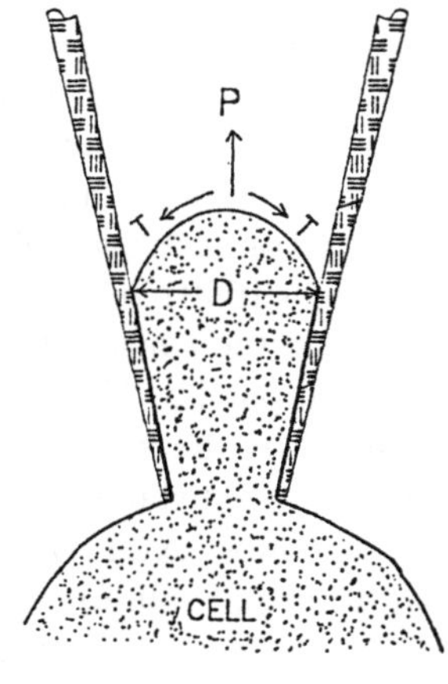

Fig. 2. Idealized image of cell-attached patch membrane. *P* Pipette pressure; *T* tension; *D* curvature of the patch

Until recently (Sokabe et al. 1991), one of the major difficulties in studying mechanosensitive ion channels was determining the magnitude of the stimulus. The absence of stimulus control has prevented us from making a quantitative investigation for the gating characteristics of the channels. A recently developed method for stimulus control (Sokabe et al. 1991) will be presented in Section 5. The choice of an appropriate physical model for gating has been confused in the literature. There have been two models, one proposing that the channel activity is an exponential function of tension squared (Guharay and Sachs 1984; Sachs 1988), and the other proposing that the activity follows an exponential function of tension (Howard and Hudspeth 1988; Howard et al. 1988). A recent analysis of the problem (Sachs and Lecar 1991) shows that we may expect the gating to be an exponential of a quadratic in tension, but that the linear terms will dominate. A simplified version of this argument will be presented in Section 4.

3 Analysis of SA Channel Gating Kinetics

3.1 Sources of Variability in Sensitivity

The activity of SA channels is generally unstable with time and is variable among patches, preventing detailed analysis of the gating kinetics. These obstacles originate from the aforementioned variability of patch geometry and possible reorganization of cytoskeletal structures during seal formation. The seal formation process is not mild and smooth as has been traditionally supposed, but rather can induce drastic changes in membrane associated structure. By using video microscopy, we observed a variety of seal formation processes (Sokabe and Sachs 1990). With gentle suction (less than 0.5 cm Hg) with small pipettes (less than 1 μm in tip diameter), the patch commonly proceeds with an expected omega shape (Sakmann and Neher 1983) that settles at around 10 μm from the pipette tip, with some variability associated with different cell types, different sized pipettes, and other unknown factors. Strong suction, the usual case when sucking by mouth, often caused rupture, vesiculation, and resealing, but in most cases creating a gigaseal. Although we do not know what happens at the molecular level during these processes, one can easily imagine that cytoskeletal structures underneath the membrane would be distorted or destroyed. We have seen evidence for these kinds of processes (Sokabe and Sachs 1990). If these structures are essential to transfer the external force to SA channels, the sensitivity of the channel may change with time and vary between different patches. The activity of membrane and cytoplasmic enzymes may also be changed by the applied forces or by calcium influx. For these reasons, it may be difficult to get stable recordings of SA channel activity.

The most extensively analyzed data has been obtained on a 35-pS SA channel (with normal saline) of cultured chick skeletal muscle cells (Guharay and Sachs 1984). Recently, a new class of cation-selective SA channel with higher voltage dependence and bigger conductance (160 pS with potassium saline) has been found in the chick muscle preparation (Hasegawa and Sokabe 1990). The gating kinetics of this channel were extensively analyzed. We will call the SA channel found by Guharay and Sachs (1984) the small SA (sSA) channel to distinguish it from the newly found large conductance SA channel (lSA channel). Results obtained from these channels will be discussed in parallel.

3.2 Pressure (Tension) Dependence

The tension dependence of SA channel activity and rate constants have traditionally been based on three assumptions. First, tension is linearly proportional to the applied pressure because the curvature of the patch has been assumed constant (Sachs 1988). Second, the increase in Helmholtz free energy (we will use Helmholtz free energy instead of Gibbs free energy since the pressure changes) of the channel gained from external force depends on the pressure squared (Guharay and Sachs 1984; Sachs 1988). Third, the SA channel molecule is a linear elastic element coupled in series with cytoskeleton that determines the area elastic properties of the membrane. Thus the rate constant, $k_{i,j}$, has the form

$$k_{i,j} = k^{o}\exp(\theta p^{2}), \tag{2}$$

where k^o is a rate constant at zero pressure and θ is a sensitivity to pressure. The first two assumptions are clearly not correct (see Sokabe and Sachs 1990; Sachs and Lecar 1991; and Sects. 4 and 5) and we have limited confidence in the last one (Sokabe et al. 1991). However, bearing in mind these limitations of the model, we can discuss the essential features of the gating mechanism of SA channels. The histograms of the open and closed times of the sSA channel were well fitted with single and triple exponential functions, respectively (Guharay and Sachs 1984). This indicates that the sSA channel has at least one open state and three closed states. Based on a simple linear kinetic model $(C_1 \underset{k_{21}}{\overset{k_{12}}{\rightleftharpoons}} C_4 \underset{k_{32}}{\overset{k_{23}}{\rightleftharpoons}} O_4 \underset{k_{43}}{\overset{k_{34}}{\rightleftharpoons}} O_4)$ six rate constants at different pressures were calculated from these data following the analysis of Colquhoun and Hawkes (1981). The traditional plot, ln $k_{i,j}$ vs. p^2, is shown in Fig. 3. The striking feature of the plot shown in Fig. 3 is that only the first forward rate constant ($k_{1,2}$, the smallest one) is pressure sensitive. A simple interpretation in terms of the rate theory is that the external force increases the free energy level of the first closed state (C_1), resulting in a decrease of the activation energy from C_1 to C_2 without affecting other parts of the energy profile of transitions. This result is independent of the assumed p^2 dependence. It is also possible that the energy profile, except for the energy minima of C_1, is lowered by the external force (Sect. 4.4).

Kinetic parameters for the lSA channel have been determined by a similar procedure (Hasegawa and Sokabe 1990). The lSA channel has one open and two closed states. As can be seen from the single channel currents in Fig. 4a, only the closed time seems to be sensitive to pressure. Calculation of the rate constants based on a linear model ($C_1 \underset{k_{21}}{\overset{k_{12}}{\rightleftarrows}} C_2 \underset{k_{32}}{\overset{k_{23}}{\rightleftarrows}} O_3$) clearly shows that, as in the sSA channel, the smallest rate constant ($k_{1,2}$) is the only one that is sensitive to pressure (Fig. 4b). The oocyte SA channel has a similar property, although if analyzed according to a linear model of ($C_1 \rightleftarrows C_2 \rightleftarrows O_3 \rightleftarrows C_4$), $k_{1,2}$ and $k_{2,3}$ are both stretch sensitive (Yang and Sachs 1990). Thus one common feature of SA channel gating is that only one or two opening rates are sensitive to external force. This is in contrast to the fact that, in most voltage-dependent channels, the electric field affects both opening and closing rates along with energy barrier between them (Ehrenstein et al. 1974; see also Sect. 3.4). The most important conclusion derived from the above results is that the physical property of the particular closed state may be quite different from that of the other states. Assuming a simple two-state model with open and closed states, the elastic constant of the closed state would be much smaller than that of the open

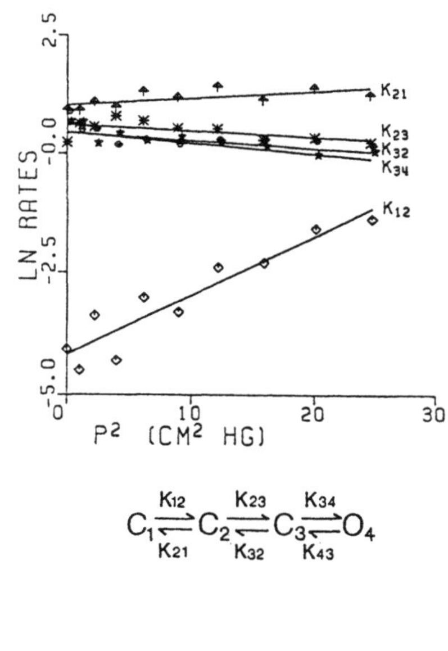

Fig. 3. Pressure dependence of the rate constants in the C_1-C_2-C_3-O_4 model in the sSA channel (Guharay and Sachs 1984). Only the smallest one (k_{12}) depends on pressure

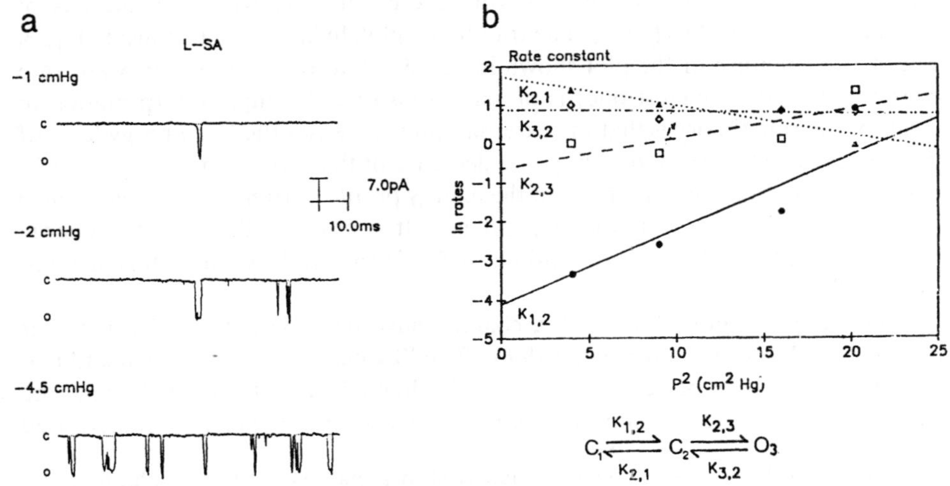

Fig. 4a. Single channel currents at different pressures observed in the lSA channel. Closed time decreases with pressure, while open time seems to be constant. *b* Dependence of the rate constants upon pressure in the C_1-C_2-O_3 model in the lSA channel. Again only the smallest rate constant ($k_{1,2}$) is pressure dependent. (Hasegawa and Sokabe, unpubl.)

state (a smaller elastic constant means more compliance). The elastic energy accumulated in the protein in the presence of tension, T, has the form

$$\Delta H = A \bullet T^2/2K_a, \tag{3}$$

where ΔH denotes Helmholtz free energy, A is the in-plane channel area and K_a is an area elastic constant (Sachs 1988). However, as will be discussed in Section 4, the physical interpretation of the kinetic results is model dependent. If we take a linear model (Howard and Hudspeth 1988), which is close to the corrected model given by Sachs and Lecar (1991), the tension dependence is interpreted in terms of the difference in the intrinsic (stress-free) size between the closed and open channel (the area of the open channel is larger than that of the closed channel). The difference in the elastic constant for the two states provides a second order correction.

3.3 Voltage Dependence

Most of mechanosensitive ion channels show little voltage dependence; only a few voltage dependent SA channels have been reported (Guharay and Sachs 1985; Methfessel et al. 1986; Martinac et al. 1987; Hasegawa and Sokabe 1990). The analysis of the voltage dependence of the gating kinetics in the sSA channel from chick skeletal muscle cells has revealed that the voltage dependence occurs in the same rate constant ($k_{1,2}$) as that responsible for the tension sensitivity (Guharay and Sachs 1985). In contrast, the voltage sensitivity of the lSA channel from the same preparation occurs in different rate constants ($k_{2,3}$ and $k_{3,2}$ in model

($C_1 \underset{\kappa_{21}}{\overset{\kappa_{12}}{\rightleftarrows}} C_2 \underset{\kappa_{32}}{\overset{\kappa_{23}}{\rightleftarrows}} O_3$) from $k_{1,2}$, the one sensitive to tension (Hasegawa and Sokabe 1990). Depolarization increases $k_{2,3}$, and decreases $k_{3,2}$ in a symmetrical way, similar to the behavior of traditional voltage-dependent channels (Ehrenstein et al. 1974). The voltage sensitivity of the sSA channel is low compared to prototypical voltage-dependent channels and is of dubious physiological value. On the other hand, the voltage sensitivity of the lSA channel is much higher, comparable to that of prototypical voltage-dependent channels, and it seems to arise in a different structure from the part that responds to tension. It can be speculated, for the lSA channel, that the pressure-dependent gate and the voltage-dependent gate are located independently in series along the ion conducting pathway.

3.4 Other Factors

There are few studies on the factors affecting SA channel gating. Calcium usually does not affect gating behavior; only in one report does a calcium-activated K channel show pressure sensitivity (Taniguchi and Guggino 1989). There is a recent report by Martinac et al. (1990) that, for the SA channels in *E. coli*, surface-active agents can modulate the midpoint of the gating curve (probability vs tension) without affecting sensitivity. They interpret their results to indicate that the agents are changing the local curvature and compressing or releasing the gating component of the channels. One result on the molecular nature of SA channels appeared in the effect of pH on the gating of the sSA channel (Guharay and Sachs 1985). Increasing extracellular pH from 7.4 to 10 increased the tension sensitivity and the voltage sensitivity in a manner suggesting titration of site(s) with a pK_a of 9.1. A single lysine or N-terminal amino acid may be responsible for modulating both the voltage and pressure responses. This is consistent with the above-mentioned result that voltage and pressure sensitivity occur in the same rate constant in the sSA channel. There are no reports of the effects of second messengers such as nucleotides, cyclic nucleotides, or phosphorylation.

4 Thermodynamic Models for SA Channel Gating

4.1 Two Contradictory Models

As mentioned above, there have been two apparently contradictory models for SA channel gating. One, called the quadratic model, (Guharay and Sachs 1984; Sachs 1988) predicts that the open probability of SA channels follows an exponential function of the square of applied tension. The other one, the linear model, predicts that the open probability follows an exponential function of applied tension (Corey and Hudspeth 1983; Howard and Hudspeth 1988; Howard et al. 1988). A close examination of the published data has shown that those data were well fitted with either model (Morris 1990). However, since no studies have been done with tension as the stimulus, only studies with pressure, adequacy of fitting is irrelevant and detailed experiments are necessary. The question of which model to use is more than an experimental issue since a physically correct model is necessary.

4.2 Quadratic Model

The model put forward by Guharay and Sachs (1984) is based on the assumption that the SA channel is an elastic molecule that can convert elastic energy to a change in state. The elastic energy in the presence of external force, T, can be obtained by an integration of the force over the area of expansion. Assuming that the expansion obeys Hook's law extended to two dimensions, the tension, T, and energy in the channel, ΔH, can be described as

$$T = Ka \bullet \Delta A/A, \tag{4}$$

$$\Delta H = A\int T_d\,(DA/A) = AT^2/2Ka, \tag{5}$$

where $\Delta A/A$ is a fractional change in the cross sectional area of the channel molecule at a given tension, T, and Ka is an area elastic constant of the molecule (recall that $T^2/2Ka$ is an energy density and needs to be multiplied by the area to obtain an energy). As the probability channel of being open (P_{op}) would obey Boltzmann's law,

$$P_{op} = P_{max}/\,[1 + K_e \exp(-\,e\,A \bullet \Delta H/k_B T^o)], \tag{6}$$

where e (< 1) denotes the efficacy of the energy conversion from the elastic to gating, K_e is an equilibrium constant controlling the tension-independent probability, P_{max} is the maximum channel activity, and $k_B T^o$ is Boltzmann's constant times the absolute temperature.

The above analysis, although apparently reasonable, does not describe a channel whose open state is of a different size from the closed state! It describes a channel whose stress-free closed state is the same size as the stress-free open state. Force acts to change the elasticity of the states. It has the unusual property that the channel opens whether you stretch it or compress it (since the energy depends upon the square of the force only). Equation (5) above, however, does represent a conceivable, although bizarre, system. Although this model might be physically realizable, it is no longer a barrier model, but a continuum model. The energy-vs-force curve for the closed state is a parabola. The respective curve for the open state is another parabola, centered with the closed-state curve (equal-size states) but located higher since the channel is normally closed. For the curves to intersect, i.e. for the open state to become more probable, the open state must be less elastic than the closed state: the parabola must be flatter. There is no activation barrier; the transition rates are predicted to become infinitely fast where the curves cross (near where the channels are open half the time). This prediction is not found to be true, so the pure quadratic model must be abandoned. However, the method of calculating the elastic energy based on Hook's law is usable when constructing a correct model (see Sect. 4.4).

4.3 Linear Model

The original linear model (Corey and Hudspeth 1983) was proposed to explain the gating property of mechanical transduction in hair cells. It was assumed that the gate was connected to a "gating spring" [possibly representing the tip link (Pickles et al. 1984) between stereocilia] that acquires elastic energy from the external force (See Fig. 5). The gate fluctuates a distance (d) between open and closed, and the gating spring (spring constant, K_g) is stretched by the additional distance d in the closed state compared to the open state.

Implicit in the analysis is the assumption that the applied force cannot distort the channel; each state is infinitely stiff; and that the elasticity of the gating spring is the same in both states. The free energy of the channel including the spring and the gate in each state (H_o, H_c) is given by

$$H_o = h_o + 1/2\, K_g\, (x - d)^2\,, \tag{7}$$

$$H_c = h_c + 1/2\, K_g\, x^2\,, \tag{8}$$

where h_o and h_c represent the strain-free energy of the open and closed channel, respectively and x is the extension of the spring by external force. The energy difference $\Delta H_{(op\text{-}c)}$ between the open and closed states can be written in the form,

$$\Delta H_{(op\text{-}c)} = H_o - H_c = (h_o - h_c) - K_g xd - d^2/2K_g = \Delta H^o - K_g xd\,. \tag{9}$$

ΔH^o is a constant (an energy difference between the states under no external force) and the energy difference depends linearly on x. In terms of force (F), instead of distance, Eq. (9) can be written,

$$\Delta H_{(op\text{-}c)} = \Delta H^o - Fd. \tag{10}$$

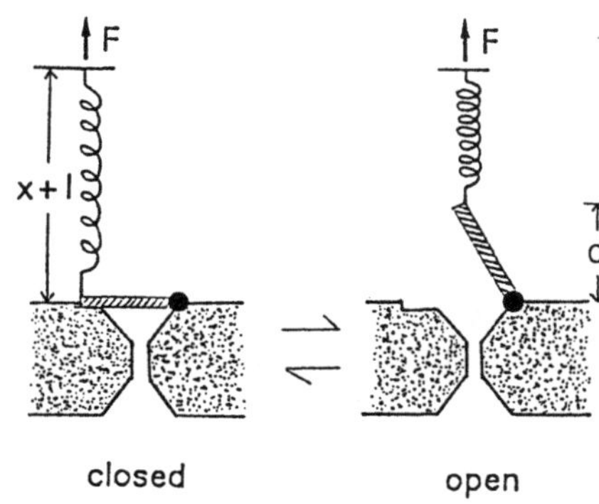

Fig. 5. A model for the transduction apparatus of the mechanosensitive ion channel in hair cells. When the channel is closed, and force *F* is applied, the gating spring is distended by x beyond its natural length (*l*), where x = F/K_g. Channel opening shortens the stretched spring by a distance *d*. (After Howard and Hudspeth 1988)

Substitution of F and d in Eq. (10) with T (tension) and area (A) yields a two dimensional version of Eq. (10) as follows,

$$\Delta H_{(op\text{-}c)} = \Delta H^{o} - T \cdot \Delta A_{(op\text{-}c)i}, \quad (11)$$

where $\Delta A_{(op\text{-}c)i}$ is an intrinsic (stress-free) difference in the cross-sectional area between the open and closed channels. The tension dependent energy term is equal to the work done on the displaced membrane when the state changes under a tension, T. This equation can explain the experimental observation that the open probability of the channel increases with external force, but does not explain the fact that, in the chick muscle SA channel, the opening rate is stretch sensitive while the return rate is not. The assumption of infinite stiffness of the channel is surely not true, but may be a reasonable approximation. The "gating spring" could be interpreted as the elasticity of the structures (cytoskeleton?) in series with the channel or the elasticity of the channel itself, assuming equal elasticity of both states.

4.4 Unified Model

Now we can clearly recognize the fundamental differences between the two models. Both the pure quadratic and the pure linear are wrong, but obviously contain elements important in a proper analysis. Sachs and Lecar (1991) have reanalyzed the problem in terms of a deformable two-state channel and showed that the system has both quadratic and linear terms. Figure 6 shows a graphical representation of the difference between the full quadratic (first and second order) and the linear model. The free energy difference between the closed and open states can be given by

$$\Delta H_{(op\text{-}c)} = \Delta H^{o} + \Delta H_{(op\text{-}c)e} + \Delta H_{(op\text{-}c)s}, \quad (12)$$

where ΔH^{o} is a stress-free energy term, $\Delta H_{(op\text{-}c)e}$ is an elastic energy term, and $\Delta H_{(op\text{-}c)s}$ is the size dependent energy term that was introduced in the linear model. The elastic energy term (introduced in the quadratic model) reflects the internal strains in the open and closed states under the applied force. In terms of the tension/area model, for small strains in which $\Delta A \ll A$,

$$\begin{aligned}\Delta H_{(op\text{-}c)e} &= A_{opi}T^2/2K_{op} - A_{ci}T^2/2K_c \\ &= T^2/2(A_{opi}/K_{op} - A_{ci}/K_c), \end{aligned} \quad (13)$$

where A_{opi} and A_{ci} are intrinsic (stress-free) areas of the open and closed channels, respectively. If the open and closed states are similar in size and equally elastic, then this term is zero. The size-dependent term is composed of two terms, the energy difference arising from the intrinsic (stress-free) size difference, $\Delta A_{(op\text{-}c)i}$ which has been introduced in Eq. (11), and that arising from the elastic size difference, $\Delta A_{(op\text{-}c)e}$. The elastic size difference is easy to calculate from Eq. (4) as follows,

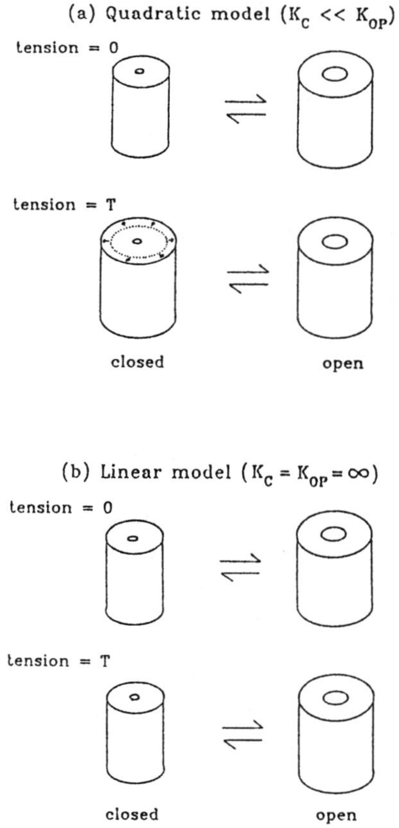

Fig. 6a,b. Schematic diagram of the change in the cross-sectional area of SA channel in two different models. Both models have an increase in intrinsic size associated with opening, but the full quadratic (unified) model (*a*) permits the closed channel to be elastically deformed without suffering a change in state. The linear model (*b*) treats each state as infinitely stiff

$$\Delta A_{(op\text{-}c)e} = A_{opi}T/K_{op} - A_{ci}T/K_c. \tag{14}$$

As the energy arising from this term is simply $T\Delta A_{(op\text{-}c)e}$, the size dependent term becomes

$$\Delta H_{(op\text{-}c)s} = -T(\Delta A_{(op\text{-}c)e} + \Delta A_{(op\text{-}c)i}). \tag{15}$$

Thus the unified equation is

$$\Delta H_{(op\text{-}c)} = \Delta H^o - T^2/2(A_{opi}/K_{op} - A_{ci}/K_c) - T\Delta A_{(op\text{-}c)i}\,. \tag{16}$$

ΔH^o again represents the force-independent structural energy difference between the open and closed channels. It would appear that if the term, $\Delta A_{(op\text{-}c)i}$, tends to zero, we can get a pure quadratic model. However, the parameters cannot be chosen arbitrarily and still preserve a barrier model, as explained above in the section on the pure quadratic model. The barrier will disappear if the quadratic terms is larger than the linear term. If $A_{opi}/K_{op} = A_{ci}/K_c$ or K_{op} and K_c are infinite, however, we can get a purely linear model for the probability of the channel being open, but *not* for the individual rate constants since they do not contain the difference term of Eq. (13).

There are two characteristics of SA channels that constrain the model: the channels are mostly closed without external force and the closing rates are independent of external force. Now let us take a look at what would happen when we introduce these experimental constraints. $\Delta H^o > 0$ to preserve the closed state occupancy in the absence of force. To preserve the asymmetry in the force dependence of the opening and closing rates, the barrier must be close to the open state. For this to be true, the energy well of the open state must be narrower than the energy well of the closed state and hence $K_{op} \gg K_c$. The second term of Eq. (16) is reduced to $A_{ci} \cdot T^2/2K_c$. Since the linear term, $T \cdot \Delta A_{(op\text{-}c)i}$ in Eq. (16) must dominate ($T \cdot \Delta A_{(op\text{-}c)i} > A_{ci}T^2/2K_c$) to suffice the observed tension dependence, this can be restated as,

$$\Delta A_{(op\text{-}c)i} > A_{ci}\, T/2K_c. \tag{17}$$

This says that the difference in the intrinsic area is larger than half of the elastic change in area of the closed state. Under these constraints the energy minima of the open state would show a larger decrease with tension than that of the closed state. Only precise experiments on SA channel kinetics will permit an estimate of the relative magnitude of the linear and quadratic terms. For a deeper understanding we will also need to know how the external force (membrane tension) is conducted to the channels. It has been suggested that force is focussed on SA channel proteins via cytoskeletal strands (see Fig. 13, Sachs 1988). Although the prediction (Guharay and Sachs 1984) was based on a flawed analysis of the energies involved, the conclusion seems to be correct (Sokabe et al. 1991).

5 Measurement of Membrane Tension

5.1 Imaging Patches

It is possible to determine the membrane tension by measuring the patch curvature and the transmural pressure according to Laplace's law [Eq. (1)]. We visualized patch membranes under a controlled pressure while recording SA channel activity (Sokabe and Sachs 1988; Sokabe et al. 1991). To obtain clear images of the patch, we used high magnification, differential interference contrast microscopy with additional video magnification. The total magnification could be increased to 15 nm/pixel. The pipettes were made from thin-wall borosilicate glass, and the tips were bent to be nearly parallel to the stage (Sokabe and Sachs 1990). Pipette pressure was applied with an oil-filled servo which had a rise time of a few milliseconds and a dynamic range of ±200 mmHg. We made patches only from the side of cells to eliminate interference arising from the underlying image of the cell itself. Under gentle suction (0.5–1 cmHg; 1 cmHg = 1.33 kPa), a bleb of cell was drawn into the pipette forming a gigaseal when the patch was 5–25 μm from the tip. Typical patch images obtained with different pressures are shown in Fig. 7. One striking feature of these images is that the radius of curvature is pressure dependent. This finding contradicts the common assumption of a constant radius of curvature (Sachs 1988) and shows that membrane tension is not simply proportional to the applied pressure. In most cases, the curvature of the patch surface was well fitted with a circular arc. Thus the tension, T, of the patch membrane was calculated from Laplace's law [Eq. (1)]. To obtain the radius of curvature and the patch area, we digitized the image and then used a nonlinear regression routine to fit the data to a geometrical model of the patch. The model consisted of two straight lines describing the inner walls of the pipette and a circular arc describing the patch (Fig. 8c). The intensity distribution about the lines was modeled with a biphasic cubic polynomial and the intensity distribution about the arc as a polynomial approximation to a Gaussian (see the original intensity distribution in Fig. 8a). Patch area was computed from the dimensions of the arc assuming rotational symmetry. We calibrated the video system in X and Y using a stage micrometer. Errors due to the cylindrical lens effect were corrected in two ways. By using a ray tracing program to model the effect of geometric distortion, we found that the geometric error was less than a few percent over most of the patch image. Then, to locate where the pipette walls were physically located relative to the intensity distribution of the inner wall, we deposited calibrated latex beads with different sizes (3–7 μm in diameter) in the pipette (Fig. 8b) and measured their diameter with a modified version of the fitting program, which just fit the intensity distribution of the pipette walls. The physical diameter of the pipette was a few percent wider than the distance between the intensity minima located over the inner walls.

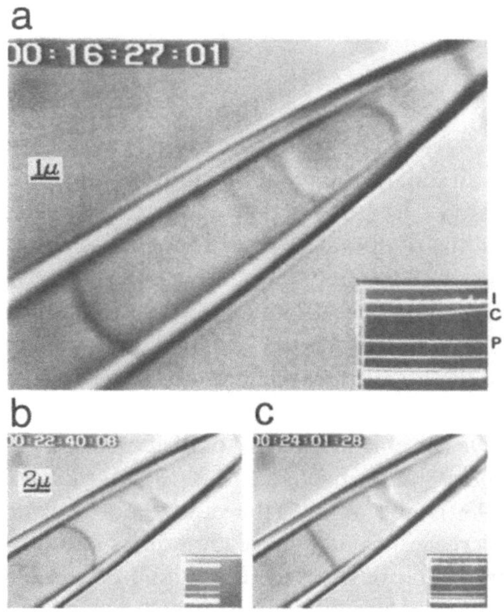

Fig. 7a-c. Typical images of a cell attached patch with different pressures. *a* –1 cmHg; *b* +1 cmHg; *c* 0 cmHg. The video time code and analog data (*I* current; *C* capacitance; *P* pressure; etc.) are superimposed on the *upper left* and *lower right*, respectively. In *a*, an SA channel opening, and the increase in capacitance associated with suction are visible. Patches were made from cultured chick skeletal muscle with normal saline in the pipette

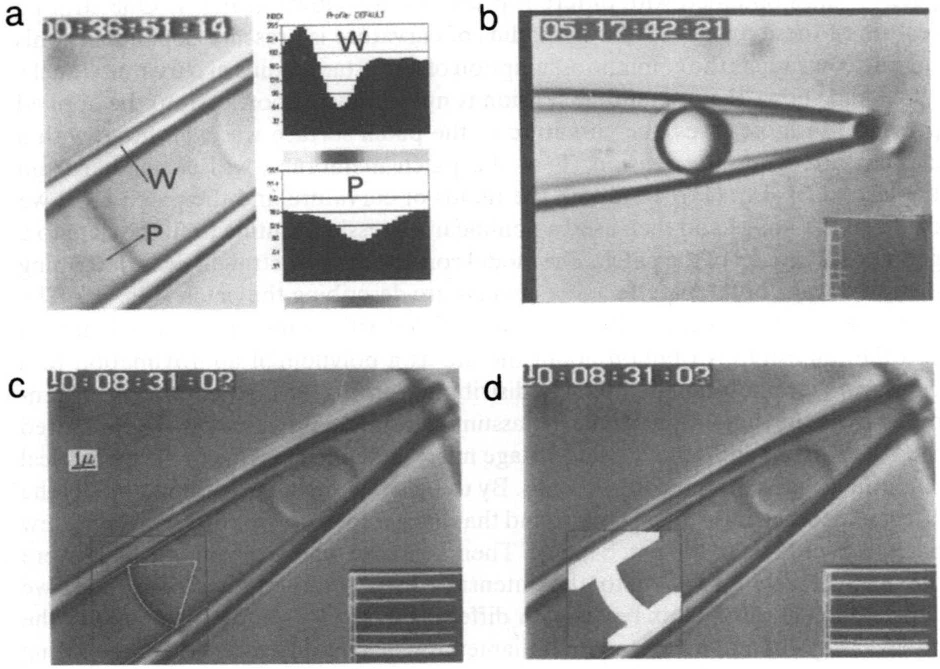

Fig. 8a-d. Calibration of images and data extraction procedures. *a* Brightness distribution across the patch (*P* Gaussian) and pipette wall (*W* biphasic). *b* An image of a calibrated latex bead (3 μm in diameter) deposited in the pipette, that was used to determine the physical interfaces of the inner walls of the pipette. *c* With a cursor, the region of interest was defined by placing two points on each inner wall of the pipette and three points on the arc of the patch. The computed straight lines and arc were superimposed on the image (*white lines*). *d* A band 10 to 15 pixels wide (12 in this case) was marked around the *white lines* in *c* forming the region of interest as shown in the panel. These data were used as input to the curve-fitting routine

As can be seen from Fig. 7, the radius of curvature of the patch is much bigger than that of the pipette, contradicting the traditional assumption that the radius of patch curvature is that of the pipette (Sachs 1988). In most cases, the minimum radius of curvature was about twice of that of the pipette over the experimental pressure range (0–50 mmHg). We did not analyze data in which the patch location changed. These data also permitted a measurement of the elastic constant of the patch membrane [see Eq. (4)]. Figure 9 shows a plot of fractional area change of a patch membrane against tension. The plot is roughly linear suggesting that patch membranes follow Hook's law extended to two dimensions. The slope of this plot gives an elastic constant of the patch membranes as about 50 dyn/cm, much less than the value expected for phospholipid/cholesterol membranes (Macintosh et al. 1989), but similar to the value reported for cultured neurons (Horie et al. 1990). The increase in patch area sometimes exceeds 10%, much larger than the 2–3% lytic limit for bilayer lysis (Kwok and Evans 1981; Wolfe et al. 1986), suggesting that we were not stressing lipids by distending the patch (see Sect. 5.3).

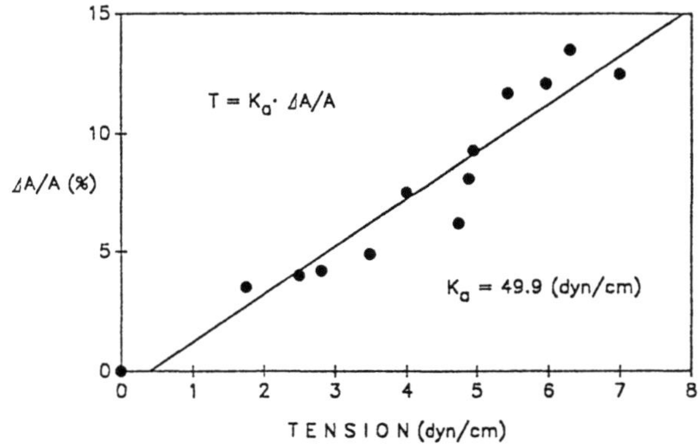

Fig. 9. Percent area change (ΔA/A) of a patch membrane as a function of tension. The plot is roughly linear, suggesting that the patch membrane follows Hook's law within the experimental pressure range (0–75 mmHg in this case). The elastic constant obtained from the slope was 50 dyn/cm. Tension was calculated by Laplace's law [Eq. (1)]

5.2 Channel Activity vs Tension

The probability of the SA channel being open, P_o, was calculated from the amplitude histogram of the channel currents by using the equation,

$$P_o = 1 - P_c^{(1/N)}, \quad (18)$$

where P_c is the fraction of area under the closed channel peak and N is the number of channels in the patch. N was estimated from the maximum observed level with the highest level of stimulation and may be underestimated. Figure 10 shows the dose response curves for SA channels from two patches with different diameters.

The filled points and the associated solid lines (linear least squares regression) represent activity as a function of membrane tension. The open symbols and associated dashed lines represent the activity as a function of applied pressure. The sensitivity of channels from different patches is nearly the same when the independent variable is tension, but not when the independent variable is pressure. This result strongly suggests that much of the variability found between different patches may represent uncertainty in the actual stimulus, and that the actual stimulus for SA channels is tension, not pressure. Although the data on P_o are plotted to emphasize a quadratic dependence on tension, the data is insufficient to comprise a test of the functional form represented by Eq. (16). More extensive measurements are in progress.

5.3 Measurement of Patch Capacitance

The increase in patch area with pressure could arise from three sources: stretching of a fixed amount of membranes (Fig. 11a), influx of new membrane (Fig. 11b), or unfolding of unresolved membrane pleats (Fig. 11c). We tested these alternatives by measuring the capacitance during changes in patch area. Capacitance was measured at 10 kHz using a phaselock amplifier, basically according to the method of Neher and Marty (1982). The patch capacitance increased following an increase in pressure and was correlated with the increase in patch area (Fig. 12). The specific

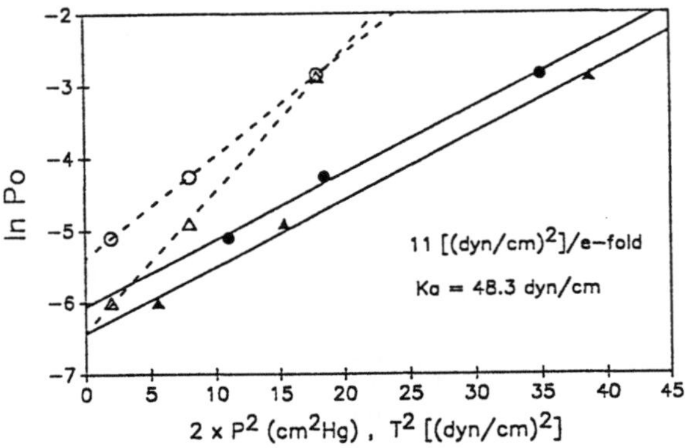

Fig. 10. Open probability of the sSA channel as a function of pressure squared (*open symbols*) and tension squared (*closed symbols*) from two patches with different diameter, 4.2 μm (*circles*) and 5.2 μm (*triangles*). Note that as a function of tension, the sensitivity of two patches is nearly identical. As a function of applied pressure, the sensitivities vary. The elastic constant of the two patches was similar (ca. 50 dyn/cm). (Sokabe et al. 1991)

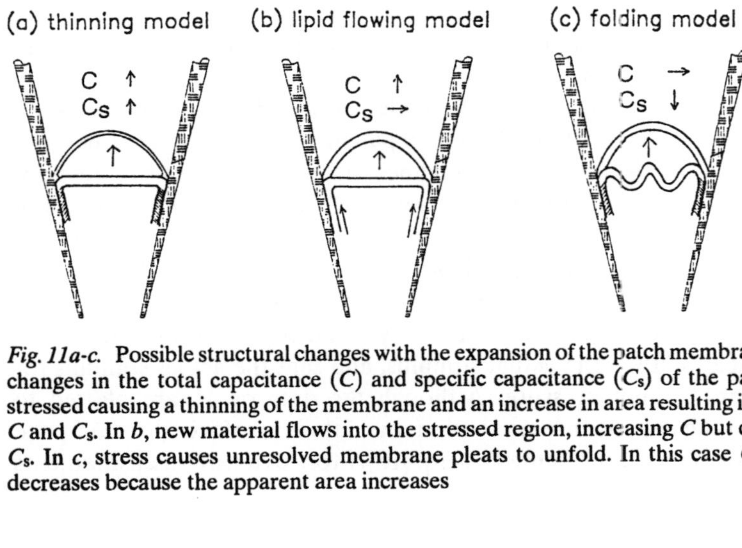

Fig. 11a-c. Possible structural changes with the expansion of the patch membrane and the expected changes in the total capacitance (C) and specific capacitance (C_s) of the patch. In *a*, lipids are stressed causing a thinning of the membrane and an increase in area resulting in an increase in both C and C_s. In *b*, new material flows into the stressed region, increasing C but causing no change in C_s. In *c*, stress causes unresolved membrane pleats to unfold. In this case C is constant but C_s decreases because the apparent area increases

capacitance (capacitance per unit area) was calculated by plotting the change in capacitance against the change in patch area. There was a linear relationship between area and capacitance irrespective of magnitude of the change or time following pressure. The slope was 0.7 μF/cm² (Sokabe et al. 1991). This result precludes the possibility of unfolding and stretching membranes. Thus the increase in patch area comes from an influx of new lipids (with perhaps some diffusible proteins) from stores along the walls of the pipette. This is consistent with the result that the increase in the patch area is far in excess (> 10%) of the strain which leads to lysis of lipid membranes. The data shown here provide strong support for the coupling

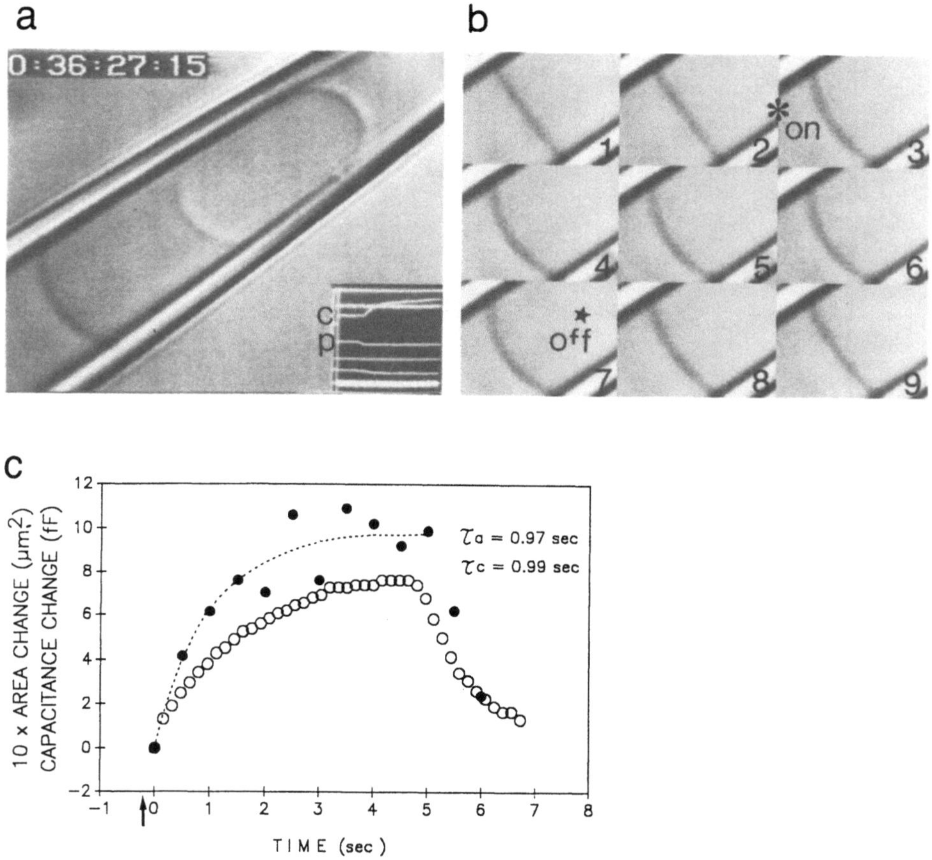

Fig. 12a-c. Simultaneous measurement of changes in patch area and capacitance. *a* A patch image taken 3.5 s after the onset of a pressure pulse. In *lower right* of the panel, the time course of capacitance (*C*) and pressure (*P*) change can be seen. *b* A time series (1 s/frame) showing the shape change of the patch following a pressure pulse (–1.44 cmHg) which started at the second frame (*) and stopped at the seventh frame. *c* Time course of the change in area (*closed circles*) and capacitance (*open circles*) of a patch membrane in response to a negative pressure pulse (–1.44 cmHg, 5-s duration) applied at –0.2 s (*arrow*). After 200 ms delay, both the area and capacitance start to increase following a single exponential (indicated by *dashed line* for area change) with similar time constants (τ_c, τ_a). Specific capacitance, C_s, was calculated from the steady state values. (Sokabe et al. 1991)

of SA channels to the cytoskeleton. Since there is little strain in the membrane lipids, then the SA channels must attach to some other tension bearing element. Since neither tubulin nor actin reagents block SA channel activation (Guharay and Sachs 1984; Hasegawa and Sokabe, unpubl.), we suspect that the spectrin/dystrophin submembrane cytoskeleton is the tension bearing element (Fig. 13).

The data also provide some information on the nature of the gigaseal. The casual flow of lipids from the membrane attached to the walls of the pipette eliminates models of seal formation that involves strong coupling of lipids to the glass

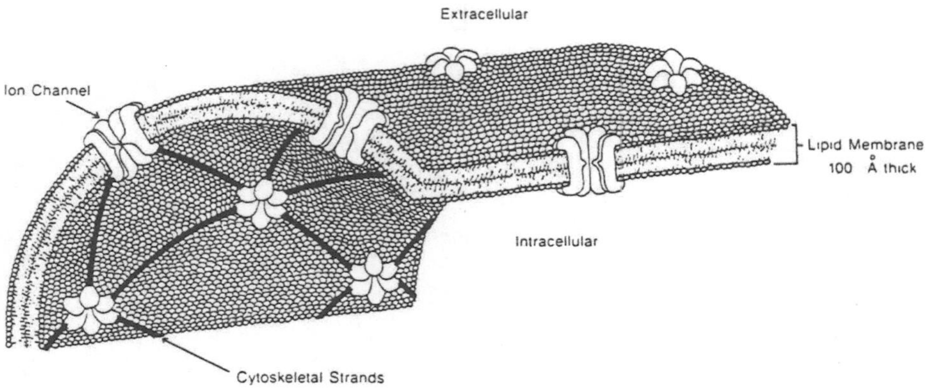

Fig. 13. Cartoon of the cell membrane showing SA channels and cytoskeletal strands in series with the channel. (Courtesy of Alan Kegler, UB Public Relations Office; Sachs 1987)

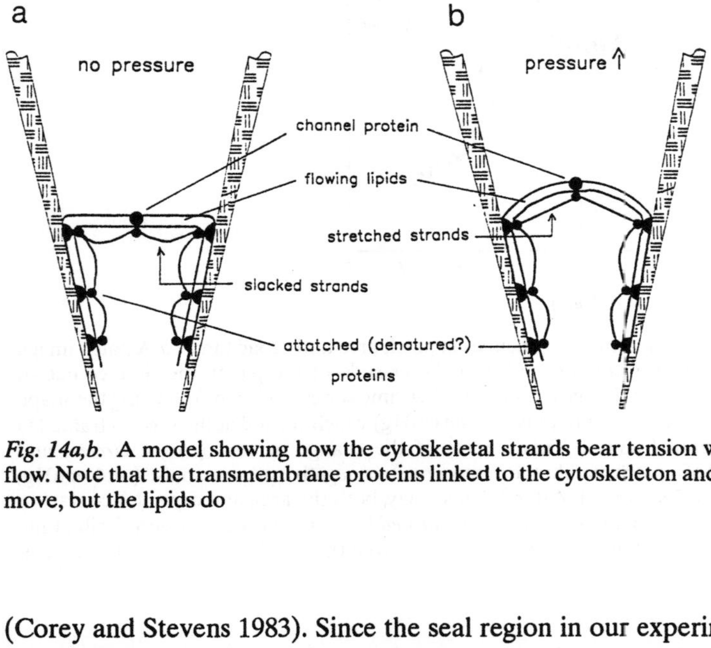

Fig. 14a,b. A model showing how the cytoskeletal strands bear tension with suction while lipids flow. Note that the transmembrane proteins linked to the cytoskeleton and the pipette wall do not move, but the lipids do

(Corey and Stevens 1983). Since the seal region in our experiments did not move and the lipids did, we must assume that the attachment to the glass involves membrane proteins. It is not hard to imagine that membrane proteins are denatured against the high energy surface of the glass, thereby anchoring the membrane structure to the pipette (Fig. 14). The lipid can then flow around these anchoring proteins like a river flowing around bridge piers. The high resistance of the seal can be accounted for if the space between the membrane and the glass is filled, like a sucrose gap, with the sugar residues of surface glycoproteins. Since gigaseals can be formed from pure lipid membranes, is there a contradiction? Clearly, no: equality of resistance does not imply equality of the conduction mechanism. In the patches obtained by the tip-dip method and by patch-clamping giant liposomes, the path length between the patch and the pipette tip may be hundreds of microns. When the membrane contains proteins, gigaseals with shorter seal length can be obtained (Sokabe, unpubl.). Membrane proteins seem to help the membrane to make a tight contact with glass walls, accelerating the formation of gigaseals with short path lengths.

6 Conclusions and Future Perspectives

Simultaneous recordings of patch images and SA channel currents permit us to calculate membrane tension, and to ascertain a reliable relationship between tension and channel activity. Coupled with the newly proposed unified thermodynamic model [Eq. (16); Sachs and Lecar 1991], we will be able to estimate thermodynamic and physical parameters of SA channel molecules and their dynamics. Measurements of patch capacitance suggested that SA channel activation couples with membrane structures other than lipids. Therefore, the mechanical property not only of SA channels, but also of membranes including extracellular matrix and cytoskeletal structures, would deeply affect the activation of SA channels. Patch images provide information on the mechanical property (elastic constant) of these structures by which we can distinguish the mechanical property of SA channels from that of membranes. Thus the interpretation of pharmacology would be greatly simplified. The work on patch images, however, provided a warning that patch-clamped membranes may become disrupted affecting the activation of SA channels (Sect. 3.1; Sokabe and Sachs 1990). We should carry out whole-cell recordings that can back up the predictions of the single-channel recordings. Reconstitution of SA channels into bilayers is another area of active investigation. By employing different degree of reconstitution, fragmented membranes, solubilized proteins, or purified proteins, we could judge which membrane component is essential for SA channel activation. The determination of the molecular structure of SA channels still seems far off. We have not yet found a specific ligand for SA channels. Screening mutants lacking SA channels may be a promising way to get the molecular structure of SA channels. Genetic studies in *E. coli* (Martinac et al. 1987), yeast (Gustin et al. 1988), and *Drosophila* (Zaggota et al. 1988) should be pursued. A plant mutant lacking geotropism may be a good tool for genetic experiments. Gadolinium, a potent blocker of certain SA channels (Gustin et al. 1988; Yang and Sachs 1989), blocks geotropism in these plants (Edwards and Pickard 1987).

Acknowledgments. This study was supported by the Japanese Ministry of Education, Science and Culture (0157004, 01880028, 02241206, 0304407), the Terumo Science Foundation, NIDDK DK-37792, the Muscular Dystrophy Association, and the USARO, 22560-LS.

References

Christensen O (1987) Mediation of cell volume regulation by Ca^{2+} influx through stretch activated channels. Nature (Lond) 330:66–68
Colquhoun D, Hawkes AG (1981) On the stochastic properties of single ion channels. Proc R Soc Lond, Ser B 211:205–235
Cooper KE, Tang JM, Rae JL, Eisenberg RS (1986) A cation channel in frog lens epithelia responsive to pressure and calcium. J Membr Biol 93:259–269
Corey DP, Hudspeth AJ (1983) Kinetics of the receptor currents in bullfrog saccular hair cells. J Neurosci 3:962–976
Corey DP, Stevens CF (1983) Science and technology of patch recording electrodes. In: Sakmann B, Neher E (eds) Single channel recording. Plenum Press, New York, pp 53–68
Deitmer JW (1981) Voltage and time characteristics of the potassium mechanoreceptor current in the ciliate *Stylonychia*. J Comp Physiol 141:173–182
Ding JP, Bowman CL, Sokabe M, Sachs F (1989) Mechanical transduction in glial cells. Biophys J 55:224a
Edwards KL, Pickard BG (1987) Detection and transduction of physical stimuli in plant. In: Wagner E, Greppin H, Millet B (eds) The cell surface in signal transduction. Springer, Berlin Heidelberg New York Tokyo, pp 41–66
Ehrenstein GR, Blumenthal R, Lattore R, Lecar H (1974) Kinetics of the opening and closing of individual excitability inducing material channels in lipid bilayer. J Gen Physiol 63:707–721
Falke LC, Misler S (1989) Activity of ion channels during volume regulation by clonal NIE115 neuroblastoma cells. Proc Natl Acad Sci USA 86:3919–3923
Falke LC, Edwards KL, Pickard BG (1988) A stretch-activated anion channel in tobacco protoplasts. FEBS Lett 237:141–144
Guharay F, Sachs F (1984) Stretch-activated single ion channel currents in tissue-cultured embryonic chick skeletal muscle. J Physiol (Lond) 352:685–701
Guharay F, Sachs F (1985) Mechanotransducer ion channels in chick skeletal muscle: The effect of extracellular pH. J Physiol (Lond) 353:119–134
Gustin MC, Zhou XL, Martinac B, Kuhn C (1988) A mechanosensitive ion channel in the yeast plasma membrane. Science 242:762–766
Hasegawa N, Sokabe M (1990) Analysis of stretch-activated channels in chick skeletal muscle. Jpn J Physiol 40 (Suppl):s237
Horie H, Ikuta S, Takenaka T (1990) Membrane elasticity of mouse dorsal root ganglion neurons decreases with aging. FEBS Lett 269:23–25
Howard J, Hudspeth AJ (1988) Compliance of the hair bundle associated with gating of mechanoelectrical transduction channels in the bullfrog saccular hair cells. Neuron 1:189–199
Howard J, Roberts WM, Hudspeth AJ (1988) Mechanoelectrical transduction by hair cells. Ann Rev Biophys Biophys Chem 17:99–124
Johnson PC (1979) The myogenic response. In: Berne RM, Sperelakis N, Geiger SR (eds) Handbook of physiology; the cardiovascular system. Am Physiol Soc, Bethesda, Maryland, pp 409–442
Kwok R, Evans E (1981) Thermoelasticity of large lecithin bilayer vesicles. Biophys J 35:637–652
Lansman JB, Hallam J, Rink TJ (1987) Single stretch activated ion channels in vascular endothelial cells as mechanotransducers? Nature (Lond) 325:811–813
Macintosh T, Simon SA, Dilger J (1989) In: Benghe G (ed) From model membranes to isolated cells: location of the water-hydrocarbon interface in lipid bilayers. CRC, Boca Raton, pp 1–13
Martinac B, Buechner M, Delcour AH, Adler J, Kuhn C (1987) Pressure sensitive ion channel in *Escherichia coli*. Proc Natl Acad Sci USA 84:1–5
Martinac B, Adler J, Kung C (1990) Mechanosensitive ion channels of *E. coli* activated by amphipaths. Nature (Lond) 348:261–263
Medina IR, Bregestovski PD (1988) Stretch-activated ion channels modulate the resting membrane potential during early embryogenesis. Proc R Soc Lond Ser B 235:95–102
Methfessel C, Witzemann V, Takahashi T, Mishina M, Numa S, Sakmann B (1986) Patch clamp measurements on *Xenopus leavis* oocyte: currents through endogenous channels and implanted acethylcholine receptor and sodium channels. Pflügers Arch Eur J Physiol 407:577–588

Morris CE (1990) Mechanosensitive ion channels. J Membr Biol 113:93–107
Morris CE, Sigurdson WJ (1989) Stretch-inactivated ion channels coexist with stretch activated ion channels. Science 243:807–809
Neher E, Marty A (1982) Discrete changes of cell membrane capacitance observed under conditions of enhanced secretion in bovine adrenal chromaffin cells. Proc Natl Acad Sci USA 79:6712–6716
Palmer RMJ, Ferrige AG, Moncada S (1987) Nitric oxide release accounts for the biological activity of endothelium-derived relaxing factor. Nature (Lond) 327:524–526
Pickles JO, Comis SD, Osborne MP (1984) Cross-links between stereocilia in the guinea pig organ of Corti, and their possible relation to sensory transduction. Hearing Res 15:103–112
Sachs F (1987) Baroreceptor mechanisms at the cellular level. Fed Proc 46:12–16
Sachs F (1988) Mechanical transduction in biological systems. Crit Rev Biomed Eng 16:141–169
Sachs F (1989) Ion channels as mechanical transducers. In: Stein WD, Bronner F (eds) Cell shape: determinants, regulation and regulatory role. Academic Press, San Diego, pp 63–92
Sachs F, Lecar H (1991) Stochastic models for mechanical transduction. Biophys J 59:1143–1145
Sackin H (1989) A stretch-activated K^+ channel sensitive to cell volume. Proc Natl Acad Sci USA 86:1731–1735
Sakmann B, Neher E (1983) Geometric parameters of pipettes and membrane patches. In: Sakmann B, Neher E (eds) Single-channel recording. Plenum Press, New York, pp 37–51
Sokabe M, Sachs F (1988) Imaging patches and simultaneous recording of stretch activated channel activity. Proc Int Congr Comp Physiol Biochem 815
Sokabe M, Sachs F (1990) The structure and dynamics of patch-clamped membranes: a study using differential interference contrast light microscopy. J Cell Biol 111:599–606
Sokabe M, Sachs F, Jing Z-G (1991) Quantitative videomicroscopy of patch clamped membranes: stress, strain, capacitance and stretch channel activation. Biophys J 59:722–728
Taniguchi J, Guggino WB (1989) Membrane stretch: a physiological stimulator of Ca^{2+}-activated K^+ channels in thick ascending limb. Am J Physiol 257:F347–F352
Ubl J, Murer H, Kolb H-A (1988) Ion channels activated by osmotic and mechanical stress in membranes of opossum kidney cells. J Membr Biol 104:223–232
Wolfe J, Dowgert MF, Steponkus PL (1986) Mechanical study of the deformation and rupture of the plasma membranes of protoplasts during osmotic expansions. J Membr Biol 93:63–74
Yang X-C, Sachs F (1989) Block of stretch-activated ion channels in *Xenopus* oocytes by gadolinium and calcium ions. Science 243:1068–1071
Yang X-C, Sachs F (1990) Characterization of stretch-activated ion channels in *Xenopus* oocytes. J Physiol (Lond) 431:103–122
Zaggota WN, Brainard MS, Aldrich RW (1988) Single-channel analysis of four distinct classes of potassium channels in *Drosophila* muscle. J Neurosci 8:4765–4779

Part II

Mechanoreceptor Organs

Chapter 5

Gravity Sensing by Higher Plants[1]

B.G. Pickard and *J. Ping Ding*[2]

Contents

[1] This chapter is dedicated to the memory of ion channel enthusiast Lee C. Falke, whose untimely death has grieved many.

[2] Biology Department, Washington University, Saint Louis, MO 63130, USA

Advances in Comparative and Environmental Physiology, Vol. 10

1 Introduction

The basic assumption of this chapter will be that the most commonly studied type of gravity sensing by higher plants depends on mechanotransductive ion channels (MCs) in the cell membrane. Quite possibly more than one kind of MC functions in gravity sensing. However, of the roughly half dozen MCs described in higher plants (Falke et al. 1988; Cosgrove and Hedrich 1990; Ding and Pickard 1990a-d; Moran 1990), one in particular has been proposed to play a role as a gravity transducer (Ding and Pickard 1990a,b,d). It is present in tissue which may well contribute heavily to gravitropic sensing in the organ in which it is found, and when patch clamped it has several properties already predicted (Hepler and Wayne 1985; Pickard 1985a,b; Edwards and Pickard 1987) for a gravity transducing channel in this type of angiosperm tissue. Most particularly, the patched channel is highly sensitive to membrane deformation, and when activated it can pass relatively large currents of calcium through the cell membrane.

Gravitationally stimulated activity of MCs present in the vacuolar membrane might complement the function of plasmalemmal MCs in certain organs that have completed normal growth. Though little is yet known about these vacuolar MCs, we will speculate on their suitability for regulating gravitropism-limiting biochemical functions.

A recent book on tropisms (Hart 1990) provides a well-rounded review and access to the literature of gravitropism, and it is complimented by several reviews (which it references) placing specific emphasis on gravity sensing. However, few data and ideas concerning the putative participation of MCs in gravitropism are as yet published (Edwards and Pickard 1987; Hart 1990). Since, from the comparative point of view of this volume, participation of MCs and the means by which energy is focused onto them is the key topic for discussion, we will briefly describe the postulated gravity transducing channel, consider the likelihood that other MCs in the cell membrane can also transduce gravitational signals, and develop some ideas about the elements of plant architecture in which gravitationally applied stress is initially expressed as strain and which lead signal energy to MCs to effect primary energy transduction. In other words, this chapter is not a review of the literature, but rather it is an attempt to explore ways in which to open up the analysis of mechanotransduction in gravitropism.

In recent years, there has been a tendency to view gravitropism as a unique sensory system coupled to the hormone system which controls growth. In contrast, we shall speculate that gravity sensing may be a delicate expression of a common mechanosensory system which operates continuously during both straight and tropic growth.

2 Mechanically Activated Channels in a Model Plant System

2.1 Overview of Possible Gravitropic Function of Mechanically Activated Channels

Mechanosensory ion channels might in principle initiate gravity responses by admission of messenger Ca^{2+}, by hyperpolarization due to limited exit of K^+, or by depolarization due to limited exit of Cl^- or entry of K^+. Shifts of internal Ca^{2+} or shifts of transmembrane potential might in principle be either symmetric or asymmetric about the cell axes (although – even in a conspicuously vacuolated, elongate plant cell – electrotonic processes will act strongly to reduce asymmetry of transmembrane potential). Furthermore, primary shifts of either might trigger secondary shifts in the other. It is also possible that the plant may have evolved multiple gravity sensing mechanisms in different organs and even within the same organ. Indeed, it is quite plausible that gravity sensing in different tissues within the same organ depends on different MCs. However, a primary role for asymmetric shifts of charge seems unlikely because shoots gravitropically stimulated after deprivation of auxin, the hormone mediating gravitropic curvature, show no lateral voltage gradients. Rather than concentrate on the entire formal array of possibilities, we shall develop a hunch based on many clues and evidences (Cormier et al. 1980; Hertel 1983; Hepler and Wayne 1985; Pickard 1985a,b; Gehring et al. 1990) that the primary phase of gravitropism of young vertical-seeking roots and stems, which over the years has proved the most appealing gravity response for experimentation, is based in considerable part on mechanoresponsive Ca^{2+} channels (MCaCs). Then, we shall make a few comparisons with gravitropism in less well-known systems.

During gravitropic stimulation, the MCaCs might be activated differentially in the upper and lower sides of the organ as a whole, or might be differentially activated in each of the sensory cells. We shall suggest that both these cases might occur, but that differential activation within each cell might be a far more sensitive and gravitropism-specific mechanism resulting from evolutionary specialization of the force-focusing characteristics of the extracellular matrix, or cell wall system. We shall alternatively promote the view that differential activation might result from gradients of hydrostatic pressure across each cell, or across each unit of hydrostatically linked cells. Then, we shall suggest that at the level of molecular architecture it may prove hard to fully separate these models either conceptually or experimentally.

If the intermediate and late steps of gravitropic response were well understood, it would greatly simplify the task of analyzing the features of force transduction. Lacking certainty about these steps, we shall nonetheless adopt some common assumptions about them in order to provide a physiological context for our discussion of MCaCs. Specifically, we shall presume that local entry of Ca^{2+} activates a membrane-associated kinase which could in turn activate auxin porters (or porter complexes) by phosphorylating them (e.g., Pickard 1985a,b; Rubery 1987; Mannan et al. 1989; Biermann and Feldman 1990; Harper et al. 1991). And, regardless of the sequence of steps leading to lateral auxin transport, there are new data bearing on where the auxin undergoing movement might act. At least one set of auxin receptors promoting cell expansion may be situated at the mural face of the cell membrane. Thus, localized export of auxin into the extracellular matrix, or cell wall, might be a bottleneck step in the regulation of growth (Rubery 1987; Barbier-Brygoo et al. 1989; Klämbt 1990; Knauth and Klämbt 1990). The implications of this evidence for the locus of channel activation will be given special attention.

2.2 Choice of Model Tissue: Distribution of Gravity Sensing

A most compelling candidate for a gravity transducing channel is an MCaC described for epidermis of the vertical-seeking leaf sheath of onion. It has also been described for the nongravitropic inner epidermis of mature onion bulb scale (Ding and Pickard 1990a-d) and for onion bulb scale parenchyma (P.M. Badot, J.P. Ding and B.G. Pickard, unpubl. data).

The presence of the MCaC in the sheath epidermis is particularly important because of the premise that this tissue must play a major role in gravitropic reception by nonwoody organs. In such organs, it is the epidermis that serves as the primary (though not exclusive) tissue governing growth (e.g., Kutschera 1987), and it is certainly the epidermis that displays the earliest and ultimately largest (e.g., Harrison and Pickard 1989) shifts in auxin levels. At least in the seedling maize shoot, the basipolar stream of auxin which must be laterally deflected to mediate tropisms has been shown to be most prominent in the region of the epidermis (Jones 1990). Insofar as measurable gravitropic response may occur in less than a minute after displacement of a maize seedling to a horizontal position (Bandurski et al. 1984), and readily visible curvature of several organs has been observed within 5–10 min (e.g., Pickard 1985a,b; Harrison and Pickard 1989), one is tempted to look for the gravitropic sensing cells most crucial for the early phase of response close to the most crucial gravitropic effector cells.

Moreover, because the epidermis appears to be the key mechanosensory tissue both for morphogenesis (Green 1987; Jesuthasan and Green 1989 and references therein) and for response to environmental factors such as wind, rain, obstructive objects, and animal activity, it is a tissue in which the ability to detect gravity might plausibly be evolutionarily refined.

In contrast, the predominant view has been that sensing occurs in a loose or tight ring of internal cells which are usually well provided with starch-filled plastids called amyloplasts. These amyloplasts have been defined as statoliths because they often settle to the lower sides of the cells, which have been defined in turn as statocytes. As will be elaborated below, in some cells amyloplasts and other heavy bodies indeed appear to serve as statoliths, and of course heavy bodies always contribute to the net weight of the cell. But for seedlings there has been no demonstration of statolithic function that compels acceptance by all plant biologists, and we anticipate strong evidence against their participation in root gravitropism except as part of the general mass and the energy reserve of the cell.

All these lines of reasoning support the hypothesis that the epidermis is an important locus for gravitropic sensing for the onion leaf sheath, and they are consistent with the hypothesis that the major contribution to sensing made by internal tissue may come from the whole population of parenchyma cells rather than simply a few cells near the vascular tissue. They appear to justify consideration of the model system.

2.3 Mechanically Activated Calcium Channels in the Model

Although most of the Ca^{2+} channels described in the literature have low or transient Ca^{2+} conductances under physiological conditions (e.g., Tsien et al. 1987; Hosey and Lazdunski 1988), the MCaC identified in protoplasted onion cells of both leaf sheath and bulb scale epidermis (Ding and Pickard 1990a,d) can admit a fairly large amount of Ca^{2+} in response to a relatively small, steadily maintained stimulus. As required for sustained gravity reception, the Ca^{2+} currents can be sustained for a long time. For example, on the basis of data such as those illustrated in Fig. 1A,C, it can be crudely but conservatively approximated that a typical patch in a pipet of 1 µm tip diameter might, if provided with 10^{-7} M "cytosolic" Ca^{2+} versus 10^{-3} M "mural" Ca^{2+}, clamped with a physiologically realistic membrane potential of –150 mV, and provided with 1 or 2 kPa suction, permit an influx of perhaps 10^5 Ca^{2+}/s. With further assumptions, it may be imagined to what extent this 1-s influx would shift the level of cytosolic Ca^{2+} in an intact cell with a similar transmembrane Ca^{2+} gradient. If the cytoplasm under the patch did not stream, if entering Ca^{2+} was not immediately bound or pumped out, if it did not diffuse away from the patch laterally during the 1-s interval, and if inward diffusion was limited to a 1- or 2-µm path (corresponding to a typical separation between cell and vacuole membranes), the flux into the cytosolic volume defined by the diffusion constraints would increase the basal concentration perhaps 1000-fold. It remains to be tested whether these protoplast channels operate in walled cells to produce the Ca^{2+} entry presumed to occur there. However, at this preliminary stage of investigation, operation in situ is a reasonable article of faith. The ability to pass large currents across a patch, based on (1) reasonably large average instantaneous conductance of the flickery channels when open; (2) persistent flickering for many minutes without noteworthy decline of conductance if suction is low; and (3) the abundance of the channels in patch after patch, is ideal for an MC presumed to detect the weak signal due to displacement of a light-weight plant in the gravitational field.

A further feature which might suit the channel system for gravity detection is linkage of multiple conductance units. Linkage cannot readily be discriminated in the flickery currents that are seen with high levels of Ca^{2+} in the patch pipet (Fig. 1A, top two traces), but is clearly seen when the channel system is provided with low to negligible levels of Ca^{2+} in the presence of high K^+. Linkage is evident in the lowest trace of Fig. 1A, but is specifically illustrated in Fig. 1B. With our experimental conditions, K^+ and Ca^{2+} currents are quite generally coincident, suggesting in the simplest interpretation that they occur through the same pores (but in the most complicated interpretation that there are physically coupled Ca^{2+} channels and Ca^{2+}-blocked K^+ channels). In any case, when the channel system is conducting

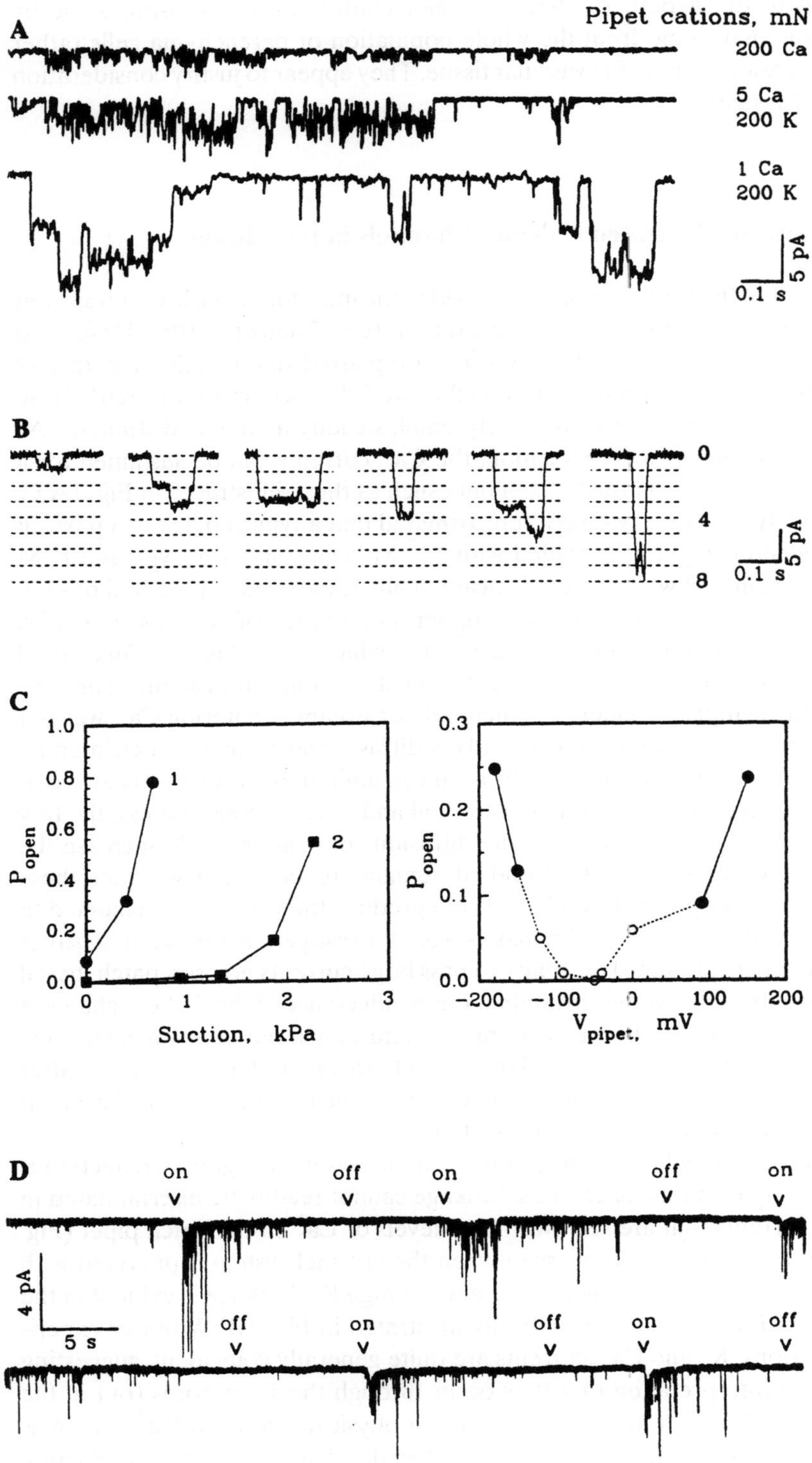
A
Pipet cations, mN
200 Ca
5 Ca
200 K
1 Ca
200 K
5 pA
0.1 s
B
0
4
8
5 pA
0.1 s
C
1.0
0.8
0.6
0.4
0.2
0.0
P_{open}
1
2
0
1
2
3
Suction, kPa
0.3
0.2
0.1
0.0
P_{open}
−200
−100
0
100
200
V_{pipet}, mV
D
on
off
on
off
on
4 pA
5 s
off
on
off
on
off

K^+ alone or in combination with more or less physiological levels of Ca^{2+}, current steps equal to as many as eight combined unitary steps are evidenced. This cooperative behavior might well enhance signal sensing.

Because gravitropic induction is strongly regulated by developmental and environmental parameters (Pickard 1985b), a gravitransductive Ca^{2+} channel might well be subject to various kinds of regulation. Thus, it is pertinent that the open time of the epidermal MC system varies markedly with patch potential when the membrane is under tension (Fig. 1C, right panel). Transmembrane voltages are typically much higher for plants than for animals, and often shift dramatically in response to both internal and external signals and circumstances, so in vivo modulation of epidermal MCaCs by transmembrane potential would seem inevitable.

When the onion epidermal cell membrane is patched, the sensitivity of its MCaCs to patch suction is at least as great as that of any yet reported. For fresh patches subjected to reasonably low levels of pipet Ca^{2+}, channel open time often increases measurably with application of fractions of a kPa of suction (Fig. 1C, left panel). It has not been assessed how suction translates to stretch for these patches; and, perhaps more importantly, it is unclear how the state of the protoplast membrane corresponds to its state before release from the cell wall. Nevertheless, exquisite sensitivity of the MCaC under patch clamp augers well for a channel postulated to function in transduction of the gravitropic signal.

The range of the channel system is large and adjustable: although patches are initially responsive to very small stimuli, when challenged with large suctions for

Fig. 1A-D. Mechanically stimulated cation currents across the cell membrane of onion epidermis protoplasts. Movement of cations out of the patch pipet is shown as *downward. A* shows currents across cell-attached patches when the pipet is provided with 200, 2, and 1 mM $CaCl_2$ with respective KCl concentrations of 0, 200, and 200 mM. Pipet voltage is clamped to +60 mV; the patch is cell-attached. At the higher concentrations of Ca^{2+}, the currents are flickery; with 1 mM $CaCl_2$ plus 200 mM KCl, a little flicker is superimposed on relatively large, long-lasting bursts. *B* illustrates the large amplitudes and long durations of currents seen when the pipet contains 200 mM potassium gluconate and essentially no Ca^{2+}; pipet voltage clamped to +60 mV; patch is cell-attached. Current bursts are selected to illustrate linkage of conductance units. The sixth panel shows a burst with opening current equivalent to that of seven or eight individual units and a return to baseline equivalent to closure of eight individual units; this is the biggest linked change seen when suction is low (here, 0.8 kPa). The *left panel* in *C*, plots of the probability that channels are open versus suction, illustrates that sensitivity of patches to suction can be high, especially when Ca^{2+} levels in the pipet are at or below 1 mN. The two plots, from different patches, illustrate that there is considerable variability of sensitivity from experiment to experiment; indeed, loss of sensitivity or "adaptation" can occur within a single experiment. Pipets contained 200 mM potassium gluconate and pipet voltage was +60 mV; patches were cell-attached. The *right panel* in *C*, a plot of the probability that channels are open versus pipet voltage, illustrates that the fraction of the time channels spend in the open state for any effective value of clamped suction (here, 1.5 kPa) depends strongly on transmembrane potential. Because the P_o is harder to assess near the reversal potential, values in this range are presented with *open circles*. In the physiological range of membrane potentials, probably corresponding to +50 to +175 mV pipet potential for this excised patch, the voltage dependence is steep. *D* illustrates "adaptation" of channel activity to constant suction, a property expected of a channel which must act in vivo with a wide stimulus range. Suction of 2.5 kPa was applied at the on markers, and at off markers suction returned to 0 kPa. Patch was excised, and pipet voltage was clamped to +100 mV

a long time they "fatigue" or "adapt" (Fig. 1D). If stimulation is not excessive, however, the system can recover rapidly. This makes good sense in vivo, for under favorable circumstances it would permit strong response to weak stimuli such as those due to gravity, while under unfavorable circumstances it would permit damped response to vigorous stimuli such as heavy wind and rain. For intact plants it is well known that although weak flexure or asymmetric friction provokes a mechanotropic re- sponse, strong flexure and friction cause a set of dramatic, symmetric, morpho- genetic responses while suppressing gravitropic sensitivity (Jaffe and Biro 1979; Pickard 1985b; Hart 1990).

2.4 Mechanoresponsive Channels in Internal Tissue

Although onion sheath epidermis has been selected as a model tissue for this discussion, similar though perhaps less delicate sensing within internal tissue has been postulated to make an important contribution to organ response. Thus, it is important that the parenchymal cells of onion bulb scale contain MCaCs closely resembling those of the epidermis (Sect. 2.2). Working with tobacco leaves, I.A. Newman has discovered at least one MC in protoplasted spongy mesophyll cells, and two MCs in protoplasted leaf rib parenchyma (unpubl. results). Some of these tobacco MCs, all of which display multiple linked conductance units, share characteristics with the onion epidermal system, though there may possibly be differences. We regret that analysis has not proceeded far enough to justify detailed comparisons. In any case, it seems likely that parenchyma cells will generally have MCaCs, and perhaps other kinds of MCs as well.

2.5 Three Possible Tests for Gravireceptive Function of Mechanosensory Channels

In order to explore whether the MCaC demonstrated in onion truly functions in gravitropic gravity reception, at least three kinds of studies can be imagined.

1. "Pharmacology". The first study is "pharmacological" comparison between channel response and gravitropism of the organ. To facilitate it, Millet and Pickard (1988a) established that Gd^{3+}, properly applied to the maize root tip, is a relatively specific inhibitor of the early phase of gravitropic reception (as contrasted with ultimate growth response). Inhibition by Gd^{3+} of a cation-selective MC – notably a K^+ channel – was promptly demonstrated in single channel experiments (Yang and Sachs 1989), and indeed, it was such experiments that led to the realization that the stretch-activated K^+ channel of frog oocytes passed small amounts of Ca^{2+}. Gd^{3+} is now commonly utilized for blockade of cation channels. In the onion leaf sheath model, unlike the maize root model in which the site of reception is thought to be separated from the site of the differential growth response by many cell lengths, the sites of reception and differential growth are thought to be axially coextensive, and specificity of any inhibition of gravity transduction would be unlikely. However, if the hypothesis under test has much generality, Gd^{3+} should inhibit onion epidermal MCaCs. It was thus gratifying to find that Gd^{3+} blocks MCaC gating when provided to outside-out patches of bulb scale epidermis at a concentration less than 2 μM (unpubl. data). The interdiction appears to occur cooperatively. Unexpectedly, at slightly lower concentrations, Gd^{3+} strongly promotes channel opening. This pharmacological action is strong evidence that MCaCs participate in gravity detection, although of course it will be desirable to seek MCs in the gravitropically sensitive maize root tip and to check their sensitivity to Gd^{3+}.

It is likely that a number of herbicides, microbial toxins, and pharmacological agents will influence the epidermal MCaC, so potential channel agonists and antagonists (e.g., Hosey and Lazdunski 1988; Jaffe 1988) should be scrutinized. Any inhibitors identified might then be screened first for their influence on gravicurvature, and then for their influence on straight growth. Considering the multiple involvements of Ca^{2+} in the complicated process of growth regulation (Trewavas 1985; Ray 1987; Li et al. 1990) few inhibitors may be expected to block gravicurvature specifically, without confounding effects on the late steps of differential growth. Nevertheless, the value of specific inhibitors justifies an effort to find them. The feasibility of such an effort is demonstrated by the fact that the first two organic compounds we subjected to test interfere fairly specifically with gravitropism and MCaC action (unpubl. data); tests of more agents are planned.

2. Presence in Diverse Gravitropic Organs. A condition for the participation of MCaCs in gravity detection is their presence in gravitropic tissues of a wide range of taxa. Although experimentation is generally carried out with seedling stems and roots, most growing organs contain tissues sensitive to gravity. Hence, a second useful test of the proposed gravitropic role of MCaCs would be demonstration by patch clamp that they are of general occurrence in the plant kingdom. Demonstration of MCaCs in the tips of primary roots should be an early priority, and, since leaves such as those of tobacco are graviresponsive, further study of MCs in their cells is warranted. Moreover, since there may well have been considerable evolutionary pressure on MCaCs, particularly in plants inhabiting soils with heavy metals which might inhibit ordinary channels, examination of the possible diversity of MCaCs and consideration whether other types of MCs might transduce gravitropic signals in special situations should prove productive.

3. Protein Studies. A third way to test participation of MCs in gravitropism might be to identify the protein components of a specific channel system and to demonstrate that a reversible shift from the agravitropic to the gravitropic state of an organ is correlated with the level of that protein. Channel proteins sensu stricto (some of which can probably be isolated and identified with organic agonist/antagonist "tags") should be present all the time, because they are postulated to fulfill multiple functions. But how might a protein of the channel complex correlated with transfer of information to subsequent effector systems be identified? Often, such identification requires tedious trial and error. We hope that we may have found a short cut.

During a sabbatical in our lab, C.E. LaMotte demonstrated such a shift from agravitropism to gravitropism in seedling roots of a maize cultivar, and worked out parameters for the shift in considerable detail. In brief, the most effective natural way to promote gravitropic capability is to illuminate the root while restricting availability of water. As shown by studies culminating in that of Leopold and LaFavre (1989), light and water stress promote the synthesis of abscisic acid (ABA) in maize root tips, and ABA in turn promotes downward gravitropism. Although the specific cultivar used by Leopold and LaFavre was reported to alternate between two fundamentally different kinds of gravitropism rather than between agravitropism and regular gravitropism, we have evidence that the action of ABA is similar in the two maize lines.

At the same time, Hong et al. (1990) have shown that HVA1 protein, characterized originally from barley aleurone as a protein synthesized (along with a homolog) in response to elevated levels of ABA (Hong et al. 1988), is also synthesized in the tip of a barley or maize root treated with ABA. Thus, its presence is at least crudely correlated with gravitropic sensitivity. Levels in the remainder of the root and in the shoot are low under the experimental conditions. Further tests are underway to see if the correlation between amount of the protein and degree of gravitropic activity holds up during rigorous variation of the parameters controlling the reversible shift between the gravitropic and agravitropic states.

It is pertinent that analysis by computer program suggests that part of the HVA1 protein could form a slightly twisted amphipathic alpha helix (Dure et al. 1989). Might helical portions of this molecule bind at the inner face of the complex to modulate transfer of information about the transduced gravity signal? The inhibition of kinases by melittin (e.g., Raynor et al. 1991) might serve as a model for such a function. On the other hand, if fully formed, this helix would have a length greater than the thickness of the cell membrane: might helical portions insert into the membrane to form channels, or join with constitutive channel-forming proteins to modify channel behavior (cf. Opsahl et al. 1990; Tosteson et al. 1990; and Sect. 3.2.1)? Or might the residual lengths of protein connect end to end to promote synchronous or near-synchronous opening of several pores? Alternatively, might the amphipathic proteins simply serve as linkers between channels, lying in the plane of the cell membrane with one side interacting with the hydrocarbon chains of the lipid and the other with the charged head groups, and termini attached to the true channel proteins? Or might HVA1 protein serve as both channel and membrane-embedded linker protein? Such questions, while perhaps simplistic, are readily subject to experimental investigation.

3 Possible Ways to Open Gravitransductive Channels

Regardless of any possible successes in pinning down MC activity responsible for gravisensing at the membrane level, gravitropism cannot be understood without knowing by what mechanism force is collected and provided to the membrane in a way recognizable by the mechanotransducers. The way in which force applied to the protoplast membrane is recognized by MC protein has not yet been investigated. Further, it is far from clear that the mechanism of activation in a protoplast membrane will be identical with the mechanism in the intact plant, where (1) gradients of hydrostatic pressure might be expressed as differential squeezing of the membrane against the extracellular matrix or cell wall system; (2) much of the mechanical stress of both internal and external origin is carried by the wall system; and (3) tremendous amplification of signals may be necessary for some sensory systems. Clearly, the general architecture of the stress focusing system in the intact plant is of special interest for gravitropism. There is of course a paucity of pertinent information about the plasmodesmata, which serve as hydraulic needle valves connecting cells within a tissue, as well as a paucity of information about the membrane, cytoskeleton, wall system, and links between them. However, technological advances hold out hope that properly posed questions on MC activation can be addressed aggressively in the near future. Therefore, we shall list some alternative proposals for mechanism at the level of the membrane and its immediate environment, and with this as a basis consider broadly how the gravitropic signal might be conveyed to the membrane.

For simplicity, we treat MCs as having simply an open and a closed condition, ignoring the generality that channels can have multiple open and closed states. Many aspects of mechanical control of channel conductance are discussed much more exhaustively by Sachs (1988) and Morris (1990).

3.1 Changing Channel Area as a Direct Response to Membrane Tension

As demonstrated by Opsahl et al. (1990), when change in state of a channel involves a change in the membrane area occupied by channel protein, simply increasing the tension on the membrane is adequate for activation. Opsahl et al. utilize as a model the toxic peptide alamethicin. It can form amphipathic alpha helices that insert into lipid bilayers as membrane-traversing clusters which can act as ion-permeant pores. Although higher conduction states are more probable at higher tensions, the precise molecular basis has not yet been shown. In this in vitro system, possible area-related determinants of conductance might be the number of clusters in the membrane, the number of helices per cluster, or the conformation of one or more peptides. Area-dependent conformational activation of membrane proteins has been demonstrated by Skou (1959) and discussed by MacGregor et al. (1989) and Morris (1990). It is noteworthy that net increase in area might in principle be either positive or negative with MC opening, correlated with "stretch-activated" or "stretch-inactivated" channels.

New evidence from Martinac et al. (1990) is consistent with the idea that simple distortion of the membrane is adequate to open MCs. These workers noted that certain small amphipathic molecules should insert preferentially into one of the two leaflets of a bilayer. Cationic amphipaths should insert predominantly from the inner surface because of the relatively greater abundance of negatively charged lipids at that face, while anionic amphipaths should insert predominantly from the outer surface due to its relatively lesser negative charge. Whether cationic or anionic amphipaths are provided to a bilayer, their insertion should tend to bend the bilayer, putting it under tension. Accordingly, Martinac et al. found that either cationic or anionic amphipaths applied to a membrane of *Escherichia coli* (see Martinac et al., this Vol.) increased the probability of MC opening.

3.2 Stretching Channel Bonds by Cytoskeletal Tug

As postulated by Guharay and Sachs in 1984 (see also Sachs 1988; Sokabe and Sachs 1990), channels might be interconnected by a spectrin-like cytoskeletal network lying under the membrane and bearing a large fraction of the tension applied to it. Cytoskeletal elements would focus energy onto critical bonds of the MC protein, bringing about conformational changes associated with MC opening.

In analogy with this hypothesis, Guharay and Sachs (1984) suggested that a submembrane network of actin with links to the membrane but without links to MCs diverts force from the mechanosensing system and regulates its sensitivity. They have found, as predicted, that sensitivity of MCs in cultured chick myoblasts is increased by cytochalasin B.

As a version of this scheme, force might be applied via inwardly directed cytoskeletal elements. This version appears to be required for models of gravitropism in which MC opening is controlled by pendular statoliths, and is not mutually exclusive with force focusing via a network in parallel with the membrane.

3.3 Stretching Channel Bonds by Direct Tug from the Extracellular Matrix

Tension experienced by the extracellular matrix or wall system might be focused on some crucially situated matrix polymer, causing it to undergo conformational change to a state in which it might pull directly on MC protein and lead to channel opening. Such a possible mechanism is of particular interest for the walled cells of plants, but also seems pertinent for animal cells when it is remembered that the extracellular matrices of young plant cells may sometimes resemble those of animals in composition and thickness (Roberts 1989). It may be a particularly useful mechanism in situations where acute sensitivity is required. This is illustrated by the mechanism of force focusing in auditory hair cells, which must surely be counted as among the most subtle of physiological sensors. Here, fine filaments link the tips of the stereocilia, where the transductory Ca^{2+} channels are located. Movement of the stereocilia stretches the filaments, which in turn stretch the protein of the channel system and cause opening (Huang and Corey 1990).

Although the direct wall-to-channel transfer could not work in isolated protoplasts, it could operate as a customary bypass of a more general mechanosensory mechanism operative under the highly stressful conditions faced by protoplasts (which rarely form in nature except during severe drought or winter desiccation). It might be particularly appropriate for gravitropism, where the signal is clearly small. Indeed, the occurrence of linked conductance units in the MCaC system of onion suggests possible juxtaposition of several channel units around a single force transmitting entity.

Such direct modes of energy transfer potentially grade into but are to be distinguished from a general transfer of energy from wall to membrane or membrane-associated elements which will be required for the opening of plasmalemmal MCs in most models invoking wall stress.

3.4 Direct Action by Compressed Wall Elements

Compression rather than tension of the wall system might be focused through some key matrix polymer onto MCs, causing them to open. Provisos about conditions of operation of this scheme are the same as those for the preceding one.

4 Distribution of Mechanical Stress and Channel Activation

In this and several following sections, we shall discuss three kinds of *mechanical stress*: (1) that imposed on the cell wall system by the weight of the tissue and presumably transferred in part to the cell membrane; (2) that imposed on the cell membrane as it is differentially pressed against the cell wall by the graded hydrostatic contribution to turgor pressure; and (3) that possibly provided to the membrane by amyloplasts acting as statoliths. But although we shall initially treat these kinds of stress somewhat formally and separately, we shall ultimately note that the difference between some of the formal ways of representing stress might possibly be mainly semantic. If at the molecular level exchanges of energy between the cell wall system and the cell membranes with their contained cytoplasm occur via integrin linkages, the formal treatments might simply represent different perspectives and different quantitative emphases on the molecular complexity of force distribution.

In any case, we shall begin by listing several possible *patterns of stress asymmetry*, assuming that whichever stress patterns may be pertinent, each gives rise to a closely similar pattern of MC activation. In order to minimize the number of geometries to be considered, we will confine the discussion to shoot axes. However, we will not simplify to the point of overlooking differences in stress distribution when during gravitropic stimulation a plant is permitted to sag or is restrained from sagging. By sagging, unless otherwise specified, we have in mind the passive downward bending of a stem or other plant axis supported at one end and bearing its own weight.

Our goal will be to lay a foundation for asking questions about the implications of the various possible stress patterns for mechanotropic response. Since it seems widely acknowledged that mechanotropic curvature depends on redistribution of auxin, this will require considering two potential means of initiating auxin translocation: (1) a primary one in which auxin porters are activated in the immediate vicinity of mechanosensory Ca^{2+} fluxes; and (2) a secondary one in which auxin migration is brought about indirectly via the creation of electrical gradients across the organ.

In view of this goal, it may be useful at the outset to note what cells of the shoot are thought to participate in auxin transport. The best information suggests that these include all the major classes of abundant cells that are not highly specialized for other purposes: the vascular parenchyma, the cortical parenchyma, and the epidermis (Rubery 1987; Jones 1990). Moreover, in a few well-studied seedling shoots, it has been demonstrated that downward auxin redistribution occurs in both the upper and lower halves of the horizontal organ.

4.1 Asymmetry Schemes

1. Cellular Asymmetry. For either hydrostatic models, cell wall models, or statolith models, it can be postulated that the gravitropically critical element of stress is distributed across the entire, nonsagging horizontal shoot axis. For hydrostatic models, it is clear that there might be pressure heads across each individual cell. For the cell wall models, one can imagine that the walls collectively perform like a complex cantilevering structure in which compression and tension are distributed to greater or lesser extent across the organ, with both the parenchymal air spaces and the middle lamellar gel of pectin in the parenchyma permitting modest slippage between walls of adjacent cells. Either tension or compression might activate MCs, but not both. For statolith models, the possibility of generating cellular asymmetry has long been recognized: heavy particles might either press via a cytoskeletal system on the lower surface or tug via cytoskeletal linkers against the upper surface of each statocyte. Insofar as it is commonly acknowledged (Jost 1907; Caspar and Pickard 1989; Hart 1990) that statoliths need not be displaced in order to act, the pressure against the lower side would also need be exerted via cytoskeletal linkers – in this case, via particularly stiff ones.

Such cellular asymmetry would have the result for cell wall models, if indeed this kind of pattern is realized in nature, that the sensory cells might be identical with the transport cells in both upper and lower sides of the horizontal organ. For statolith models, it would have the result that the statocytes, generally located in a narrow ring around the core containing the bulk of the vascular tissue, might have to communicate with key auxin transporting and curvature effecting cells via some long-range secondary mechanism.

2. Organ Asymmetry, or "Half-Organ Activation". If the upper half of a horizontal organ were to experience tension while the lower half experienced compression, as is certainly the case when a shoot axis sags, cell wall models would require that MCs be activated more or less omnidirectionally within each cell, but only in cells of one half of the plant. That is, either compression or tension could activate MCs, but not both. Statolith models would require that the radially inner and outer portions of statocyte membranes be different – one sensitive, the other insensitive, to statolith action. They would further require secondary communication with a large fraction of the cells thought to participate in creating a lateral auxin gradient. Hydrostatic models permit no predictions about effects of sagging without precise specification of molecular mechanisms; while the hydrostatic pressure gradients would be little changed, the net turgor pressure within each cell would change in a complicated way.

Cellular asymmetry and "half-organ" asymmetry schemes need not be mutually exclusive, and it seems likely that in many situations the two kinds of asymmetry might overlap. The effects of sag might probably enhance gravitropic stimulation in some situations, but in others might tend to suppress the rate at which cellular polarity could be developed.

4.2 Testing the Schemes

1. Discrimination Against Half-Organ Activation Schemes for "Simple" Gravitropism. Hild and Hertel (1972) have performed an experiment which tends to discriminate against half-organ activation schemes for gravitropically responding shoots which do not sag much. It should be emphasized at the outset that while the experimental designs are not absolutely definitive, they are worthy of considerable thought and could serve as basis for further interesting experimentation. Using young coleoptiles of maize, Hild and Hertel alternated 2-min stimuli of opposing vector for 1 h. Half of the plants were then left horizontal in the plane of prestimulation: kinetics of their gravitropic response were quite different from kinetics of controls which had not been pretreated. Half the plants were swiveled 90° before horizontal placement: these bent with essentially the same kinetics as did controls. In the simplest interpretation either the lower side, the upper side, or the upper and lower sides, but not the lateral sides, of the cells are stimulated during pretreatment and, in a loose sense of the word, adapt.

Because of their importance, it would be worth while to extend these experiments, probably using dicot stems in addition to coleoptiles, and shortening the length both of the individual alternating prestimuli and of their summed duration. There should be careful comparison between control stimuli at right angles to the alternating prestimuli and those within the plane of prestimulation. Further, plants grown under other conditions can express "adaption" in other ways than in the experiments of Hild and Hertel, and this suggests other useful comparisons.

2. Comparing Schemes for Tropism by Sagging, Bowed, and Rubbed Plants. If a vertical, basally anchored stem is gently pulled into an arc or bow for a few minutes, it will typically respond with a tropic curvature away from the concave side (Clifford et al. 1982; Pickard 1985b; Hart 1990). For precision and clarity, we shall henceforth refer to this kind of bending stimulus as "bowing". Bowing of woody trunks and branches also results in curvature away from the concave side (Timell 1986). Here, although features such as the (probable) localization of sensing cells to the cambium plus its immediate derivatives and the final means of achieving curvature by reaction wood formation are not the same as in seedlings, the mediational mechanisms appear to be comparable (Yamaguchi 1990). Since cells in bowed organs are compressed on the concave side of the organ while cells are stretched on the other, one expects that mechanical strain should be characterized by a "half-organ scheme".

The experimental design of Hild and Hertel described in Section 4.2.1 above should be extended to give information about sagging plants, for this could provide a new kind of mechanotropic data which seems lacking in the literature. Sagging per se should be mimicked by bowing; controlled bowing in opposite directions could be substituted for opposing horizontal placements during the prestimulation period. If our reasoning is correct, this pretreatment should cause "gravitropic adaptation" in the plane perpendicular to bowing as well as in the plane of bowing. Of course, it should cause "adaptation" to further unidirectional bowing stimuli as well.

As a final variant, it might be instructive to substitute gentle opposing rubs for gravitropic pretreatment. The responses of a wide range of plants to delicate stroking with a cork rod have many properties in common with gravitropism (Stark 1916, 1919, 1921, 1922). Furthermore, the spatiotemporal pattern of thigmotropic curvature of maize roots closely resembles that of gravitropism (Millet and Pickard 1988b). Probably, "adaptation" to rubbing would occur in both the prestimulation plane and the perpendicular plane, though it might not be as great as for bowing.

5 Activation of Auxin Porters

5.1 Primary Activation of Auxin Transport

If the cellular asymmetry schemes of Section 4.1 (in which MCs are gravitropically activated at one side of each cell) are combined with the assumption of Section 2.1 that local entry of Ca^{2+} produces local activation of auxin porters, it follows that auxin porters are also activated at one side of each cell. According to accepted theory of auxin transport (Rubery 1987), such cellular asymmetry of porters will rapidly give rise to an asymmetry of auxin across the whole shoot axis. The "half-organ" schemes of Section 4.2 have a somewhat different implication: porters are activated omnilaterally in each cell, but only in cells on one side of the shoot axis. Of course, the resulting expulsion of auxin on that side of the axis would have no net vector. On the other hand, the pool of auxin in the cell wall space would increase on that side. According to the new theory that auxin receptors at the mural face of the cell membrane play a key role in growth regulation (Sect. 2.1), that side would elongate preferentially. If it was the concave side of a sagging shoot that was activated, the shoot would begin to curve upward. If instead it was the concave side of a bowed shoot that was activated, the shoot would simply begin to develop a growth curvature away from that side.

5.2 Possible Secondary Electrical Effects on Channels and on Lateral Transport

Adding to the complexity of understanding the pattern of channel and auxin transport activation, electrical feedbacks might rapidly confound the primary asymmetries. While gravitropic stimulation of a shoot does not produce electrical asymmetry in the absence of auxin (e.g., Schrank 1959; Newman 1963; Pickard 1985a,b), in a normal plant the gravitropically developed asymmetries of mural auxin lead to asymmetries of proton secretion which lead in turn to electrical asymmetries. These charge asymmetries develop more rapidly than do pH asymmetries, because the early-secreted protons exchange with wall-bound Ca^{2+} (Ryan et al. 1990; Newman pers. comm.).

1. Possible Feedback on Transduction. Such secondary pH or charge asymmetries might influence the rate of transduction by MCaCs. This could indeed be the case if proton or other ion shifts directly or indirectly influenced the transmembrane potentials of sensory cells (cf. Senn and Goldsmith 1988), for the pressure-dependent activity of MCaCs in a patched membrane poised within a physiologically realistic range of membrane potential is strongly promoted by hyperpolarization and suppressed by depolarization (Fig. 1C).

2. Possible Feedback on Auxin Porters. Secondary electrical asymmetry might also reinforce the initial auxin asymmetry by causing electrophoretic migration of auxin porters within the plasmalemma of each transporting cell. It has been proposed for axial auxin transport and for establishment of a polar auxin transport system in embryos that electrophoresis should occur from the negative to the positive end of the organ, and evidence has been marshalled to support this idea (Raven 1979; Nuccitelli 1988; Rathore et al. 1988; Rathore and Robinson 1989). Secondary electrophoretic migration of porters might well be consistent with the spatiotemporal patterns of tropic curvature on which so many of our inferences about the geometry of force distribution and transduction depend, but unfortunately there are few data relevant to lateral auxin redistribution.

We can, nevertheless, glean some evidence on electrical effects from a set of experiments by Schrank and coworkers on oat coleoptiles (Schrank 1944, 1951, 1957, 1959; Schrank and Backus 1951; Webster and Schrank 1953), from a set by Stark with oat coleoptiles (1916, 1919, 1921, 1922), and from one experiment on dandelion flower stalks by Clifford et al. (1982). These will be summarized briefly as follows: (1) A horizontal coleoptile, presumably not undergoing much sag, will develop a transverse electrical field with the lower side positive to the upper. More or less simultaneously, it will undergo upward gravitropic curvature. Similarly, (2) a unilaterally illuminated vertical coleoptile will develop a net positive charge on the shaded side and curve phototropically toward the light. (3) A coleoptile that is rubbed gently with a cork rod will curve thigmotropically in the direction of the rubbing, but there is no evidence on the distribution of charge in this case. (4) A dandelion stalk which is bowed for a few minutes curves mechanotropically toward the side maintained concave, but no electrical data are available. It may be pertinent that curvature develops biphasically, with a long early phase of low-amplitude curvature and a later phase in which bending occurs at about the same rate as in gravitropically stimulated controls. (5) A coleoptile which is rapidly tapped by a unilaterally positioned vibrator becomes relatively positive on the side opposite the vibrator, and bends toward the vibrator. Unfortunately, the effects of tapping are difficult to analyze: they include application of local pressure but also rapid bowing and rebounding.

Finally, effects of applied currents have been assessed. The most careful work on lateral currents appears to be from the lab of Schrank (cited above); these several papers also review a fairly extensive early literature on electrotropism (e.g., Jost 1907; Stern 1924; Lund 1947). In order to understand the effect of transverse current, it is necessary to understand the effects of longitudinally applied current. Shrank reported that small longitudinal currents inhibit polar auxin transport, regardless of which way the current flow is oriented. (This work was been confirmed and extended by Morris in 1980 for pea stems.) Schrank then showed that transverse flow of current causes auxin-dependent curvature, which he attributed partly (but not totally) to lateral transport of auxin. Schrank's experiments were well executed and interpreted for their time, but from a modern perspective three factors suggest that the laterally imposed current was merely inhibiting polar flow of auxin in the vicinity of the negative salt bridge, with resultant damming of auxin upstream. First, the curvatures toward the positive side of the plant began well above the level of the transversely placed salt bridges, and migrated downward. Second, after a considerable lag, curvature developed below the salt bridges – but in the opposite direction from that above. Third, the placement of the bridges was critical: for strong effect, the bridges had to be well below the apical region, in which the bulk of gravitropic lateral transport is known to occur (cf. Pickard 1985b). The nature of Schrank's effect is mysterious, but it appears to be somewhat unnatural in terms of gravitropic polarization. Artificial application of electric fields across a plant presents complex analytic problems due to the geometry and differing electrical properties of the cuticle and the several tissues with their different kinds of walls, their different symplastic connectivities (van der Schoot and van Bel 1990), and the variety of ions which can carry current, be exchanged, released, and compartmentalized. Further, differential poisoning by current-carrying electrode products must always be excluded. Thus, if variables could be effectively separated, one might learn a great deal about how small mechanical stimuli establish large asymmetries of auxin by studying imposed electrical fields, but severe experimental problems are to be expected. Further complicating implications of the development of charge asymmetry are discussed by Pickard (1985b).

6 Load-Bearing at the Cell Wall Level

In order to better understand how stress is focused in the mechanotropisms, including gravitropism, it will be helpful to consider some architectural features of the walls and membranes. This is usefully done in the present context of load-bearing by the wall system, but the information is equally applicable to evaluation of statolith and hydrostatic hypotheses discussed below.

6.1 The Wall System

The wall system of the plant, made up of a variety of large polymers, is less well explored than most other plant components because it is has presented a major challenge to both separation and visualization technologies. The arrangement of its cellulose, hemicellulose, pectin, methylated pectin, minor carbohydrate constituents, and proteins is only beginning to be understood (Roberts 1989). The proteins are diverse (Cassab and Varner 1988; Roberts 1989), and their distributions are probably very heterogeneous. One may speculate that the somewhat amorphous-looking plant cell walls, having been subject to heavy evolutionary pressures, feature a variety of fine-scale architectural devices specially suited to distribute stress effectively for control of plant form and integrity.

Although the walls are formed by individual cells, and vary from one cell type to another, they may be considered as a system because they are cemented together by the pectin-rich "middle lamella". In the epidermis, adjacent cells may lie in close contact and adhere to each other uniformly, but in parenchyma the zones of cell-to-cell adhesion are interspersed with air spaces. In regions of contact between like cell types, plasmodesmata connect the cytoplasms of the adjacent cells. Considering all this architectural heterogeneity at both cellular and molecular levels, it is obvious that a seedling displaced from the vertical will experience heterogeneous distribution of stress. Just as the stronger elements of a building, such as pillars and beams, bear most of its weight and experience strain as they do so, the strong elements of the seedling bear its weight and experience strain. Just as the strain deformation of a building beam is not readily noticed without accurate measurement, it would require sensitive observation to demonstrate the strain deformation of the key structural elements of the cell wall. Just as the substrate-supported building is designed for minimal sag as its rigid members continuously experience strain deformation, the seedling axis has evolved to resist sag (both longitudinal and radial) and continuously experiences strain deformation whether or not it is supported by soil or surrounding objects. And just as it is usually practical to distribute strain in a cantilevered section of a building as widely as possible in the vertical direction, rather than, for example, concentrating it all on the lowest rigid element, it seems evident that some of the strain in a root or stem must be distributed throughout the tissue. In support of this point of view, it may be noted that in the absence of reasonable distribution of support, a horizontally placed plant would subside to a flat sheet. Moreover, it should be noted, turgid longitudinal halves or quarters of roots and shoots of reasonable size will tend to support their own weight when set horizontal.

We have discussed above that both small nonwoody plants and large woody ones execute curvatures in response to bulk tension and compression and to localized friction. The problem of force distribution is just as difficult to resolve for the rub, touch, tap, or bowing that initiates thigmotropism as it is in the case of either unsupported or well supported, horizontally placed organs. The distribution of stress would appear to be quite different for superficial pressure and friction than for flexure. On the other hand, a given stimulus, such as rubbing, is apt to involve both local friction and global flexure; and even for a delicate touch, the "honeycomb" of cell walls might transmit force in complicated patterns across the tissue. This could help to explain why, although shoots generally bend toward a touch, occasionally they bend away from it.

One specific example with roots will suffice to illustrate the complexity. At the outset, we must note that roots generally react to a stimulus in the opposite direction from shoots, owing to the poise of growth regulators including auxin. In the case of gravitropism, this permits roots to seek support and water while shoots seek sunlight; for thigmotropism, this permits roots to avoid obstacles in the soil while shoots avoid displacement from the upright by transient stimuli, or seek support in the case of certain vines, and so forth. Millet and Pickard made the curious observation (1988b) that when root caps are stroked on one side without removing their secreted mucigel, half the population of roots bends strongly toward the stroked side and half bends strongly away. However, when all the external mucigel is soaked off, essentially all roots bend away from the stroked side. Perhaps force is distributed in a different way in the two cases; and perhaps there is a comparatively larger component of flexure in the soft meristematic zone just behind the relatively rigid cap if the cap is laden with the viscous mucigel. On the other hand, some all- or-none effect of a second, polarizing step influenced by mucilage cannot be excluded without further experiment.

How do these ordinary mechanical stresses compare with those of gravitropism? What does all this suggest about the possibilities mentioned in Sections 3 and 4? At the minimum, it seems that many patterns of wall stress distribution might be adequate to activate MCaCs, and when activation is asymmetric the organ might curve. More speculatively, it seems possible that while transmittal of stress to the MCaCs might occur in part by somewhat unspecialized paths in those kinds of mechanical response for which the stimulus is strong, gravitropic stress might be focused by a path specially evolved to facilitate detection of and response to the very weak gravitropic signal.

6.2 Wall-to-Membrane Transmittal of Force

Assuming that the force borne by the wall system plays an important role in at least some varieties of mechanotropism, it is important to ask how the mural force could be focused onto the MCs embedded in the cell membrane just inside the wall. We have recently provided evidence for onion bulb scale epidermis that the wall is attached to the membrane by a macromolecule which reacts with an antibody prepared against hydroxyproline-rich glycoprotein (HRGP) isolated from cell walls of soybean (Pont-Lezica and Pickard 1990). Soybean HRGP has been well characterized, and is known to form a long, relatively rigid rod with an alpha-helical core of protein wrapped round with arabinogalactan oligomers (Cassab and Varner 1988). It is extracted from the wall only with harsh treatments. The HRGP-like antigen of onion, like soybean HRGP itself, is located in the wall, and is not found on the cell membrane if the membrane is pulled away from the wall by plasmolysis. However, if the wall is enzymatically digested, the antigen is abundant on the protoplast membrane. In the same onion cells, it is clear that during plasmolysis by strong salt (especially $CaCl_2$) the withdrawing membrane remains pinned to the wall by numerous fine tubules which are drawn out from regularly spaced sites of attachment; these sites are independent of the laterally positioned fields of plasmodesmata. The sites also appear to function as junctions for a subplasmalemmal network of cytoskeleton (Pickard and Suthar 1988; Palevitz 1989). Numerous clues suggest that the sites are functionally equivalent to integrin, and it may be at these sites that the HRGP-like antigen binds wall to membrane (Pont-Lezica and Pickard 1990).

On the basis of the linkage, or cooperativity, of MCaC conductance units illustrated in Fig. 1B, we are tempted to suggest a tentative, simplistic model. Several (as many as eight) conductance units might be grouped around each integrin. In the patch, stretching of the membrane itself might lead to either independent or linked activation of the physically attached units. Subplasmalemmal cytoskeleton might link integrin molecules, so that with more vigorous stretching large numbers of units might be activated at nearly the same time (as is observed under some conditions: see Ding and Pickard 1990c). However, in cells surrounded by extracellular matrix in the intact plant, activation might under some circumstances be the result of tugs on the integrin-channel system from the HRGP-like antigens linking integrin to wall. Or, viewed from a different perspective, it might be the result of membranes subject to imbalanced forces pressing and/or tugging against the walls to which they are linked by integrins.

In summary, there appears to be a set of uniformly distributed connecting linkers suitable for transmitting stress from wall to membrane, but there is no information on whether the stress might be conveyed to the MCs as localized shear of the bilayer (Sect. 3.1), as localized distortion of a cytoskeletal net (Sects. 3.2), or specific distortion of the channel system per se (Sects. 3.3 and 3.4). Perhaps all these things can happen to onion epidermal MCaCs, depending on the specific circumstances of perturbation. Progress toward identifying wall-to-membrane linkers is important even if it does not immediately resolve the issue of how force is distributed in any specific situation, because it can permit exploration of key stress-bearing elements at the molecular level. A variety of techniques make it possible to chemically link a macromolecule to its neighbors and identify them; it might soon be possible to investigate whether and how the HRGP-like antigen is associated with (for example) integrin, actin, other cytoskeletal elements, or channel protein. Thus, characterization of the wall-to-membrane linkers and their neighbors may provide an opportunity to better discriminate between the several possible mechanisms of activating channels laid out in Section 3.

7 Is Gravitropic Stress-Sensing an Expression of a More General Growth-Regulating System?

Evidently, gravitropic stimuli and a variety of mechanical perturbations regulate whether growth will be symmetric or asymmetric. The evidence is consistent with the possibility that transduction is accomplished by MCaCs. Could gravitropism, and other mechanotropisms, constitute a modest evolutionary optimization or specialization of a very general growth-coordinating system?

We speculate that the most frequent use the plant makes of MCaCs is coordination of local wall expansion. As an organ grows, the walls of its cells "soften" and yield to the osmotically maintained turgor pressure exerted by protoplasm within. Such softening is thought to be coordinated in a general way by growth hormones, most notably auxin (Ray 1987). However, expansion must surely be fine-tuned on a local basis, or else the cells would tend to tug and rip each other apart; whereas, in fact, adjoining walls appear to keep in register and maintain a uniform thickness, and the plasmodesmata that connect neighbor cells tend to maintain alignment. Part of this coordination doubtless results from passive exercise of mechanical restraint by the less extensible regions of the wall system: just because a wall is loosened does not mean that it will fully yield to internal turgor. Yet, it seems likely that there is an active mechanism of coordination as well. If MCaCs are locally activated by shearing of one cell's walls by its neighbors, locally admitted Ca^{2+} might trigger local wall loosening. An additional, attractive possibility is that this loosening might be mediated by movement of auxin into the wall space, where it could locally bind to plasma membrane receptors and thereby locally induce a variety of biochemical wall-loosening events. It must be emphasized that the concept that a class of auxin receptors controlling growth is situated in the plasmalemma with binding sites at the mural face is a relatively recent one (Barbier-Brygoo et al. 1989; Klämbt 1990; Knauth and Klämbt 1990) which has attracted considerable controversy and remains to be fully tested.

Moreover, such action in coordination of growth is compatible with a morphogenetic role in determining the key focal points of internally generated epidermal strain which Green (1987) has championed as determinants of pattern generation. It is also compatible with a role, perhaps in coordination with other kinds of MCs, in mechanical determination of the orientation of cell division (Lintilhac and Vesecky 1984).

In summary, it is plausible that MCaCs operate more often in response to the stimuli generated internally during straight growth and normal morphogenesis than to external stimuli, but serve as well during correction of orientational displacement and other responses to the environment.

8 Hydrostatic Pressure

All of the above considerations of molecular architecture and auxin regulation are compatible with a possible dependence of gravisensing on hydrostatic pressure. Indeed, if the weight of the cell rather than merely that of statolithic components constitutes the mass sensed by gravity detectors, and this weight is largely due to protoplasm, it is reasonable to ask whether the head of hydrostatic pressure across each cell (or hydraulically connected set of cells) might be the factor detected by the transducers in the membrane. This basic idea was proposed by Czapek in 1898, and again by Gillespie (1959) and by Pickard and Thimann (1966). The total turgor in the cell is high, and almost uniformly distributed around the cytosolic face of the cell membrane except for the gravitation-dependent hydrostatic head. This high absolute pressure prestresses the membrane by forcing it against the restraining wall, perhaps establishing it in a state to recognize the small hydrostatic asymmetries. If, for example, the membrane is forced against the lower wall harder than against the upper wall, the molecules in the bilayer at the lower side will rearrange a little so as to tend to thin the membrane. The opposite will occur at the upper side. These rearrangements, though small, might possibly include changes in protein conformation or in the extent of protein insertion. They might be analogous to control of activity of an enzyme in a membrane on a Langmuir trough by small changes in surface tension, first demonstrated by Skou (1959). They might also be analogous to the control by membrane tension of alamethicin channel conductance demonstrated by Opsahl et al. (1990). In other words, the hydrostatic pressure hypothesis may be considered as a variant of the scheme of Section 3.1.

In any case, the weight of the protoplasm is transmitted to the wall system via the membrane. Moreover, since the membrane seems to be pinned to the wall with specific linkers, as described in Section 6.2, these might well play some role in stress transmission. From a molecular point of view, the reciprocal distribution of force between cytosol, plasmalemma and wall system is challenging to model. In fact, the precise molecular geometries become crucial, and use of macroscopic formulations tends to become ambiguous. Without further experimental delimitation, molecular possibilities for how load might be distributed throughout the cells and how force might be focused on channels are very numerous.

The concept of readily deformed but strategically placed structural members under heavy load may be illustrated with a simple macroscopic model. Suppose a structure such as a tower were to be built with fitted metal rods connected at each end with a film of adhesive rubber. Where would the weight be borne? The metal must bear stress for the tower to stand, and the tower could not be built with the rubber films alone; but nevertheless the rubber transmits the weight too.

Recently, experiments on gravitational influence on rotational cytoplasmic streaming of giant cells of Characean algae have been presented as evidence that plant cells can detect gravity by means of their internal hydrostatic pressure differential (Staves et al. 1990; Wayne et al. 1990). As first described at the turn of the century (Ewart 1903), streaming is faster with than against the gravity vector. This is not simply due to the differential work involved in moving uphill, for Wayne et al. can control the "polar ratio" of downward and upward streaming speeds with the level of Ca^{2+} in the medium. Polarity is absent without external Ca^{2+}, is "reversed" at levels somewhat less than micromolar, and "normal" at levels greater than micromolar. Nifedipine, a class of Ca^{2+} channel blocker, reversibly inhibits gravity response, while the "super calcium ion" La^{3+} establishes inverted polarity in the presence of relatively high external Ca^{2+}. "Normal" polarity is reversed by application of neutral red, known to activate certain K^+ channels, but this effect is not seen in the presence of tetraethylammonium chloride, ordinarily used as a K^+ channel blocker. The polarity ratio can be controlled by external imposition of hydrostatic pressure, and by varying solutes in the external medium (Staves et al. 1990 and pers. comm.).

As evidenced by their intensive continuation of experiments (pers. comm.), the authors recognize that the case for hydrostatic control of this gravity effect is not closed. Moreover, each of the individual experiments to date is subject to some ambiguity of interpretation. Nevertheless, the case is impressive. Further, it is being built for a type of cell which is potentially among the most prominent candidates for statolith action, owing to the numerous, large, dense plastids that are fixed along the cell membrane and lever against it by cytoskeletal connections. (Indeed, in our opinion, the case for participation of pendular plastid statoliths has not yet been ruled out, since any agent that influences membrane properties or behavior might readily influence the way in which pendulum-focused stress acts on MCs in the membrane.) Further, this case is being built for an effect which must be an incidental and almost inconsequential outcome of the basic properties of the cell, rather than one developed in response to evolutionary pressure. Thus, mechanisms with at least this level of sensitivity seem likely to have evolved to aid walled cells in behaviors critical for organismic survival.

Early botanists reported the same gravitation-dependent polar ratio of rotational streaming in certain cells of *Vallisneria* and *Elodea*, two higher plants that live in fresh water. Thus, conclusions based on Characean algae appear to be important for our present discussion of higher plant gravitropism.

Predictions of macromodels based on hydrostatic pressure overlap those based on cell wall stress, and for micromodels the distinction between the two kinds of models is blurred. Only experimentation at the molecular level can establish a precise account of how stress is focused during responses to gravity and other mechanical stimuli. Meanwhile, since the role of the cell wall appears critical in hydrostatic pressure models, might it be useful to view them as a form of architectural stress-focusing model?

9 Statoliths

There are some situations in which statoliths have a distinct advantage over other force-focusing mechanisms. Most notably, an aquatic plant is buoyed up by its watery medium, and its wall system bears only a small fraction of the load it would bear in air. It is probably no accident that the first unequivocal demonstration of statolith action was with the aquatic alga Chara (Buder 1961; Sievers and Volkmann 1979). Its rhizoids make large, dense, readily displaced particles by packing crystalloid bodies with $BaSO_4$ (Sievers and Volkmann 1979). When these bodies are permitted to settle into the mature region of the rhizoid, gravitropic sensitivity is lost, but when they are permitted to fall back to the growing tip, sensitivity returns. While there have been no demonstrations in higher plants quite so convincing, one may look forward to them. (Another option for water plants, of course, is simply to increase the density of the sensing organ. This might be achieved by minimizing the vacuole in favor of denser cytoplasm, making relatively large dense cells, and depositing starch, protein, or other heavy materials in various "storage" bodies.)

Statoliths might be quite useful in initiating a variety of morphogenetic responses for which a vector is unimportant or of tropic processes for which the kinetics are not quite so critical as in the primary phase of seedling gravitropism. For example, suppose that gravitropic correction of an organ which has stopped extending depends on renewed availability of hormone after displacement in the gravitational field. Mobile amyloplasts in this tissue might deform the vacuole membrane as they shifted, thereby activating either mechanosensory enzymes which could hydrolyze the hormone conjugates stored in the vacuole or, more interesting from the present point of view, activating MCs and thus permitting transmembrane ionic shifts which might control breakdown of conjugates (cf. Clifford et al. 1989). In fact, MCs have been discovered in what appear possibly to be vacuolar membranes of aleurone cells (D.S. Bush and R.E. Jones, pers. comm.).

One of several organs which might seem to be a candidate for such a vacuolar bombardment mechanism is the mature stem of cereal grasses (see Kaufman et al. 1987). Certain regions of the stem contain many "statocytes" with particularly heavy, conspicuous amyloplasts. When, for example, an oat stem is blown to the ground by strong wind, its pulvini resume growth and right it. The upward bending is correlated with the appearance of an absolute increase of free auxin, with considerably more measured on the lower side than on the upper side. If statoliths were to cause vacuolar release of auxin, it would be necessary to suppose that the auxin could be secondarily distributed across the stem. While it was formerly believed that this redistribution did not occur (Wright 1986), new, still preliminary, evidence (Brock et al. 1989) suggests that it may.

Another possibility for action of statoliths which needs to be explored is that they make a secondary contribution to the gravitropism of organs that putatively depend, more generally, on architectural stress focusing. For example, the gravitropic adjustments made by a stem may be prolonged, and are much more complicated than one would imagine from studying merely the early phase of gravitropic curvature (Pickard 1985b). Perhaps the "starch sheaths" that ring the vascular tissue of many stems are indeed composed of statocytes, and perhaps their action adds to the effect of a more general kind of gravity sensing. This review simply does not stress the role of amyloplasts because there is some evidence against their participation in certain cases of seedling gravitropism and no evidence strongly supporting their participation in the early phase of seedling gravitropism. Clearly, the possible role of amyloplasts in other situations deserves critical experimentation.

10 Overview

In concluding, we would like to emphasize that this chapter is not a review of the literature in an ordinary sense. It is an expression of our faith that mechanosensory Ca^{2+} channels are an excellent candidate for transducers of the physical stresses that regulate many critical plant processes such as growth and gravitropic modification of growth, and that reexamination of mechanosensitive processes with such channels in mind can produce new and testable ways of modeling them. It is also an expression of our fascination with the ancient mystery that tiny seedling plants are able to amplify the extraordinarily weak gravitational signal with such rapidity and versatility. We ask our readers to view all the ideas in this essay with the same level of skepticism that we do! With the benefit of present technologies, these ideas should be subject to rapid revision, refinement and (we hope) extension.

Acknowledgments. Support for preparation of this review was provided by the Innovative Research Program of NASA via Grant NAGW-1782. We are grateful to Ian A. Newman and William F. Pickard for critical reading of the manuscript and stimulating suggestions. We are also grateful for stimulating discussions with Randy Wayne. We acknowledge inspiration from S. Krawczyk (1978), who published the first recordings of currents through single ion channels in plant cell membranes.

Addendum. This chapter was completed in October 1990. So many relevant new reports are appearing that updating is impractical; from our lab, however, these include the following: Ding JP, Pickard BG (1991) Mechanosensory calcium channels in plants. Plant J (in press) and Pont Lezica, McNally, Pickard. Wall-to-membrane linkers in onion epidermis (submitted).

References

Bandurski RS, Schulze A, Dayanandan P, Kaufman PB (1984) Response to gravity by *Zea mays* seedlings. I. Time course of the response. Plant Physiol 74:284–288

Barbier-Brygoo H, Ephritikhine G, Klämbt D, Ghislain M, Guern J (1989) Functional evidence for an auxin receptor at the plasmalemma of tobacco mesophyll protoplasts. Proc Natl Acad Sci USA 86:891–895

Biermann B, Feldman L (1990) Expression of a maize cDNA homologous to the catalytic domain of second messenger-dependent protein kinases. Plant Physiol 93 (Suppl):79

Brock TG, Kapan U, Ghosheh NS, Kaufman PB (1989) The effect of gravistimulation on auxin transport in oat pulvini. ASGSB Bull 3:127

Buder J (1961) Der Geotropismus der Characeenrhizoide. Ber Dtsch Bot Ges 74:14–23

Caspar T, Pickard BG (1989) Gravitropism by a starchless mutant of *Arabidopsis*: implications for the starch-statolith theory. Planta 177:185–197

Cassab GI, Varner JE (1988) Cell wall proteins. Annu Rev Plant Physiol Plant Molec Biol 39:321–353

Clifford PE, Fensom DS, Munt BI, McDowell WE (1982) Lateral stress initiates bending responses in dandelion peduncles: a clue to geotropism? Can J Bot 60:2671–2673

Clifford PE, Douglas S, McCartney GW (1989) Amyloplast sedimentation in shoot statocytes having a large central vacuole – further interpretation from electron microscopy. J Exp Bot 40:1341–1346

Cormier MJ, Anderson JM, Charbonneau H, Jones HP, McCann RO (1980) Plant and fungal calmodulin and the regulation of plant NAD kinase. In: Cheung WY (ed) Calcium and cell function, vol 1. Academic Press, London, pp 201–218

Cosgrove DJ, Hedrich R (1990) Stretch-activated cation, anion, and calcium channels coexist in guard cells. Plant Physiol 93(Suppl):17

Czapek F (1898) Weitere Beiträge zur Kentniss der geotropischen Reizbewegungen. Jahrb Wiss Bot 32:175–407

Ding JP, Pickard BG (1990a) Ca^{2+} channel (from onion epidermis) sensitive to stretch and voltage, and regulated by mural Ca^{2+} levels. 10th Int Biophys Congr, Vancouver (Abstr), p 359

Ding JP, Pickard BG (1990b) Epidermal, "high-conductance" Ca^{2+} channel responsive to stretch, voltage, and mural but not cytosolic Ca^{2+} is likely transducer for gravitational stimuli, osmotic shifts, touch, and the intracellular shears occurring during normal growth and development. Plant Physiol 93(Suppl):78

Ding JP, Pickard BG (1990c) Pulsed bursting of stretch and voltage activated Ca^{2+} channels: model and possible basis for several ultradian phenomena. Plant Physiol 93(Suppl):78

Ding JP, Pickard BG (1990d) Clustering of conductance units amplifies signal-to-force ratio for a putative gravitropic transducer channel. ASGSB Bull 4:106

Dure III L, Crouch M, Harada J, Ho T-HD, Mundy J, Quatrano R, Thomas T, Sung ZR (1989) Common amino acid sequence domains among the LEA proteins of higher plants. Plant Molec Biol 12:475–486

Edwards KL, Pickard BG (1987) Detection and transduction of physical stimuli in plants. In: Wagner E, Greppin H, Millet B (eds) The cell surface and signal transduction. Springer, Berlin Heidelberg New York, pp 41–66

Ewart AJ (1903) On the physics and physiology of protoplasmic streaming in plants. Clarendon Press, Oxford

Falke L, Edwards KL, Pickard BG, Misler S (1988) A stretch-activated anion channel in cultured tobacco cells. FEBS Lett 237:141–144

Friedmann M, Mannan RM, Poovaiah RW (1990) Calmodulin-binding proteins and protein phosphorylation in corn root tips. Plant Physiol 93 (Suppl):79

Gehring CA, Williams DA, Cody SH, Parish RW (1990) Phototropism and geotropism in maize coleoptiles are spatially correlated with increases in cytosolic free calcium. Nature (Lond) 345:528–530

Gillespie BC (1959) An hypothesis for the geotropic curvature of plants. MA Thesis, Stanford Univ, Stanford

Green PB (1987) Inheritance of pattern: analysis from phenotype to gene. Am Zool 27:657–573

Guharay F, Sachs F (1984) Stretch-activated single ion channel currents in tissue-cultured embryonic chick skeletal muscle. J Physiol 352:685–701

Harper JF, Schaller GE, Sussman MR, Putnam-Evans C, Charbonneau H, Harmon AC (1991) A calcium-dependent protein kinase with a regulatory domain that is similar to calmodulin. Science 252:951–954

Harrison MA, Pickard BG (1989) Auxin asymmetry during gravitropism by tomato hypocotyls. Plant Physiol 89:652–657

Hart JW (1990) Plant tropisms and other growth movements. Unwin Hyman, London

Hepler PK, Wayne RO (1985) Calcium and plant development. Annu Rev Plant Physiol 36:397–439

Hertel R (1983) The mechanism of auxin transport as a model for auxin action. Z Pflanzenphysiol 112:53–67

Hild V, Hertel R (1972) Initial phases of gravity-induced lateral auxin transport and geotropic curvature in corn coleoptiles. Planta 108:245–258

Hong B, Uknes S, Ho T-HD (1988) Cloning and characterization of a cDNA encoding a mRNA rapidly-induced by ABA in barley aleurone layers. Plant Molec Biol 11:495–506

Hong B, Pickard BG, Ho T-HD (1990) Is force focused on putative gravitransducer channels by an ABA-induced amphipathic protein? ASGSB Bull 4:100

Hosey MM, Lazdunski M (1988) Calcium channels: molecular pharmacology, structure and regulation. J Membr Biol 104:81–105

Huang PL, Corey DP (1990) Calcium influx into hair cell stereocilia: further evidence for transduction channels at the tips. Biophys J 57:530a

Jaffe MJ (1988) Calcium as a messenger in the thigmoperception of tendrils. Plant Physiol 86 (Suppl):41

Jaffe MJ, Biro R (1979) Thigmomorphogenesis: the effect of mechanical perturbation on the growth of plants, with special reference to anatomical changes, the role of ethylene, and

interaction with other environmental stresses. In: Mussell H, Staples RC (eds) Stress physiology in crop plants. Wiley, New York, pp 25–69
Jesuthasan S, Green PB (1989) On the mechanism of decussate phyllotaxis: biophysical studies on the tunica layer of *Vinca major*. Am J Bot 76:1152–1166
Jones AM (1990) Location of transported auxin in etiolated maize shoots using 5-azindoindole-3-acetic acid. Plant Physiol 93:1154–1161
Jost L; translated by Gibson RJH (1907) Lectures on plant physiology. Clarendon Press, Oxford
Kaufman PB, Brock TB, Song I, Rho YB (1987) How cereal grass shoots perceive and respond to gravity. Am J Bot 74:1446–1457
Klämbt D (1990) A view about the function of auxin-binding proteins at plasma-membranes. Plant Molec Biol 14:1045–1050
Knauth B, Klämbt D (1990) Is cell elongation regulated by extracellular auxin. Bot Acta 103:103–106
Krawczyk S (1978) Ionic channel formation in a living cell membrane. Nature (Lond) 273:56–57
Kutschera U (1987) Cooperation between outer and inner tissues in auxin-mediated plant organ growth. In: Cosgrove DJ, Knievel DP (eds) Physiology of cell expansion during plant growth. Am Soc Plant Physiol, Rockville, MD, USA, pp 215–226
Leopold AC, LaFavre AK (1989) Interactions between red light, abscisic acid and calcium in gravitropism. Plant Physiol 89:875–878
Li Y, Hagen G, Guilfoyle T (1990) Auxin induced genes and gravitropism. Plant Physiol 93 (Suppl):78
Lintilhac PM, Vesecky TB (1984) Stress-induced alignment of division plane in plant tissues grown in vitro. Nature (Lond) 307:363–364
Lund EJ (1947) Bioelectric fields and growth. Univ Texas Press, Austin
MacGregor RD, Sachs F, Hunt CA (1989) Mechanisms for opening stretch-activated ion channels. Biophys J 55:245a
Mannan DM, Jena PK, Poovaiah BW (1989) Role of calmodulin and calmodulin-dependent multifunctional protein kinase in gravitropism. ASGSB Bull 3:131
Martinac B, Adler J, Kung C (1990) Mechanosensory ion channels of *Escherichia coli* activated by amphipaths. Nature (Lond) 348:261–263
Millet B, Pickard BG (1988a) Gadolinium ion is an inhibitor for testing the putative role of stretch-activated ion channels in geotropism and thigmotropism. Biophys J 53:155a
Millet B, Pickard BG (1988b) Reproducible thigmotropic response by maize roots. Plant Physiol 86 (Suppl):29
Moran N (1990) Stretch-activated channels in plasmalemma of pulvinar motor cells. Plant Physiol (Suppl) 93:17
Morris CE (1990) Mechanosensitive ion channels. J Membr Biol 113:93–107
Morris DA (1980) The influence of small direct electric currents on the transport of auxin in intact plants. Planta 150:431–434
Newman IA (1963) Electric potentials and auxin translocation in *Avena*. Aust J Biol Sci 16:629–646
Nuccitelli R (1988) Physiological electric fields can influence cell motility, growth and polarity. Adv Cell Biol 2:213–233
Opsahl LR, Mak D-OD, Webb WW (1990) Membrane tension sensitivity of the alamethicin channel. 10th Int Biophysics Congr, Vancouver, (Abstr) p 398
Palevitz BA (1989) The cytoskeleton as a mediator between environmental signals and cell development. ASGSB Bull 3:25
Pickard BG (1985a) Early events in geotropism of seedling shoots. Annu Rev Plant Physiol 36:55–75
Pickard BG (1985b) Roles of hormones, protons and calcium in geotropism. In: Pharis RP, Reid DM (eds) Hormonal regulation of development III. Role of environmental factors. Encyclopedia of plant physiology. New Series, vol 11. Springer, Berlin Heidelberg New York, pp 193–281
Pickard BG, Suthar MR (1988) Interaction between wall, membrane and cytoskeleton in onion epidermis. ASGSB Bull 2:43
Pickard BG, Thimann KV (1966) Geotropic response of wheat coleoptiles in absence of amyloplast starch. J Gen Physiol 49:1065–1086
Pont-Lezica RF, Pickard BG (1990) Visualization with HRGP antibodies of wall-to-membrane linkers in epidermis. ASGSB Bull 4:107

Rathore KS, Robinson KR (1989) Ionic currents around developing embryos of higher plants in culture. Biol Bull 176(S):46–48

Rathore KS, Hodges TK, Robinson KR (1988) Ionic basis of currents in somatic embryos of *Daucus carota*. Planta 175:280–289

Raven JA (1979) The possible role of membrane electrophores is in the polar transport of IAA and other solutes in plant tissues. New Phytol 82:285–291

Ray PM (1987) Principles of plant growth. In: Cosgrove DJ, Knievel DP (eds) Physiology of cell expansion during plant growth. Am Soc Plant Physiol, Rockville, MD, USA, pp 1–17

Raynor RL, Zheng B, Kuo JF (1990) Membrane interactions of amphiphilic polypeptides mastoparan, melittin, polymyxin B, and cardiotoxin. J Biol Chem 266:2753–2758

Roberts K (1989) The plant extracellular matrix. Curr Opinion Cell Biol 1:1020–1027

Rubery PH (1987) Auxin transport. In Davies PJ (ed) Plant hormones and their role in plant growth and development. Martinus Nijhoff, Boston, pp 341–362

Ryan P, Newman I, Arif I (1990) Fluxes of Ca^{2+}, H^+ and K^+, and ion exchange in the cell wall of *Chara*. Plant Physiol (Suppl) 93:50

Sachs F (1988) Mechanical transduction in biological systems. CRC Crit Rev Biomed Eng 16:141–169

Schrank AR (1944) Relation between electrical and curvature responses in the *Avena* coleoptile to mechanical stimuli. Plant Physiol 19:198–211

Schrank AR (1951) Electrical polarity and auxins. In: Skoog F (ed) Plant growth substances. Univ Wisconsin Press, Madison, pp 123–140

Schrank AR (1957) Bioelectric implications in plant tropisms. Symp Soc Exp Biol 11:95–117

Schrank AR (1959) Electronasty and electrotropism. In: Ruhland W (ed) Encyclopedia of plant physiology, vol XVII, part I. Springer, Berlin Heidelberg New York, pp 148–163

Schrank AR, Backus GE (1951) The relationship of auxin to electrically induced growth responses in the *Avena* coleoptile. J Cell Comp Physiol 38:361–376

Senn AP, Goldsmith MHM (1988) Regulation of electrogenic proton pumping by auxin and fusicoccin as related to the growth of *Avena* coleoptiles. Plant Physiol 88:131–138

Sievers A, Volkmann D (1979) Gravitropism in single cells. In: Haupt W, Feinleib ME (eds) Encyclopedia of plant physiology. New Series, vol 7. Physiology and movements. Springer, Berlin Heidelberg New York, pp 567–572

Skou JC (1959) Studies on the influence of the degree of unfolding and the orientation of the side chains on the activity of a surface spread enzyme. Biochim Biophys Acta 31:1–10

Sokabe M, Sachs F (1990) The structure and dynamics of patch-clamped membranes: a study using differential interference contrast light microscopy. J Cell Biol 111:599–606

Stark P (1916) Experimentelle Untersuchungen über das Wesen und die Verbreitung der Kontaktreizbarkeit. Jahrb Wiss Bot 57:189–320

Stark P (1919) Das Resultantengesetz beim Haptotropismus. Jahrb Wiss Bot 58:201–207

Stark P (1921) Studien über traumatotrope und haptotrope Reizleitungsvorgänge mit besonderer Berücksichtigung der Reizübertragung auf fremde Arten und Gattungen. Jahrb Wiss Bot 60:67–134

Stark P (1922) Weitere Untersuchungen über das Resultantengesetz beim Haptotropismus. Jahrb Wiss Bot 61:126–167

Staves MP, Wayne R, Leopold AC (1990) Hydrostatic factors affect the gravity-induced polarity of cytoplasmic streaming in Characean algae. Plant Physiol 93 (Suppl):78

Stern K (1924) Elektrophysiologie der Pflanzen. Springer, Berlin

Timell TE (1986) Compression wood in gymnosperms, vol 2. Springer, Berlin Heidelberg New York

Tosteson MT, Alvarez O, Hubbell W, Bieganski RM, Attenbach C, Caporales LH, Levy JJ, Nutt RF, Rosenblatt M, Tosteson DC (1990) Biophys J 58:1367–1375

Trewavas AJ (ed) (1985) Molecular and cellular aspects of calcium in plant development. Plenum Press, New York

Tsien RW, Hess P, McCleskey EW, Rosenberg RL (1987) Calcium channels: mechanisms of selectivity, permeation and block. Annu Rev Biophys Biophys Chem 16:265–290

van der Schoot C, van Bel AJE (1990) Mapping membrane potential differences and dye-coupling in internodal tissues of tomato (*Solanum lycopersicum* L.). Planta 182:9–21

Wayne R, Staves MP, Leopold AC (1990) Gravity-dependent polarity of cytoplasmic streaming in *Nitellopsis*. Protoplasma 155:43–57

Webster Jr. WW, Schrank AR (1953) Electrical induction of lateral transport of 3-indoleacetic acid in the *Avena* coleoptile. Arch Biochem Biophys 47:107–118

White RG, Hyde GJ, Overall RL (1990) Microtubule arrays in regenerating *Mougeotia* protoplasts may be oriented by electric fields. Protoplasma 158:73–85

Wright M (1986) The acquisition of gravisensitivity during the development of nodes of *Avena fatua*. J Plant Growth Reg 5:37–48

Yamaguchi K (1990) The effects of auxin transport inhibitors on the reaction wood formation. Thesis, Kyoto Univ, Kyoto

Yang X-C, Sachs F (1989) Block of stretch-activated ion channels in *Xenopus* oocytes by gadolinium and calcium ions. Science 243:1068–1071

Chapter 6

Mechanoreceptors in Respiratory Systems

J.G. Widdicombe[1] and *G. Sant'Ambrogio*[2]

Contents

[1] Department of Physiology, St George's Hospital Medical School, London, SW17 0RE, UK
[2] Department of Physiology and Biophysics, University of Texas Medical Branch, Galveston, TX 77550, USA

Advances in Comparative and Environmental Physiology, Vol. 10

1 Introduction

The respiratory system can be divided into the respiratory tract and lungs, and the extrinsic respiratory muscles which are responsible for ventilating the lungs. The main extrinsic respiratory muscles are the diaphragm, intercostals and abdominal muscles, largely responsible for inspiratory and expiratory efforts. Many other muscles have an accessory respiratory function, including those in the neck and the perineum. In general, mechanoreception in the extrinsic respiratory muscles is similar to that in other skeletal muscles, the main differences being quantitative rather than qualitative, especially for the diaphragm.

The respiratory tract and lungs are essentially a tubular cul-de-sac, with quite different structural features in its various parts. For example, the main part of the nose has no extrinsic skeletal muscles, these being found only at the nostrils and palate. Its bony wall makes it a nondistensible cavity, and mechanoreceptors can only be found in the mucosa. On the other hand, the pharynx and larynx have conspicuous striated muscles in their walls, and mechanoreception can therefore reside both in the mucosa and in these intrinsic muscles. The lower respiratory tract, below the larynx, has smooth muscle in its walls, and mechanoreceptors reside both in this smooth muscle and in the mucosa. Finally, the alveoli lack any conspicuous contractile tissues, but nonetheless have mechanoreceptors in the alveolar walls that respond to tissue changes.

2 Upper Airway Receptors

2.1 Afferent Innervation of the Upper Airway

The afferent innervation to the upper airway is supplied by the trigeminal, glossopharyngeal and vagus nerves.

2.1.1 Receptors in the Nose

The ophthalmic and maxillary divisions of the trigeminal nerve contain fibres for the nonolfactory modalities of the nasal mucosa; the corresponding cell bodies are in the trigeminal ganglion. The ophthalmic branch leads to the anterior and posterior ethmoidal nerves that reach the upper and anterior regions of the lateral wall of the nasal cavity and the septum. The maxillary branch provides fibres to the inferior and middle turbinates and the corresponding septal region, the lower anterior portion of the septum and the incisive canal, the anterior part of the floor of the nose and the inferior meatus.

The presence of nonmyelinated fibres, presumed to be afferents, that terminate as free endings has been described in several studies (e.g. Cauna 1982). Differentiated nervous receptors have not been found within the nasal mucosa. Substance P-containing nerve fibres are abundant within and under the nasal epithelium as well as in the trigeminal ganglion (Uddman and Sundler 1986). Pretreatment with high doses of capsaicin depletes these fibres of substance P and causes a loss of response to various irritants. Other neuropeptides, like calcitonin gene-related peptide (CGRP), galanin and neurokinins have also been demonstrated in the nose (Uddman and Sundler 1986). These neuropeptides in association with the afferent nerves of the nose are presumably involved in axon reflexes, local reflexes through the trigeminal ganglion, as well as central reflexes.

2.1.2 Receptors in the Pharynx and Epipharynx

The nongustatory afferent innervation of the pharynx is provided by the lingual branch of the mandibular division of the trigeminal nerve for the anterior two-thirds of the tongue, and by the pharyngo-oesophageal branch of the glossopharyngeal nerve for the posterior third of the tongue and part of the pharynx through its contribution to the pharyngeal plexus.

Receptors with free endings have been described under and within the squamous cells of the epithelium of the epipharynx; they could represent either the rapidly or the slowly adapting receptors that have been characterized by recording single-unit activity in the pharyngo-oesophageal branch of the glossopharyngeal nerve (Nail et al. 1969; Hwang et al. 1984). At present no morphological data are available on the receptors presumably present in the muscles of the pharyngeal wall: muscle spindles, Golgi tendon organs etc. Such receptors could also account for some of the activity recorded from glossopharyngeal fibres.

2.1.3 Receptors in the Larynx

The vagus nerve carries the afferent supply to the larynx through its laryngeal branches. The superior laryngeal nerve (SLN) supplies most of the afferent fibres distributed to the ipsilateral and supraglottal areas of the larynx. The recurrent laryngeal nerve (RLN) and the external branch of the SLN that supply the motor innervation for the laryngeal muscles, have only a minor role; their afferent fibres are mainly distributed ipsilaterally and subglottically. The SLN has a considerable number of fibres in relation to the relatively small structure it innervates. In the cat only 22% of the total fibres (efferent and afferent) of the SLN are myelinated at birth, increase to 48% at 2 months (Miller and Dunmire 1976) and to about 90% in the adult (Miller and Loizzi 1974). The myelination process is accompanied by an increase in diameter and conduction velocity of the nerve fibres (Miller and Loizzi 1974; Sasaki 1979).

Laryngeal receptors have been classified according to their location as "mucosal", "articular" and "muscular" (Wyke and Kirchner 1976). A prevalence of undifferentiated, free-endings type receptors has been found in all three locations. "Corpuscular" receptors with a defined structure have been described in submucosal layers and in the fibrous capsules of the intercartilaginous joints. Endings similar to muscle spindles and Golgi tendon organs have been described within the intrinsic laryngeal muscles of some species. Indeed, fibre calibre spectra of laryngeal nerves of humans and cats are consistent with the presence of muscle proprioceptors (Wyke and Kirchner 1976). Although no data are available on the receptor types in the newborn, some inferences on muscle proprioceptors can be made from the studies of Sasaki et al. (1977) and Sasaki (1979) on the changes in diameter and myelination of the recurrent nerve of dogs during development. Most of the fibres are nonmyelinated at birth and it is only at 1 year of age that the RLN acquires the adult composition. Together with these changes in myelination there is a progressive increase in the diameter of the myelinated fibres which attain their adult values at about 8 weeks. It seems reasonable to speculate that corresponding to these changes in nerve composition and size exist some functional limitations during the early stages of development.

2.2 Functional Characteristics

2.2.1 Trigeminal Afferents

Changes in afferent activity have been documented in the trigeminal nerve in response to respiratory-related stimuli, such as airflow and transmural pressure as well as chemical and mechanical irritant stimuli (Dawson 1962; But and Klimova-Cherkasova 1967; Ulrich et al. 1972; Glebovsky and Bayev 1984; Tsubone 1989, 1990). Single and multifibre recordings from the ethmoidal branch of the trigeminal nerve have clearly demonstrated the presence of specific cold receptors responding to the cooling effect of airflow (Tsubone 1989). The presence within the ethmoidal nerve of units activated either by tactile stimuli or changes in transmural pressure (Tsubone 1990) has been clearly established. Some of the units that responded to light touch could also be recruited by noxious odourants such as ammonia. The vast majority of the receptors activated by pressure responded only to collapsing pressure (Fig. 1), and only a few could also be activated by distending pressure. Their response to maintained pressure indicates a slow rate of adaptation. The pressure threshold for their recruitment is higher than that of laryngeal pressure receptors to an extent that precludes, in most cases, their activation during normal, unobstructed breathing. Considering that the pressure endings are located within the mucosa or the submucosa overlying the rigid, bony structure of the nasal cavity, it must be conceived that relatively small distortions of the receptor microenvironment are sufficient for its activation. The lingual nerve, another branch of the trigeminal

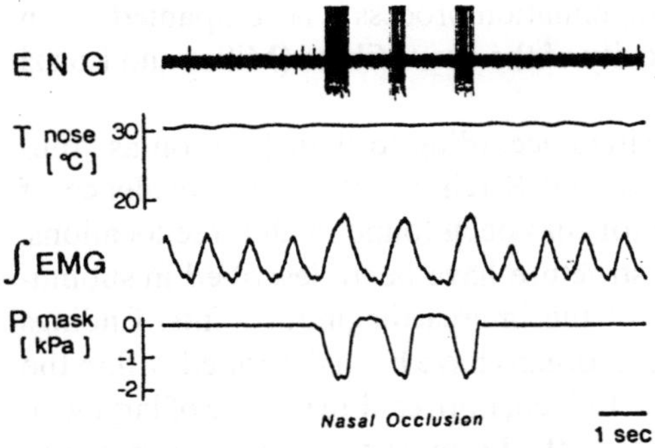

Fig. 1. Single-unit action potentials recorded from a nasal pressure receptor with the corresponding fibre in the ethmoidal nerve. Traces from *above*: *ENG* electroneurogram; *Tnose* temperature of the nasal cavity; *EMG* integrated electromyogram of an external intercostal muscle; *Pmask* pressure in the upper airway. (Tsubone 1990)

nerve, supplies the anterior two-thirds of the tongue and has afferent fibres that correspond to tactile, nociceptive and thermoceptive (warm and cold) modalities (Zotterman 1936).

2.2.2 Glossopharyngeal Afferents

Only a few studies are available concerning glossopharyngeal afferents. Since these studies were conducted in paralyzed and artificially ventilated animals, it is difficult to characterize the behaviour of the corresponding receptors in terms of the events normally occurring during spontaneous breathing. Several of the endings studied were silent during the respiratory cycle and were recruited, exhibiting a rapid adapting discharge, by mechanical stimuli (Nail et al. 1969). Some of them could be activated by airflow and pharyngeal distension as well. Recently, Hwang et al. (1984) recorded from afferent units of the glossopharyngeal nerve (pharyngo-oesophageal branch) that were activated by collapsing and/or distending pressure with either a slowly or a rapidly adapting discharge. Changes in transmural pressure in the nasopharyngeal and/or oropharyngeal regions could activate these units.

2.2.3 Laryngeal Afferents of the Vagus Nerve

Laryngeal endings have been studied in much greater detail than trigeminal and glossopharyngeal afferents. The overall afferent activity recorded from the superior laryngeal nerve reveals a strong respiratory modulation. At a single-unit level, this respiratory related activity can be attributed to specialized receptors activated by changes in pressure, temperature and laryngeal motion; this latter stimulation can be attributed either to the contraction of intrinsic laryngeal muscles or to transmitted tracheal tug (G. Sant'Ambrogio et al. 1983).

An ending stimulated by negative transmural pressure is recruited during inspiration through the upper airway, provided that its pressure threshold is reached as in the course of an inspiratory effort against an occluded upper airway. Such an ending remains silent during tracheostomy breathing and tracheal occlusions when the larynx is not exposed to any pressure change (Fig. 2). Receptors responding to

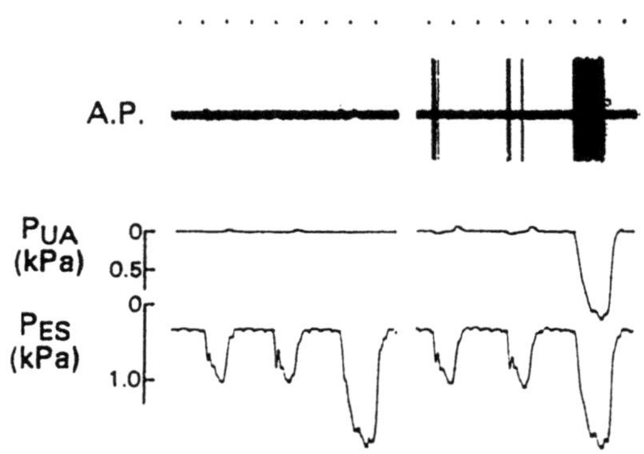

Fig. 2. Behaviour of a laryngeal pressure receptor during tracheostomy breathing and tracheal occlusion (*left panel*), during upper airway breathing and upper airway occlusion (*right panel*). Traces from *above*: time marker in seconds; *A.P.* action potentials; *Pua* pressure in the upper airway in kPa; *Pes* oesophageal pressure in kPa. Note that the receptor is active only in the presence of negative pressure in the upper airway (*right panel*)

pressure are the most numerous among the laryngeal endings (G. Sant'Ambrogio et al. 1983).

Some degree of pressure responsiveness is shown also by those receptors activated by laryngeal motion, but this stimulus does not represent the sole factor for their respiratory modulation. In fact these receptors are activated by laryngeal muscle contraction (e.g. the posterior cricoarytenoid muscle in Fig. 3) and/or tracheal tug, both conditions still present during tracheostomy breathing and during tracheal occlusion when the larynx is bypassed and not subjected to any change in pressure and temperature. This type of receptors has been defined as "drive", since it reflects the magnitude of the respiratory output (G. Sant'Ambrogio et al. 1983).

Specific cold receptors have been demonstrated on the vocal cords at the junction between the arytenoid process and the vocal ligament (G. Sant'Ambrogio et al. 1985a). Their activation during inspiration is uniquely due to the cooling effect of the inspired air: when the temperature and/or the humidity of the inspired air is such that no heat loss is permitted from the laryngeal mucosa their activation, all other factors being equal, ceases.

Another category of laryngeal endings remains silent or has a random spontaneous activity, generally unrelated to the respiratory cycle, and can strongly be activated by mechanical or chemical irritants (Fig. 4). By analogy with a similar group of tracheobronchial receptors, we may define these endings as laryngeal irritant receptors. "Drive" receptors are activated by the contraction of intrinsic laryn-

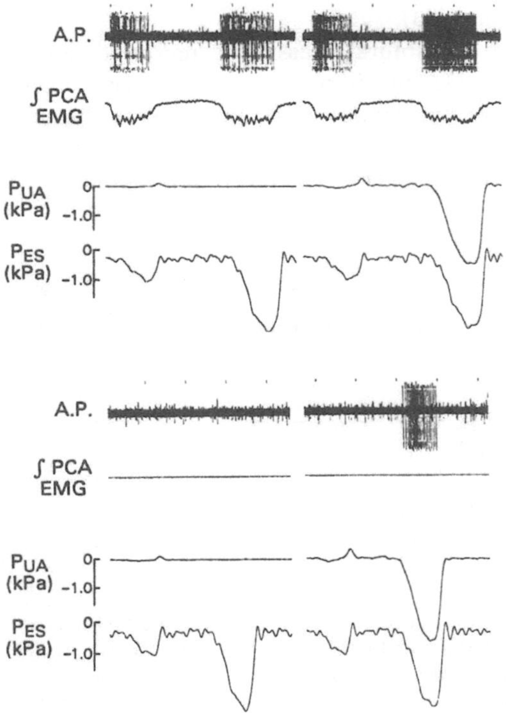

Fig. 3. Behaviour of a "drive" receptor during tracheal breathing and tracheal occlusion (*left panels*), upper airway breathing and upper airway occlusion (*right panels*). Paralysis of the posterior cricoarytenoid muscle, induced by cold block of the recurrent laryngeal nerves, eliminates or markedly reduces the receptor discharge (*lower panels*). The receptor is now activated by negative pressure in the upper airway when the upper airway is occluded (*lower right panel*). Traces from *above: A.P.* action potentials; *PCA EMG* electromyogram of the posterior cricoarytenoid muscle; *Pua* pressure in the upper airway; *Pes* pressure in the oesophagus. (Widdicombe et al. 1988)

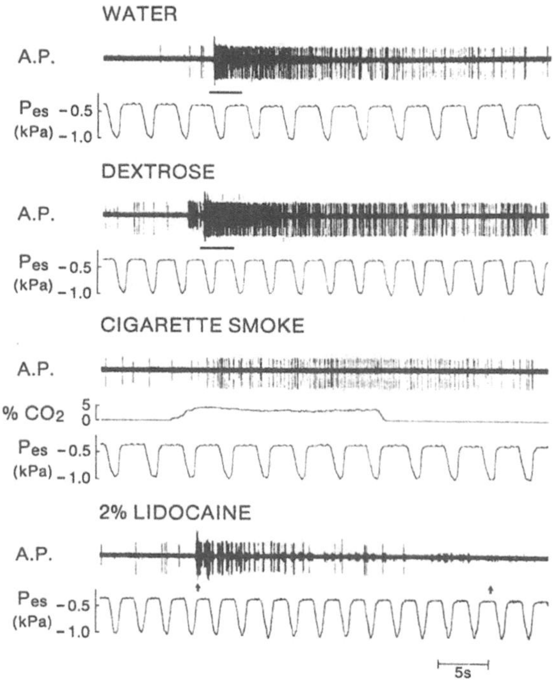

Fig. 4. Response of a nonmodulated laryngeal receptor to tussigenic agents. Exposure of the upper airway to smoke is indicated by the CO_2 signal. The response of the receptor to mechanical stimulation (*bottom panel*) was abolished by lidocaine in 21 s. This stimulation was performed (between *arrows*) with a cotton applicator soaked in a 2% solution of lidocaine.

geal muscles or the passive pull transmitted to the larynx through the trachea (tracheal tug). Selective paralysis of laryngeal muscles, by cold block of either the RLN or the external branch of the SLN, eliminates the respiratory modulation of about 40% of the "drive" receptors. In 15% of these receptors, the respiratory modulation could be abolished by fixation of the trachea, whereas a mixed situation was ascertained for the remaining "drive" receptors for which both muscle paralysis and blocking of the tracheal movement were necessary (G. Sant'Ambrogio et al. 1985b).

The observation that the contraction of laryngeal muscles modifies the activity of several of the "drive" receptors cannot be considered as an indication of a muscular location. In fact, even a mucosal ending could be stimulated by the same distortions resulting from muscular activity or tracheal tug.

Cooling of the laryngeal lumen, besides stimulating specific cold receptors, has a marked depressive effect on laryngeal mechanosensitive endings. This effect is much greater than that observed for tracheobronchial stretch receptors: a decrease of 10 °C in laryngeal luminal temperature reduces the activity of laryngeal mechanoreceptors by a factor of 2.5, while that of tracheal mechanoreceptors is reduced by a factor of 1.25 (F.B. Sant'Ambrogio et al. 1986; G. Sant'Ambrogio et al. 1986).

All types of laryngeal receptors that show a respiratory modulation, as well as those responsive to irritant stimuli, have fibres that, on the basis of their blocking temperature, can be classified as myelinated (Widdicombe et al. 1988).

2.2.3.1 Responses of Laryngeal Mechanoreceptors to Changes in Osmolality, Ionic Composition and Carbon Dioxide

Of great interest are the observations that several of the receptor types recognized for their mechanoresponsiveness can be influenced by changes in osmolality and/or ionic composition of the surface liquid overlying the laryngeal mucosa. Since these changes could occur following environmental exposures and/or modifications of airway secretions, we can envisage the possibility of significant alterations in receptor functions and related reflex responses.

The response of laryngeal endings to water has been studied extensively in newborn and adults of several species (Boushey et al. 1974; Harding et al. 1978; Storey and Johnson 1975; Boggs and Bartlett 1982). Receptors from adult and newborn mammals exhibit similar responses to water and various solutions and their stimulation was found to depend on the lack of chloride or other permeant anions. The water-responsive endings are not identifiable with taste buds, but represent presumably a distinct group of myelinated fibres with free endings located within the laryngeal mucosa (Boushey et al. 1974; Harding et al. 1978). Harding et al. (1978) demonstrated that the water responsive laryngeal endings do not belong to a homogeneous category, but exhibit differences in latency and duration of response, mechanosensitivity etc. Recently, the presence of two distinct types of responses has been confirmed and further characterized (Fig. 5): a short latency, short duration response and a long latency, long duration response (Anderson et al. 1990a). The short latency, short duration response depends on the lack of chloride anions, whereas the long latency, long duration response depends on the hypoosmolality of the solutions tested (Anderson et al. 1990a). Receptors activated by hypoosmolar solutions have a clear respiratory modulation and are generally identifiable with "drive" and pressure laryngeal endings. Those stimulated by the lack of chloride anions are either silent or randomly active during the breathing cycle and are identifiable with the laryngeal "irritant" receptors.

Reflex responses involved in the regulation of upper airway muscles and the maintenance of upper airway patency depend on pressure-responsive laryngeal endings (Mathew et al. 1982; F.B. Sant'Ambrogio et al. 1985). Considering that a large proportion of these receptors respond also to water, we may speculate that changes in osmolality of the fluid overlying the laryngeal mucosa could modify the responses to transmural pressure and thus affect the mechanisms for the preservation of upper airway patency. On the other hand, considering that both low chloride solutions and several tussigenic stimuli affect the nonmodulated irritant laryngeal receptors, we may also speculate that changes in the ionic composition of the laryngeal surface liquid may influence the responsiveness of these endings, thereby modifying the threshold or the strength of the cough reflex.

Studies on laryngeal receptors and reflexes have clearly demonstrated the presence in the larynx of endings responsive to carbon dioxide (Boushey et al. 1974; Boushey and Richardson 1973). In fact, carbon dioxide has an excitatory effect on some receptors and an inhibitory influence on others. Most of the receptors found to be stimulated belong to the type of receptors either silent or randomly active during the breathing cycle that respond vigorously to irritant stimuli and to solutions lacking chloride anions, i.e. laryngeal irritant receptors (Anderson et al.

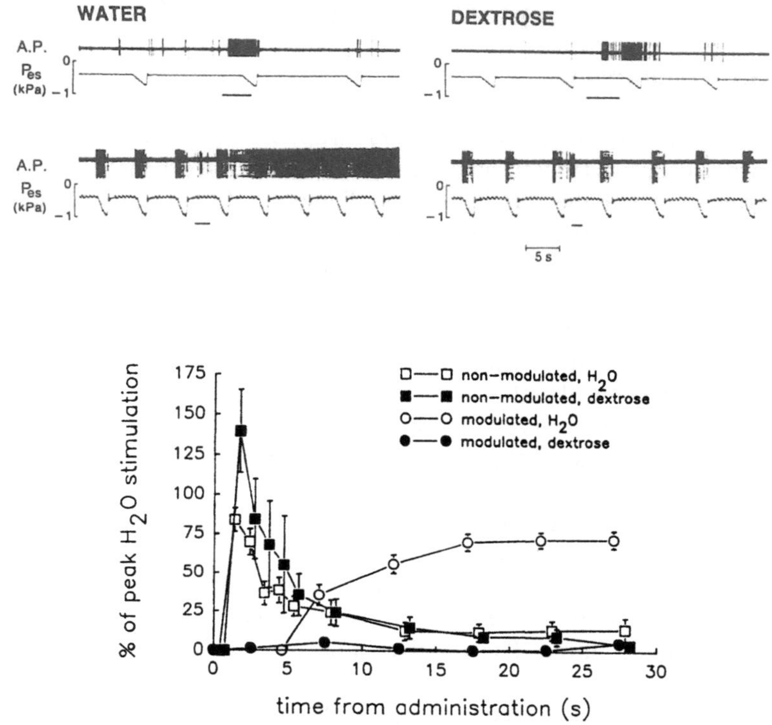

Fig. 5. The behaviour of two different types of water-responsive receptors is represented in the *upper* panels. In each panel traces are: action potentials (*A.P.*) and oesophageal pressure (*Pes*) in kPa. Application of the test solutions is indicated by *thick lines*. The nonrespiratory-modulated receptor (*upper traces*) was stimulated by both water and dextrose, while the respiratory-modulated receptor (*lower traces*) was only stimulated by water. The time course of the response of the two types of laryngeal receptors is represented in the *lower* panel. Note that the nonmodulated receptors (n = 12) responded with a short-delay, short-duration discharge to both water and dextrose, while the modulated receptors (n = 21) responded only to water with a long-delay, long-duration discharge. (Anderson et al. 1990a)

1990b). On the other hand, most of the receptors inhibited by carbon dioxide show a regular respiratory modulation and belong to the types of receptors described as "drive" or pressure endings (Anderson et al. 1990b). Similarly to the considerations presented above concerning the changes in osmolality and ionic composition of the surface liquid of the larynx, we may speculate that changes in carbon dioxide concentration within the upper airway lumen could modify the responses of upper airway muscles to collapsing pressure, as well as the cough reflex mediated by laryngeal pressure and irritant receptors, respectively.

3 Mechanoreceptors of the Respiratory Muscles and Joints

3.1 Diaphragm

On histological analysis the diaphragm appears to contain a scant number of muscle proprioceptors (muscle spindles and Golgi tendon organs). These observations are consistent with the scarcity of large-diameter afferent (group I) and small-diameter efferent fibres (representing the fusimotor gamma supply) in the phrenic nerve. The ratio of spindles to tendon organs is particularly low in comparison with other skeletal muscles; only four to five muscle spindles have been found in each hemidiaphragm of the cat, mostly located in the vertebral region (Duron 1981). The other receptors present are undifferentiated free-endings, Pacinian corpuscles and encapsulated peritoneal end-organs on the abdominal surface.

As in other nerves the nonmyelinated fibres largely exceed in number the myelinated ones: in the cat 1900 vs 770. However, they represent only a small proportion of the total cross-sectional area of the phrenic nerve with a good correspondence to their contribution to the composite evoked action potential from the whole nerve (Duron 1981).

Recording from the afferent fibres of the phrenic nerve during quiet breathing reveals only a few active fibres whose conduction velocity is consistent with group I myelinated fibres. The relative importance of group Ia (muscle spindles) and group Ib (Golgi tendon organs) fibres is, however, still uncertain due to the difficulty of identifying these receptors within the diaphragm. The majority of the endings that could be recognized as tendon organs increased their activity during inspiration, being silent or firing at lower rate in expiration. When single shocks were given to the ventral root corresponding to the dorsal root site of the recorded activity, brief bursts of action potentials could be evoked, indicating the "in series" arrangement of these endings (Fig. 6). Corda et al. (1965) determined that the receptors usually active during eupnoea are Golgi tendon organs which thus have a relatively low threshold of activation. Essentially two patterns of activity have been recorded during eupnoea from diaphragmatic receptors identified as muscle spindles (Corda et al. 1965). In most cases they increase their activity in expiration, behaving as "passive" stretch receptors in response to the expiratory stretching of the diaphragm. A selective block of the fusimotor fibres, obtained by applying lignocaine onto the phrenic nerve, lowers the frequency at which the spindle discharges, indicating some degree of fusimotor drive (Fig. 7). Only a few spindles have an inspiratory augmentation of their discharge, indicating the presence of a strong inspiratory drive that overrides the effect of the inspiratory diaphragmatic shortening. In fact, blocking of fusimotor fibres changes the behaviour of these spindles that become "passive" in type. Spindle-type afferents with an intermediate behaviour have also been described.

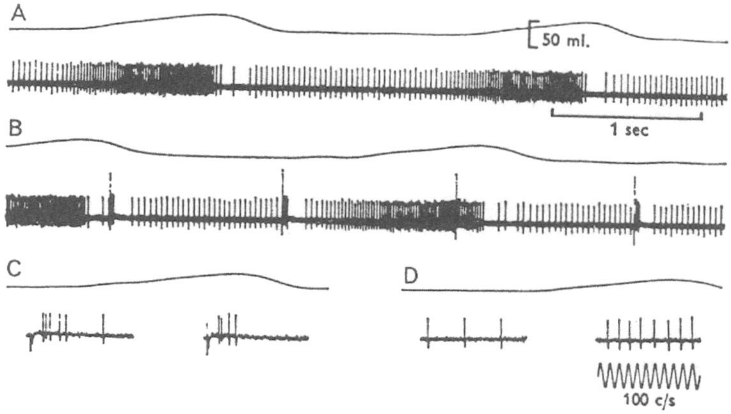

Fig. 6A-D. Discharge from a tendon organ located in the crural part of the diaphragm. Recorded from C5 dorsal root, together with tidal volume (*upper tracings*). *A* Spontaneous breathing; *B* single shocks to the ventral root of C5 2db above threshold for diaphragmatic contraction; *C* same on sweeps; *D* serves as control for C. (Corda et al. 1965)

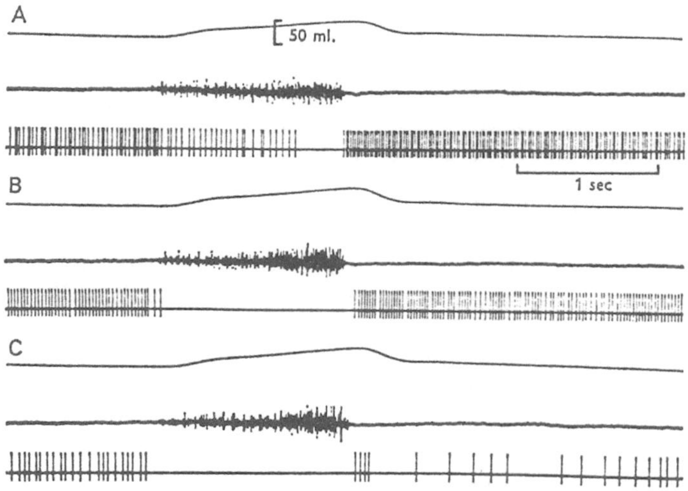

Fig. 7A-C. Discharge from a diaphragmatic muscle spindle of the "passive" type. Spindle located in the pars costalis of the diaphragm, and recorded from C5 dorsal root together with tidal volume (*top tracing*, upward deflection denotes inspiration) and the electromyogram of the diaphragm (*middle tracing*). Spontaneous breathing of air. *A* Control; *B* 3 min; *C* 4–5 min after application of 0.4% lignocaine solution to the phrenic nerve. Lignocaine blockade of the fusimotor fibres led to a general decrease in firing rate. Note, no effect on EMG or tidal volume. (Corda et al. 1965)

3.2 Intercostal Muscles

The number of muscle proprioceptors is greater in the intercostal muscles than in the diaphragm, but it varies considerably among them. In general, the muscles of the more cranial intercostal spaces have a greater number of spindles than those of the more caudal interspaces. The external intercostals have the highest density of spindles and the interchondral portion of the internal intercostals the lowest density, with the remaining internal intercostals having an intermediate value. The fibre spectrum of the corresponding intercostal nerves reflects these relative values. As a general conclusion we may say that all the cranial external and internal intercostal muscles have, for the high concentration of spindles, characteristics of postural muscles. On the other hand, the diaphragm and interchondral intercostals, that have a more specialized inspiratory function, are characterized by a scant spindles density. At variance with the diaphragm, muscle spindles are more numerous than Golgi tendon organs in the intercostal muscles (Duron 1981).

Intercostal muscle spindles have been studied to a greater extent than Golgi tendon organs (Critchlow and von Euler 1963; Sears 1963). Virtually no information exists concerning the functional properties of other receptors type, like the Pacinian corpuscles. The vast majority of intercostal spindles show a respiratory modulation: inspiratory augmenting for the external intercostals and expiratory augmenting for the internal intercostals. These types of modulation are present in both primary and secondary endings. A selective block of the small fusimotor (gamma) fibres changes the pattern of spindle discharge, indicating the presence of a strong fusimotor drive during the phase of respiration in which the corresponding muscle contracts and shortens (Fig. 8). Experimental results have in fact demonstrated the presence of a phasic inspiratory fusimotor drive to inspiratory intercostal spindles that not only offsets the unloading of the receptor, due to shortening of the main muscle, but actually increases its discharge during the corresponding phase of the breathing cycle. An analogous behaviour is seen in expiratory intercostal spindles which in fact increase their activity in expiration. However, not all the fusimotor fibres have a phasic pattern of discharge. Some have a "tonic" activation without a clear co-activation with the alpha motoneurones. "Tonic" fusimotor activity seems to be more promptly affected by cerebellar influences and postural reflexes.

3.3 Abdominal Muscles

No direct data are available on abdominal muscle proprioceptors. However, experimental evidence derived by stimulation of afferent fibres from abdominal muscles indicates that this activity has characteristics similar to that of internal intercostal afferents.

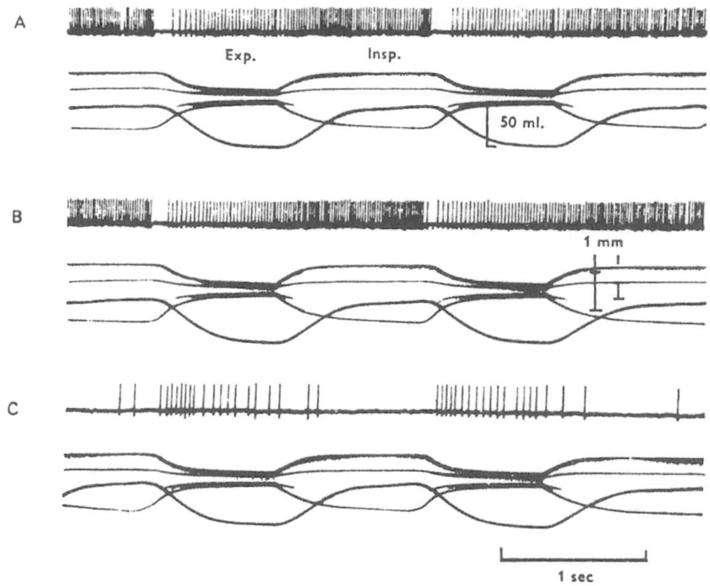

Fig. 8A-C. Afferent discharge from a muscle spindle of an inspiratory intercostal muscle. *A* control; *B* 2.5 min after application of lidocaine solution 0.25% to the intercostal nerve; *C* 3.5 min after the lidocaine application. Alpha-motoneurone conduction appeared unimpaired at 4.2 min after cocainization as judged from the "myographic" record of the response to ventral root stimulation. Note the "reversal" of the discharge pattern. Intercostal width (*middle traces*) and tidal volume (*lowest trace*) recorded together with the neurogram. (Critchlow and von Euler 1963)

3.4 Costovertebral Joints

Unitary action potentials originating from costovertebral joints have been recorded in thoracic dorsal roots of cats and rabbits (Godwin-Austin 1969). The corresponding endings are localized within the capsule of the costotransverse joints. These receptors are stimulated by movements of the rib that determines an increase in tension in the fibres of the capsule in which they are located. Those receptors placed within the rostral aspect of the capsule are stimulated by a caudal (expiratory) movement of the rib, while those within the caudal part of the capsule are stimulated by rostral (inspiratory) displacement of the ribs. Approximately three-quarters of the receptors increase their activity with expiratory displacements (Fig. 9). The majority of the endings studied have characteristics of slowly adapting receptors and have a dynamic response directly related to the velocity of the movement. A few endings with characteristics of rapidly adapting receptors have also been found: they respond to rib displacement with a brief burst of action potentials.

In summary, we can conclude that receptors in the costovertebral joints relay information on the position of the ribs, and on the direction and velocity of their movements.

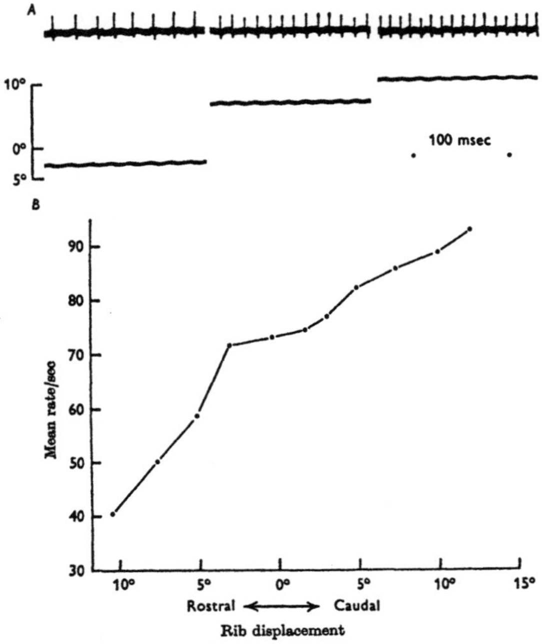

Fig. 9A,B. Relationship between rib position and joint receptor discharge rate. *A* Upper traces show the unitary discharge of a slowly adapting joint mechanoreceptor at three rib positions taken from the complete series plotted in B. Rib position is recorded in the *lower trace*. *B* A graph of mean discharge rate in impulses/s plotted against rib position. (Godwin-Austen 1969)

4 Lower Respiratory Tract and Lungs

Mechanoreception from the lower respiratory tract and lungs is predominantly vagal. A few studies have described some symnpathetic pathways for bronchopulmonary afferent fibres but, in general, if the vagus nerves are cut, all mechanosensitive reflexes from the lungs disappear. Like the other nerves supplying the respiratory tract, the vagi consist largely of nonmyelinated fibres, and the same is true of the pulmonary branches of the vagus nerves (Sant'Ambrogio, 1982; J.C.G. Coleridge and H.M. Coleridge 1984; H.M. Coleridge and J.C.G. Coleridge 1986).

4.1 Slowly Adapting Pulmonary Stretch Receptors

The existence of sensory receptors responding to lung inflation was deduced by Hering and Breuer in 1868 when they described the reflex that bears their name. They showed that inflation of the lung cut short inspiratory muscle activity and prolonged expiration, and that this reflex was mediated by a vagal pathway. The subsequent studies of Adrian (1933), who recorded activity in single vagal nerve fibres from slowly adapting stretch receptors, established the nature of this vagal pathway and showed many of the properties of the receptors. They have since been abundantly studied.

Single-fibre recording shows that the receptors are often of low-volume threshold, and many have a spontaneous discharge at functional residual capacity.

Inflation of the lungs stimulates them or increases their activity, and maintained inflation causes a slowly adapting regular discharge, the properties of which fit well with the Hering-Breuer inflation reflex. Thus inflation both recruits new receptors and increases the discharge of those firing. The vagal fibres are myelinated, with conduction velocities in the range 15–60 m s^{-1} (Saint'Ambrogio 1982; Coleridge and Coleridge 1986).

4.1.1 Histology

Indirect evidence has long suggested that the slowly adapting receptors are in the airway smooth muscle. They are found most frequently in the airway wall which has much smooth muscle; they are rather insensitive to luminal mechanical stimulation; and their discharge patterns are affected by airway smooth muscle contraction. However, direct evidence for their histological identification had to await the work of Krauhs (1984). With sequential electron microscopy, she mapped out receptors in the airway smooth muscle, and showed that they were complex structures with ramifying nerve terminals and somewhat resembling carotid sinus baroreceptors (Fig. 10). The ramifications are nonmyelinated and their general orientation is in the direction of the smooth muscle fibres. Electron microscopic studies showed that the receptor nerve terminals were closely associated with collagen that appeared to be attached to their membranes. How the receptors are stimulated is not clear. Presumably the collagen attachment pulls on the nerve terminal membranes, setting up a generator potential which spreads through the ramifications and summates eventually to cause an action potential in the afferent nerve.

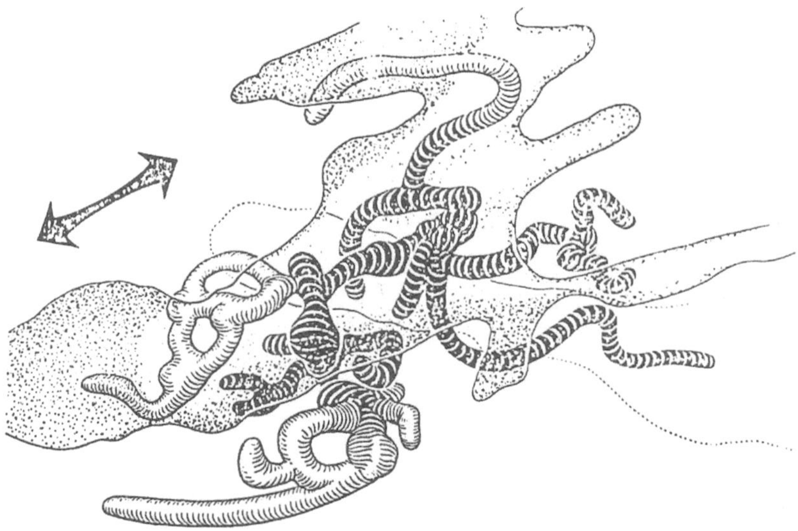

Fig. 10. Reconstruction from electron micrograph of a slowly adapting stretch receptor from the trachea of a dog. The *stippled* area represents smooth muscle, and the *tubular* structure the receptor complex. The *arrows* show the longitudinal direction of contraction of the muscle. (Krauhs 1984)

4.1.2 Functional Characteristics

The slowly adapting receptors are sensitive to stretching of the airway wall, as happens on inflation of the lungs (Fig. 11), and as already described. As lungs and airways inflate, the total discharge from receptors builds up partly be recruitment of new receptors. Deflation of the lungs inhibits some receptors but stimulates others. The lung volume/receptor discharge relationship is fairly complex, at least two different types of receptor having been identified on this basis (Sant'Ambrogio 1982). Whether the receptors with different discharge characteristics also have different reflex actions, quantitatively or qualitatively, is not known. Although they are called slowly adapting, in fact they have a wide range of adaptation rates (Widdicombe 1954). Indeed it has been suggested that the slowly and rapidly adapting receptors (see below) are part of the same series and have the same reflex actions (Paintal 1983); however, there is much evidence against this view (Coleridge and Coleridge 1986; Widdicombe 1986).

The slowly adapting receptors are not very sensitive to intraluminal stimuli, but respond readily to mechanical probing of the airway smooth muscle, and they are stimulated or sensitized by contraction of the smooth muscle; this last observation is consistent with their lying in series with smooth muscle elements (Coleridge and Coleridge 1986). Thus, potentially smooth muscle contraction can modulate the strength of the Hering-Breuer inflation reflex, although such modulation has not been shown to have physiological importance. The receptors are relatively insensitive to chemical agents, although those such as histamine and acetylcholine, which cause smooth muscle contraction and change lung mechanics, will indirectly affect the frequency of discharge of the receptors. The only physiological chemical change that acts on the receptors seems to be a reduction in the concentration of inhaled carbon dioxide, which stimulates receptor activity. Conversely, an increase in CO_2 tension from low values to physiological or higher values will inhibit the receptors (Sant'Ambrogio 1982; Coleridge and Coleridge 1986). It has been suggested that this mechanism is one component of the respiratory response to inhaled CO_2, the relative inhibition of receptors tending to stimulate breathing. At a more pathological level, high concentrations of sulphur dioxide will completely inhibit slowly adapting receptors, at least in the rabbit, and this property has been used to study

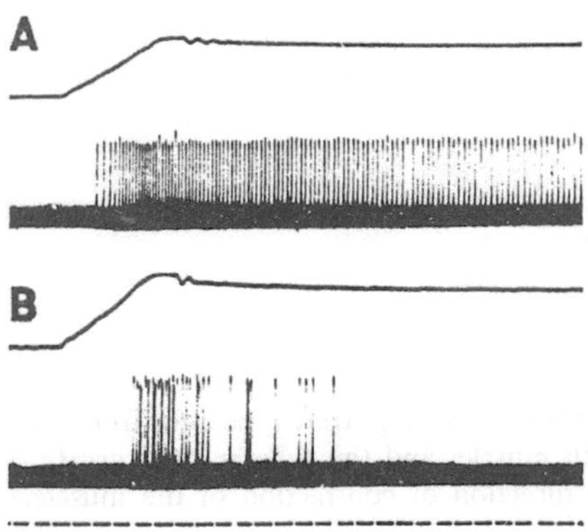

Fig. 11A,B. Responses to inflation of the lungs of two kinds of afferent fibre in the vagus of a cat. The first receptor (*A*) adapts slowly, the second (*B*) very rapidly to maintained inflation of the lungs. *Upper traces* intratracheal pressure; *lower traces* action potentials; time in 0.1 s. (Knowlton and Larrabee 1946)

pulmonary reflexes in this species by the selective abolition of the Hering-Breuer inflation reflex (Davies et al. 1978).

4.1.3 Reflex Actions

In terms of reflexes, the slowly adapting receptors cut short inspiration and prolong expiration. These phenomena have been extensively studied and related to the neuronal circuitry of the medullary respiratory rhythm generator (von Euler 1986). Slowly adapting receptor activity also reflexly inhibits airway smooth muscle, and causes reflex tachycardia (Daly 1986). In man the respiratory effects of the reflex are weak compared with many other mammalian species (Widdicombe 1961), and possibly the reflex actions on the autonomic nervous system are more important. The receptors in human airways have been identified histologically, and functionally by vagal fibre recording.

4.2 Rapidly Adapting Stretch Receptors

In 1929, Keller and Loeser identified a second group of pulmonary stretch receptors, which adapted rapidly to maintained inflations and deflations of the lungs. They concluded that they might mediate the cough reflex. This view has been well substantiated. The receptors were characterized by single-fibre recording by Knowlton and Larrabee in 1946 (Fig. 11).

4.2.1 Histology

Rapidly adapting receptors can be identified during single-fibre recording by gentle mechanical probing of the airway mucosa (Widdicombe 1954). This method shows that, at least in the cat, they are concentrated in the larger airways and especially at the points of airway branching. Electron microscopic studies of these areas show nonmyelinated fibres in the epithelium, both at its basal layer and close to the tight junctions near the lumen (Das et al. 1978) (Fig.12). Light microscopy shows that many of these epithelial nerves are connected to myelinated fibres passing to the vagus nerves, and the fibres from rapidly adapting receptors fit this picture, having conduction velocities in the range 15–40 m s^{-1}. The epithelial nerves can be shown to be afferent or sensory, since they degenerate on midcervical vagotomy, which would leave parasympathetic postganglionic nerve fibres intact (Das et al. 1979). Although the epithelial nerves seem to be part of the rapidly adapting receptor complex, the full extent of the receptor has not been mapped out. However, in the dog they may extend to a range 1 cm in diameter, with branches both near the mucosal surface and deep in the smooth muscle (Sant'Ambrogio et al. 1978). The receptors have been identified histologically in most mammalian species, including man. Histologically the mouse has no epithelial nerves (Pack et al. 1984), and the ferret none in the trachea and only a few in the bronchi (Robinson et al. 1986). Interestingly, these two species lack a cough reflex from the tracheobronchial tree.

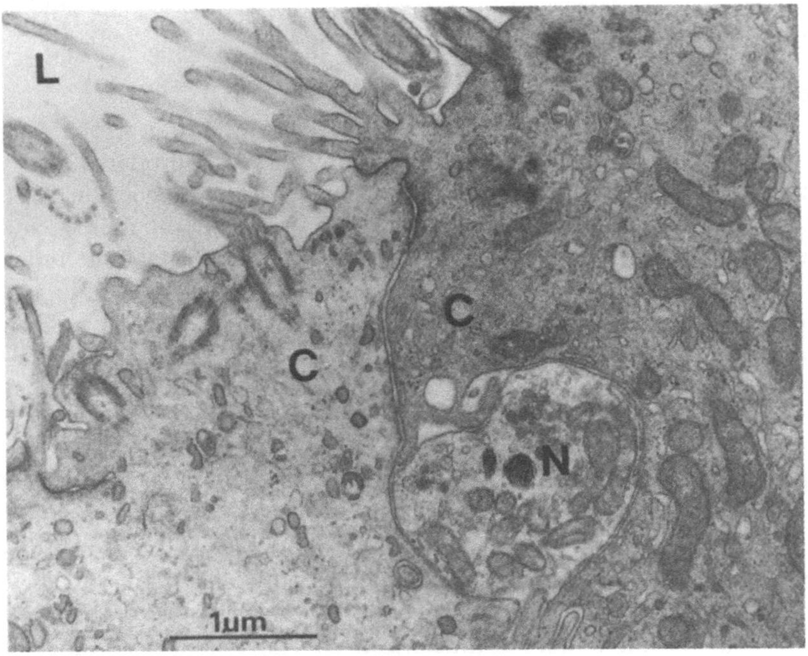

Fig. 12. An electron micrograph of normal tracheal epithelium. An axon profile (*N*) is located between ciliated (*C*) cells close to the airway lumen (*L*). The profile contains many small agranular vesicles and some mitochondria. (Laitinen 1986)

4.2.2 Functional Characteristics

Maintained inflation and deflation of the airways causes a rapidly adapting discharge at a highly irregular, probably random frequency (Fig. 11). The range of adaptation varies, the most rapidly adapting receptors only responding during change of stretch, and these also show a pronounced off-response rather like a Pacinian corpuscle (Widdicombe 1954). Others respond with a lower adaptation rate and may individually be stimulated predominantly either by inflation or by deflation of the airways. All receptors seem to be very sensitive to intraluminal stimulation, consistent with their terminal fibres being in the epithelium. Light mechanical probing of the mucosa, the inhalation of inert dust, and probably even the movement of mucus will stimulate them (Coleridge and Coleridge 1986). In addition, they respond to a very wide range of chemical irritants and inflammatory mediators, and for this reason have sometimes been called "irritant receptors" (Mills et al. 1970). Histamine, acetylcholine, 5-hydroxytryptamine and some prostaglandins will stimulate them (Fig. 13), whether given into the vascular bed or by aerosol. Inhaled irritants that excite the rapidly adapting receptors include sulphur dioxide, ammonia, ethyl ether vapour and cigarette smoke. In some instances these stimuli cause maintained discharges, unlike the rapidly adapting response to a me-

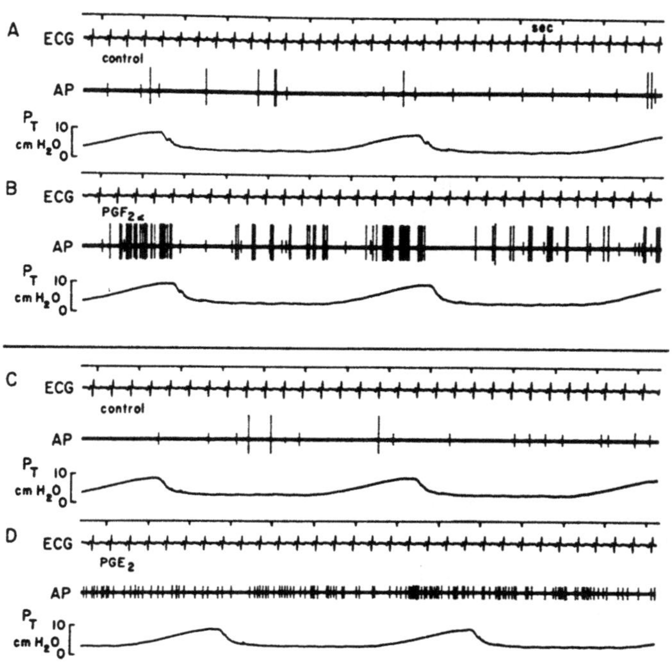

Fig. 13A-D. Comparison of the effects of prostaglandin (PG) F_2 and PGE_2 on an "irritant" receptor (*larger spikes*) and a pulmonary C-fibre (*smaller spikes*) in a dog with open chest and on positive-pressure ventilation. Both endings were located in the lower lobe of the left lung. *A* Before, and *B*, 16 s after injection of 4 µg kg^{-1} PGF_2 into the right atrium. The interval between B and C was 6 min. *C* Before, and *D*, 42 s after injection of 20 µg kg^{-1} PGE_2 into the right atrium. From *above downward* in each record: 1-s time trace; electrocardiogram (*ECG*); action potentials (*AP*) recorded from an afferent slip of the left vagus nerve; tracheal pressure (P_T). (Coleridge and Coleridge 1986)

chanical stimulus. Thus the receptors appear to be polymodal, a term used to describe some somatic nociceptors.

Contraction of airway smooth muscle underlying the receptors appears to stimulate or sensitize them, and the action of some agents such as histamine may be partly due to this indirect effect as well as to a direct action on the receptor (Coleridge and Coleridge 1986). Recent work has indicated that the receptors are sensitized by oedema in the airway walls (Kappagoda et al. 1987), so inflammatory mediators that produce oedema may affect the receptors by this third mechanism as well.

In view of their patterns of stimulation it is not surprising that the rapidly adapting receptors can be activated in a variety of experimental lung diseases. These include anaphylactic reactions, microembolism, atalectesis, pulmonary congestion and oedema, and pneumothorax (Mills et al. 1969, 1970). In all of these conditions one would expect either mechanical changes in the lungs or the release of inflammatory and other mediators which would stimulate or sensitize the rapidly adapting receptors.

4.2.3 Reflex Actions

As already implied, the main reflex response from the receptors seems to be the cough. However, these receptors may have a more physiological role. As concluded by Larrabee and Knowlton (1946), their activation by lung inflation sets up a deeper inspiration or augmented breath, the deep sighs that occasionally take place in mammals, presumably to open up collapsing parts of the lungs. Selective stimulation of rapidly adapting receptors by brief pressure pulses triggers an augmented breath, and also cuts short the expiratory pause (Davies and Roumy 1982). Thus their actions seem to be the opposite of those of slowly adapting receptors which cut short inspiration and prolong expiration. The similarity between an augmented breath and the preliminary deep inspiration of the cough relfex is striking, but the two have not been established as mechanistically the same.

Stimulation of rapidly adapting receptors reflexly contracts airway smooth muscle and causes secretion of mucus (Coleridge and Coleridge 1986). Since both these events can in turn stimulate or sensitize the rapidly adapting receptors, the possibility of positive feedback exists. Reflex actions of the rapidly adapting receptors on other autonomic outputs such as those to the cardiovascular system have not been established, but have possibly not been sought.

4.3 Pulmonary and Bronchial C-Fibre Receptors

These receptors, by definition, have nonmyelinated afferent fibres. Those in the lung parenchyma were originally studied by Paintal (1955, 1969), who described them as J-receptors (juxtapulmonary capillary receptors) a term now usually replaced by pulmonary C-fibre receptors. These should be distinguished from the bronchial C-fibre receptors (Coleridge and Coleridge 1984). Although both have many reflex characteristics in common, experimentally the former are affected by agents put into the pulmonary vascular bed and the latter by agents in the systemic (bronchial) circulation. There are differences in their relative sensitivity to inflammatory and other mediators.

4.3.1 Histology

C-fibres have been identified in the alveolar wall, both in experimental animals and man (Meyrick and Read 1971; Fox et al. 1980). It is likely that these are parts of the C-fibre receptor complex. As described above, nonmyelinated nerves have been identified in the airway epithelium. Although indirect evidence suggests that most or many of these are parts of the rapidly adapting receptor complex, some could be terminals of the C-fibre receptors. For both the pulmonary and bronchial C-fibre receptors, however, no detailed receptor maps have been worked out and the extent of the terminal complex of a single receptor is not known.

4.3.2 Functional Characteristics

C-fibre receptors respond to almost the same range of stimuli as the rapidly adapting receptors, but in general are more sensitive to chemical and less sensitive to mechanical stimuli (Coleridge and Coleridge 1984). Thus they can be identified by mechanical probing, respond to large lung inflations but not deflations, and are sensitive to a variety of chemical irritants. These include agents such as phosgene that cause pulmonary oedema and, at least for the pulmonary C-fibre receptors, increased interstitial liquid is thought to be one of the most important stimuli (Paintal 1969). Inflammatory mediators that activate them include histamine, 5-hydroxytryptamine, acetylcholine and some prostaglandins (Fig. 13). Thus, like the rapidly adapting receptors, the C-fibre receptors are polymodal. It is therefore not surprising that they are activated in experimental pathological conditions such as pulmonary oedema, pneumonia, microembolism and anaphylactic reactions.

4.3.3 Reflex Actions

Reflexly the C-fibre receptors cause apnoea followed by rapid shallow breathing when stimulated vigorously, and rapid shallow breathing when the stimulus is smaller (Coleridge and Coleridge 1984; Widdicombe 1986). There has been some dispute as to whether the C-fibre receptors also can induce cough (Paintal 1986). This has never been described during their selective activation under experimental conditions, and indeed some recent evidence shows that the pulmonary C-fibre receptors may inhibit cough (Tatar et al. 1988). However, the same stimuli that cause coughing will usually also activate both C-fibre receptors and rapidly adapting receptors, and the reflexes set up from the former may well modulate the strength and pattern of the cough. C-fibre receptors cause powerful secretion of mucus in the lower airways and bronchoconstriction, and a range of other autonomic nervous reflexes such as bradycardia and vasodilatation. The reflex vasodilatation includes the vascular beds of the nose and lower airways (Lung and Widdicombe 1987; Sahin et al. 1987). In experimental animals the receptors also inhibit spinal reflexes, although the importance of this in humans has not been determined.

4.3.4 Axon Reflex Actions

The receptors probably underlie the axon reflex inflammatory changes seen in airways disease ("neurogenic inflammation"; McDonald 1990). These include vasodilatation and oedema, mucus secretion and, at least in the guinea pig, smooth muscle contraction. The axon reflexes are mediated by neuropeptides such as substance P, neurokinin A and calcitonin gene-related peptide. These neuropeptides have been identified in sensory nerves in the airway mucosa and epithelium (Uddman and Sundler 1986; Lundberg 1990). The extent to which they are found in either or both of the C-fibre and rapidly adapting receptors is uncertain, but both types of receptor could mediate the "neurogenic inflammation" associated with airway axon reflexes.

It will be apparent that rapidly adapting receptors and C-fibre receptors have many features in common, both being polymodal and responding to a variety of irritant and inflammatory stimuli. This is not surprising because other tissues, such as the skin and skeletal muscle, contain two groups of "nociceptive" endings with myelinated and nonmyelinated fibres, respectively, and both groups contribute to the responses, both axon and central reflex, resulting from pathological changes in the tissue. In general the C-fibre receptors seem to have a more profound action on the autonomic nervous system and somatic reflexes than do the rapidly adapting receptors, whereas the latter cause prompt respiratory reflexes such as an augmented breath and coughing. The interaction between the two groups of reflexes has been little studied.

4.4 Neuroepithelial Body Receptors

Neuroepithelial bodies (NEBs) are groups of paracrine cells found in the airway epithelium, especially in species such as the rabbit and in fetal and neonatal animals (Lauweryns and Goddeeris 1974; Pack and Widdicombe 1984). They have an afferent innervation, and possibly also a motor one. The NEBs are stimulated by hypoxia to release their active peptides, which include bombesin and either calcitonin or calcitonin gene-related peptide. The latter is also found in the nerves to the NEBs (Shimosegawa and Said 1991). It has been suggested that these peptides have local actions, possibly on the pulmonary vasculature. Activation of the afferent nerves related to the NEBs presumably will set up reflexes, but if so these reflexes have not yet been identified. The whole question of the role of the NEBs is still an intriguing enigma.

5 Conclusions

The walls of the airways contain a variety of nervous receptors, the characteristics of which vary from nose down to alveoli. However, virtually all parts of the respiratory tract contain more than one type of receptor. The majority of receptors respond both to mechanical stimuli and to various chemical stimuli, and some also respond to temperature. They have been studied mainly by isolating receptor mechanisms, for example by single-fibre recording, and by trying to apply selective stimuli. This is to introduce most unreal conditions. In the whole body any mechanical change in the airways, disease there, or inhalation of chemical irritants and agents, is bound to activate a large number of receptors, both at different airway sites and at the same site. The resultant reflex changes will be highly complex depending on the interactions of the afferent pathways, which have been little studied. In addition, the receptors may set up local changes, either by axon or by central nervous reflexes, which in turn will modulate their activity. Contractions of respiratory muscles will in turn have a respiratory feedback. The whole picture of the physiological and pathological mechanisms of respiratory receptors and reflexes is at present extreme sketchy, in spite of intensive research.

References

Adrian ED (1933) Afferent impulses in the vagus and their effect on respiration. J Physiol 79:332–358
Anderson JW, Sant'Ambrogio FB, Mathew OP, Sant'Ambrogio G (1990a) Water-responsive laryngeal receptors in the dog are not specialized endings. Respir Physiol 79:33–44
Anderson JW, Sant'Ambrogio FB, Orani GP, Sant'Ambrogio G, Mathew OP (1990b) Carbon dioxide-responsive laryngeal receptors in the dog. Respir Physiol 82:217–226
Boggs CF, Bartlett D (1982) Chemical specificity of a laryngeal apneic reflex in puppies. J Appl Physiol 53:455–462
Boushey HA, Richardson PS (1973) The reflex effects of intralaryngeal carbon dioxide on the pattern of breathing. J Physiol 228:181–191
Boushey HA, Richardson PS, Widdicombe JG, Wise JCM (1974) The response of laryngeal afferent fibres to mechanical and chemical stimuli. J Physiol 240:153–175
But VI, Klimova-Cherkasova VI (1967) Afferent fibres from the upper respiratory tract in branches of the trigeminal nerve. Bull Exp Biol Med 64:1275–1277
Cauna N (1982) Blood and nerve supply of the nasal lining. In: Proctor DF, Andersen IB (eds) The nose: upper airway physiology and the atmospheric environment. Elsevier Biomedical, Amsterdam, pp 45–69
Coleridge HM, Coleridge JCG (1986) Reflexes evoked from the tracheobronchial tree and lungs. In: Cherniack NS, Widdicombe JG (eds) Handbook of physiology, 3. The respiratory system, vol II. Control of breathing. American Physiological Society, Bethesda, pp 395–429
Coleridge JCG, Coleridge HM (1984) Afferent vagal C-fiber innervation of the lungs and airways and its functional significance. Rev Physiol Biochem Pharmacol 99:1–110
Corda M, Eklund G, von Euler C (1965) External intercostal and phrenic alpha motor responses to changes in respiratory load. Acta Physiol Scand 63:391–400
Critchlow V, von Euler C (1963) Intercostal muscle spindle activity and its gamma motor control. J Physiol 168:820–847
Daly M deB (1986) Interactions between respiration and circulation. In: Cherniack NS, Widdicombe JG (eds) Handbook of physiology, 3. The respiratory system, vol II. Control of breathing. American Physiological Society, Bethesda, pp 529–594
Das RM, Jeffery PK, Widdicombe JG (1978) The epithelial innervation of the lower respiratory tract of the cat. J Anat 126:123–131
Das RM, Jeffery PK, Widdicombe JG (1979) Experimental degeneration of intraepithelial nerve fibres in cat airways. J Anat 128:259–263
Davies A, Roumy M (1982) The effect of transient stimulation of lung irritant receptors on the pattern of breathing in rabbits. J Physiol 324:389–401
Davies A, Dixon M, Callanan D, Huszczuk A, Widdicombe JG, Wise JCM (1978) Lung reflexes in rabbits during pulmonary stretch receptor block by sulphur dioxide. Respir Physiol 34:83–101
Dawson WW (1962) Chemical stimulation of peripheral trigeminal nerve. Nature (Lond) 196:341–345
Duron B (1981) Intercostal and diaphragmatic muscle endings and afferents. In: Hornbein TF (ed) Regulation of breathing. Dekker, New York, pp 473–540
Euler von C (1986) Brain stem mechanisms for generation and control of breathing pattern. In: Cherniack NS, Widdicombe JG (eds) Handbook of physiology, 3. The respiratory system, vol II. Control of breathing. American Physiological Society, Bethesda, pp 1–67
Fox B, Bull TB, Guz A (1980) Innervation of alveolar walls in the human lung: an electron microscopic study. J Anat 131:683–692
Glebovsky VD, Bayev AV (1984) Stimulation of nasal cavity mucosa trigeminal receptors with respiratory airflows. Sechenov Physiol J USSR 70:1534–1541
Godwin-Austen RB (1969) The mechanoreceptors of the costo-vertebral joints. J Physiol 202:737–753
Harding R, Johnson P, McClelland MA (1978) Liquid-sensitive laryngeal receptors in the developing sheep, cat and monkey. J Physiol 277:409–422
Hering E, Breuer J (1868) Die Selbststeuerung der Athmung durch den Nervus vagus. Sitzungsber Akad Wiss Wien 57:672–677

Hwang JC, St. John WM, Bartlett D (1984) Afferent pathways for hypoglossal and phrenic responses to changes in upper airway pressure. Respir Physiol 55:341–354
Kappagoda CT, Man GCW, Teo KK (1987) Behaviour of canine pulmonary vagal afferent receptors during sustained acute pulmonary venous pressure elevation. J Physiol 394:249–265
Keller CJ, Loeser A (1929) Der zentripetale Lungenvagus. Z Biol 89:373–395
Knowlton GC, Larrabee MG (1946) A unitary analysis of pulmonary volume receptors. Am J Physiol 147:100–114
Krauhs JM (1984) Morphology of presumptive slowly adapting receptors in dog trachea. Anat Res 210:73–85
Laitinen L (1986) Detailed analysis of the neural elements in human airways. In: Kaliner M, Barnes PJ (eds) Neural regulation of the airways in health and disease. Dekker, New York, pp 35–56
Larrabee MG, Knowlton GC (1946) Excitation and inhibition of phrenic motoneurones by inflation of the lungs. Am J Physiol 147:90–99
Lauweryns JM, Goddeeris P (1975) Neuroepithelial bodies in the human child and adult lungs. Am Rev Respir Dis 111:469–476
Lundberg JM (1990) Peptide and classical transmitter mechanisms in the autonomic nervous system. Arch Int Pharmacodyn 303:9–19
Lung MA, Widdicombe JG (1987) Lung reflexes and nasal vascular resistance in the anaesthetized dog. J Physiol 386:465–474
Mathew OP, Abu-Osba YK, Thach BT (1982) Genioglossus muscle responses to upper airway pressure changes: afferent pathways. J Appl Physiol 52:445–450
McDonald DM (1990) The ultrastructure and permeability of tracheobronchial blood vessels in health and disease. Eur Respir J 3 (Suppl 12):572S-585S
Meyrick B, Reid L (1971) Nerves in the rat intra-acinar alveoli: an electron microscopic study. Respir Physiol 11:367–377
Miller AJ, Dunmire CR (1976) Characterization of the postnatal development of superior laryngeal nerve fibres in the postnatal kitten. J Neurobiol 7:483–494
Miller AJ, Loizzi RF (1974) Anatomical and functional differentiation of superior laryngeal nerve fibres affecting swallowing and respiration. Exp Neurol 42:369–387
Mills JE, Sellick H, Widdicombe JG (1969) Activity of lung irritant receptors in pulmonary microembolism, anaphylaxis and drug-induced bronchoconstriction. J Physiol 203:337–357
Mills JE, Sellick H, Widdicombe JG (1970) Epithelial irritant receptors in the lungs. In: Porter R (ed) Breathing: Hering-Breuer centenary symposium. Churchill, London, pp 77–92
Nail BS, Sterling GM, Widdicombe JG (1969) Epipharyngeal receptors responding to mechanical stimulation. J Physiol 204:91–98
Pack RJ, Widdicombe JG (1984) Amine-containing cells of the lung. Eur J Respir Dis 65:559–578
Pack RJ, Al-Ugaily LH, Widdicombe JG (1984) The innervation of the trachea and extrapulmonary bronchi of the mouse. Cell Tissue Res 238:61–68
Paintal AS (1955) Impulses in vagal afferent fibres from specific pulmonary deflation receptors. The response of these receptors to phenyl diguanide, potato starch, 5-hydroxytryptamine and nicotine, and their role in respiratory and cardiovascular reflexes. Q J Exp Physiol 40:89–111
Paintal AS (1969) Mechanism of stimulation of type J pulmonary receptors. J Physiol 203:511–532
Paintal AS (1983) Lung and airway receptors. In: Pallot DJ (ed) Control of respiration. Croom Helm, London, pp 78–107
Paintal AS (1986) The visceral sensations – some basic mechanisms. In: Cervero F, Morrisson JFB (eds) Visceral sensation. Elsevier, Amsterdam, pp 3–19
Robinson NP, Venning L, Kyle H, Widdicombe JG (1986) Quantitation of the secretory cells of the ferret tracheobronchial tree. J Anat 145:173–188
Sahin G, Webber SE, Widdicombe JG (1987) Lung and cardiac reflex actions on the tracheal vasculature in anaesthetized dogs. J Physiol 387:47–57
Sant'Ambrogio FB, Mathew OP, Clark WD, Sant'Ambrogio G (1985) Laryngeal influences on breathing pattern and posterior cricoarytenoid muscle activity. J Appl Physiol 58:1298–1304
Sant'Ambrogio FB, Sant'Ambrogio G, Mathew OP (1986) Effects of airway cooling on tracheal stretch receptors. Respir Physiol 66:205–214
Sant'Ambrogio G (1982) Information arising from the tracheobronchial tree of mammals. Physiol Rev 62:531–569

Sant'Ambrogio G, Remmers JE, De Groot WA, Callas G, Mortola JP (1978) Localization of rapidly adapting receptors in the trachea and main stem bronchus of the dog. Respir Physiol 33:359–366
Sant'Ambrogio G, Mathew OP, Fisher JT, Sant'Ambrogio FB (1983) Laryngeal receptors responding to transmural pressure, airflow and local muscle activity. Respir Physiol 54:317–330
Sant'Ambrogio G, Mathew OP, Sant'Ambrogio FB (1985a) Laryngeal cold receptors. Respir Physiol 59:35–44
Sant'Ambrogio G, Mathew OP, Sant'Ambrogio FB (1985b) Role of intrinsic muscle and tracheal motion in modulating laryngeal receptors. Respir Physiol 61:289–300
Sant'Ambrogio G, Brambilla-Sant'Ambrogio F, Mathew OP (1986) Effect of cold air on laryngeal mechanoreceptors in the dog. Respir Physiol 64:45–56
Sasaki CT (1979) Development of laryngeal function: etiologic significance in the sudden infant death syndrome. Laryngoscope 89:1964–1982
Sasaki CT, Suzuki M, Horiuchi M, Kirchner JA (1977) The effect of tracheostomy on the laryngeal closure reflex. Laryngoscope 87:1428–1433
Sears TA (1963) Activity of fusimotor fibres innervating muscle spindles in the intercostal muscles of the cat. Nature (Lond) 197:1013–1014
Shimosegawa T, Said SI (1991) Pulmonary calcitonin gene-related peptide immunoreactivity: nerve-endocrine cell interrelationships. Am J Respir Cell Mol Biol 4:126–134
Storey AT, Johnson P (1975) Laryngeal water receptors initiating apnea in the lamb. Exp Neurol 47:42–55
Tatar M, Webber SE, Widdicombe JG (1988) Lung C-fibre receptor activation and defensive reflexes in anaesthetized cats. J Physiol 402:411–420
Tsubone H (1989) Nasal 'flow' receptors of the rat. Respir Physiol 75:51–64
Tsubone H (1990) Nasal 'pressure' receptors. Jpn J Vet Sci 52:225–232
Uddman R, Sundler F (1986) Innervation of the upper airways. In: Widdicombe JH (ed) Clinics in chest medicine, vol 7. The upper airways. WB Saunders, Philadelphia, pp 201–209
Ulrich CE, Haddock MP, Alarie Y (1972) Airborne chemical irritants, role of the trigeminal nerve. Arch Environ Health 24:37–42
Widdicombe JG (1954) Receptors in the trachea and bronchi of the cat. J Physiol 123:71–104
Widdicombe JG (1961) Respiratory reflexes in man and other mammalian species. Clin Sci 21:163–170
Widdicombe JG (1986) Sensory innervation of the lungs and airways. In: Cervero F, Morrisson JFB (eds) Progress in brain research, vol. 67. Visceral sensation. Elsevier, Amsterdam, pp 49–64
Widdicombe JG, Sant'Ambrogio G, Mathew OP (1988) Nervous receptors of the upper airway. In: Mathew OP, Sant'Ambrogio G (eds) Respiratory function of the upper airway. Dekker, New York, pp 193–231
Wyke BD, Kirchner JA (1976) Neurology of the larynx. In: Hinchcliffe R, Harrison D (eds) Scientific foundations of otolaryngology. Year Book Medical Publishers, Chicago, pp 546–574
Zotterman Y (1936) Specific action potentials in the lingual nerve of cat. Skand Arch Physiol 75:105–119

Chapter 7

Cardiovascular Mechanoreceptors

M.W. Chapleau[1]

Contents

[1]Department of Internal Medicine, Cardiovascular Center, University of Iowa College of Medicine and Department of Veterans Affairs Medical Center, Iowa City, Iowa 52242, USA

Advances in Comparative and Environmental Physiology, Vol. 10

1 Introduction

Cardiovascular mechanoreceptors are sensory nerve endings located throughout the cardiovascular system. These sensory endings are activated by mechanical deformation and transmit impulses to the central nervous system that provide information regarding stretch or filling of visceral organs, e.g., the heart, and changes in vascular pressure or volume.

This chapter focuses on mechanisms of activation and function of mechanoreceptors located in systemic arteries that are sensitive to changes in arterial blood pressure. These receptors are commonly referred to as *baroreceptors*. The reader is referred to other reviews for information on mechanoreceptors located in other areas of the cardiovascular system (Paintal 1973; Bishop et al. 1983; Coleridge and Coleridge 1980).

2 Location and General Function of Arterial Baroreceptors

2.1 Location

Arterial baroreceptors are located primarily in the vessel walls of the carotid sinuses and aortic arch (Fig. 1). These regions are innervated by the carotid sinus and aortic depressor nerves, respectively. A description of the anatomy and morphology of baroreceptors and the vessels they innervate, along with the classification of different afferent fiber types (e.g., myelinated vs nonmyelinated) can be found in other reviews (Kirchheim 1976; Brown 1980).

2.2 General Function

Baroreceptors are activated primarily by mechanical deformation during increases in arterial pressure. Increased afferent nerve activity is transmitted to the central nervous system and triggers reflex adjustments in sympathetic and parasympathetic nerve activities that buffer the rise in arterial pressure through changes in heart rate and vascular resistance (Fig. 1). Conversely, decreases in pressure reduce baroreceptor activity which produces reflex adjustments that minimize the fall in pressure. Thus, the system functions in a negative-feedback manner to regulate the level of arterial pressure.

Several excellent reviews describe the discharge characteristics and the pressure-activity relation of baroreceptors (Kirchheim 1976; Brown 1980), the central nervous system pathways that participate in the baroreflex (Spyer 1981), and the multiple reflex responses to changes in baroreceptor activity (Kirchheim 1976; Sagawa 1983; Chapleau et al. 1989a).

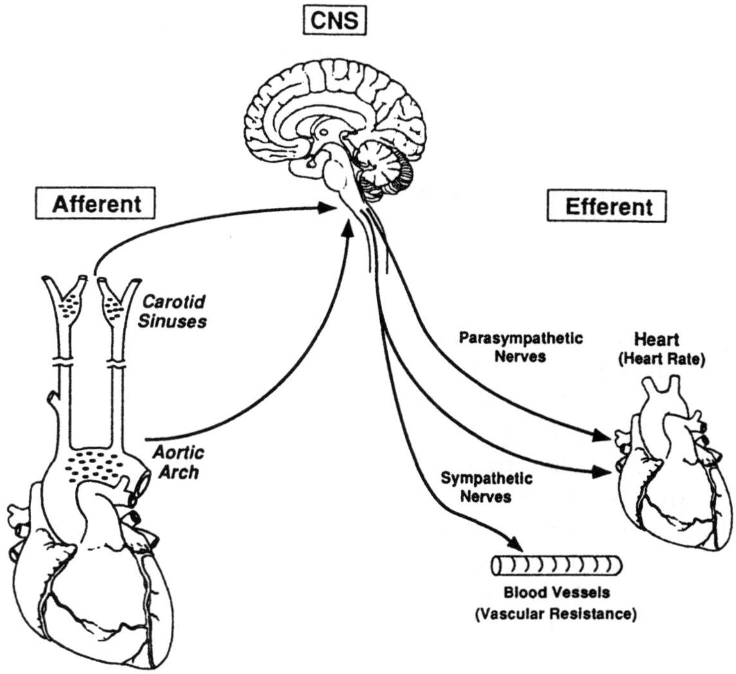

Fig. 1. Location of arterial baroreceptors in carotid sinuses and aortic arch and afferent and efferent baroreflex pathways. Increased frequency of impulses transmitted to the central nervous system (CNS) during increased arterial pressure triggers reflex excitation of parasympathetic and inhibition of sympathetic neurons, reducing heart rate and vascular resistance to buffer the rise in pressure

3 Mechanisms of Activation of Baroreceptors

3.1 General Transduction Processes of Mechanoreceptors

Our understanding of the processes that mediate the transduction of changes in arterial pressure, to changes in frequency of action potentials transmitted in baroreceptor neurons, is limited. The small size of the nerve endings and complex coupling of the endings to the vessel wall have prevented measurement of deformation or tension at the level of the nerve endings, and membrane potential or ionic events at the receptor level. The fundamental mechanisms of activation may be similar to those demonstrated in other slowly adapting mechanoreceptors (e.g., crustacean stretch receptors and muscle spindles) that are more amenable to study at the cellular level (Katz 1950; Eyzaguirre and Kuffler 1955; Nakajima and Takahashi 1966; Nakajima and Onodera 1969a,b; Hunt et al. 1978; Edwards 1983; Ito et al. 1981, 1982, 1985).

The classical model of mechano-electrical transduction in mechanoreceptors includes: (1) a mechanical filter such as the wall of a blood vessel; (2) the nerve ending that depolarizes in a graded manner with increased deformation; and (3) a

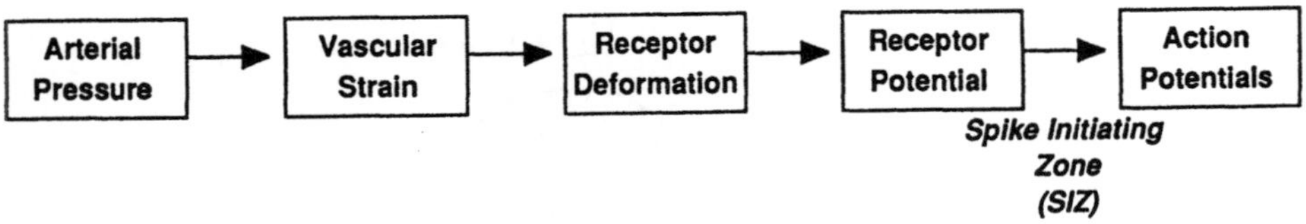

Fig. 2. Mechanoelectrical transduction processes involved in transduction of changes in arterial pressure to changes in frequency of action potentials transmitted to the central nervous system

specialized area of the nerve membrane termed the "spike initiating zone" or "encoder region" where the magnitude of depolarization is encoded into frequency of action potential discharge (Katz 1950; Eyzaguirre and Kuffler 1955; Loewenstein 1971; Grigg 1986; Fig. 2). Increased arterial pressure produces vessel distension or increased vascular strain, which leads to deformation of the receptor endings. The deformation may generate an inward ionic current that results in a graded depolarization or receptor potential. The receptor potential conducts passively along the receptor membrane, decaying with time and distance. The magnitude of depolarization at the "spike initiating zone" determines the frequency of action potentials transmitted to the central nervous system. Voltage- or ion-gated channels or an electrogenic membrane pump may modulate the degree of depolarization. Modulation of baroreceptor activity may occur at any of the multiple transduction steps.

3.2 Mechanical Deformation as the Primary Stimulus

3.2.1 Vessel Strain Rather than Pressure or Wall Tension Determines Baroreceptor Activity

Baroreceptors are not activated even at high vascular pressures if vessel distension is prevented, e.g. by encasing the carotid sinus in a plaster cast (Hauss et al. 1949) or by increasing extravascular pressure (Angell-James 1971a). Therefore, increased intravascular pressure and wall tension do not activate baroreceptors directly. Increases in baroreceptor activity are correlated more closely with increases in vessel wall strain than with pressure (Andresen et al. 1978, 1980; Brown 1980). The wall strain is a function of the transmural pressure gradient across the vessel wall and the distensibility or compliance of the wall. Therefore, vasoactive neurotransmitters or hormones and pathological changes in the composition of the vessel wall may alter baroreceptor activity through changes in vascular distensibility.

3.2.2 How Does Deformation Depolarize the Receptor Ending?

The mechanism by which a cell may sense changes in stretch or tension remained controversial until the discovery of stretch-activated (SA) ion channels on various types of cells (Guharay and Sachs 1984; Sachs 1987; Lansman et al. 1987; see chapter by Sokabe and Sachs, this Vol.). SA-channels are sensitive specifically to stretch, which increases the probability of channel opening rather than the size or conductance of each channel. The ion selectivity of the SA-channel may vary in different cell types; in general the channels are permeable in a nonselective manner to cations, e.g., Na^+ and Ca^{2+}. Consistent with this view, the receptor potential recorded from crustacean stretch receptors and muscle spindles is dependent on the concentration of cations in the extracellular fluid (Hunt et al. 1978; Edwards 1983).

SA-channels are blocked in a relatively selective manner by gadolinium (Gd^{3+}) (Gustin et al. 1988; Yang and Sachs 1989). Gd^{3+} is the most potent blocker of SA-channels known, and is often used as a way to identify SA-channels.

Preliminary results from our laboratory indicate that Gd^{3+} also prevents activation of arterial baroreceptors. Exposure of the vascularly isolated carotid sinus in anesthetized rabbits to Gd^{3+} inhibits baroreceptor activity in a dose-dependent manner without a change in vessel diameter, and the inhibition of activity is reversible after removal of Gd^{3+} (Hajduczok et al. 1990). The inhibition of baroreceptor activity by Gd^{3+} is specific for *mechanical* activation since the receptors can still be activated *chemically* with veratrine. These findings support the concept that the basic mechanism of activation of baroreceptors involves SA-channels.

3.3 Modulation of Nerve Activity Via Ionic Mechanisms

3.3.1 Sensitivity of Baroreceptors to Extracellular Ions

The frequency of action potentials in baroreceptor neurons is influenced by factors other than the degree of vascular stretch and deformation. Changes in the ionic environment of baroreceptors can dramatically alter baroreceptor discharge in the absence of changes in vascular strain. Reductions of extracellular Na^+ concentration increase the pressure required to activate baroreceptors (pressure threshold) and decrease the slope of the pressure-activity relation (Saum et al. 1977; Thoren et al. 1982; Andresen and Kunze 1987). A decrease in Na^+ of only 5% in the isolated carotid sinus reduces baroreceptor activity, and triggers a reflex increase in systemic arterial pressure and changes in renal function that would tend to restore Na^+ concentration to normal (Kunze and Brown 1978). Thus, baroreceptors may regulate not only arterial pressure, but also sodium and body fluid balance.

Increases in extracellular K^+ decrease the baroreceptor pressure threshold and increase the slope of the pressure-activity curve (Saum et al. 1977; Thoren et al. 1982) most likely by decreasing resting membrane potential or by altering the gating of SA channels (Guharay and Sachs 1984).

An inward Ca^{2+} current (in addition to Na^+) may contribute to the receptor potential in mechanoreceptors (Edwards 1983; Hunt et al. 1978). One might predict that an increase in the extracellular concentration of Ca^{2+} would therefore increase baroreceptor activity. However, increased Ca^{2+} increases the pressure threshold for action potential generation and suppresses nerve activity (Thoren et al. 1982; Andresen and Kunze 1987). A decrease in Ca^{2+} causes opposite responses. The inhibitory influence of Ca^{2+} on baroreceptor activity may result from: (1) interference with generation of action potentials at the "spike initiating zone" through alteration of surface charge (Hille et al. 1975); (2) blockage of SA-channels (Yang and Sachs 1989); and/or (3) activation of Ca^{2+}-dependent K^+ channels (Edwards 1983; Rudy 1988).

A related finding is that blockade of voltage-dependent Ca^{2+} channels with nifedipine increases the sensitivity of carotid baroreceptors in dogs (Heesch et al. 1983). Nifedipine is used to treat various cardiovascular disorders including hypertension (high blood pressure) and may cause its beneficial effects in part by sensitizing baroreceptors and reflexly inhibiting sympathetic nerve activity. Others, however, have failed to observe an effect of nifedipine on aortic baroreceptors in rats (Kunze et al. 1986).

3.3.2 Role of the Na^+, K^+ Pump

Depolarization of baroreceptors and increased intracellular Na^+ during vascular stretch may stimulate an electrogenic Na^+, K^+-ATPase that pumps Na^+ out of and K^+ into the cell (Nakajima and Takahashi 1966; Sokolove and Cooke 1971). The Na^+, K^+ pump serves to reestablish normal ionic gradients across the cell membrane. Evidence suggests that a Na^+ pump is present on baroreceptor neurons and that it is an important modulator of baroreceptor activity (Quest and Gillis 1974; Saum et al. 1976; Heesch et al. 1984a). Inhibition of the pump, e.g., with ouabain, low concentrations of extracellular K^+, or replacement of extracellular Na^+ with lithium, decreases the baroreceptor pressure threshold and increases the slope of the pressure-activity curve. Thus, activation of the pump exerts an inhibitory influence on baroreceptors.

3.3.3 Role of K^+ Channels

Activation of voltage-dependent and/or Ca^{2+}-dependent K^+ channels on neurons may contribute to repolarization of the action potential and spike after-hyperpolarization, and may regulate discharge frequency by modulating the rate of membrane depolarization (Rudy 1988). A *transient K^+ channel* that operates over a narrow subthreshold range of membrane potential generates an outward current (A-current) which limits the rate of depolarization, suppresses the generation of action potentials, and regulates the minimum discharge frequency of neurons (Rogawski 1985).

The activity of the majority of myelinated baroreceptor neurons is triggered abruptly as a pressure threshold is reached, with a slow ramp increase in nonpulsatile pressure (Kirchheim 1976; Brown 1980; Chapleau and Abboud 1987). These neurons are incapable of firing at lower frequencies during stimulation with nonpulsatile pressure below the pressure threshold. It has been suggested that the A-current suppresses activity at low pressures and is thus responsible for the characteristic abrupt activation of these fibers at threshold (Seagard et al. 1990; van Brederode et al. 1990). 4-Aminopyridine, a relatively specific blocker of the A-current (Rogawski 1985), reduces pressure threshold and converts the abrupt relation between pressure and baroreceptor activity to a graded sigmoid relation with low levels of activity at low pressures (van Brederode et al. 1990).

K^+ channels also play a role in determining the activity of other types of sensory mechanoreceptor neurons (Hunt et al. 1978; Ito et al. 1981, 1982; Ottoson and Swerup 1982; Edwards 1983; Gastrelius and Grampp 1983).

3.4 Neurohumoral Modulation of Baroreceptor Activity

3.4.1 Influence of Changes in Vasomotor Tone

Vasoactive agents, either present endogenously (e.g., hormones, neurotransmitters) or administered exogenously (e.g., drugs for treatment of hypertension), may influence the activity of baroreceptors through their vasoconstrictor or dilator actions (Landgren 1952a; Kirchheim 1976; Munch and Brown 1987; Munch et al. 1987; Yang and Andresen 1990). The net effect of a vasoactive substance on baroreceptor activity depends on both the local change in vascular tone and the change in arterial pressure that may occur, particularly during systemic administration of a drug. For example, a vasoconstrictor agent may contract vascular smooth muscle in the carotid sinus but also elevate arterial pressure. The local increase in vascular tone will tend to reduce the diameter of the sinus and decrease baroreceptor activity, whereas the increased arterial pressure will tend to distend the sinus and increase nerve activity. The net effect on vessel diameter and baroreceptor activity is difficult to predict.

When a comparison is made at an equivalent pressure, a vasoconstrictor such as the hormone angiotensin II will decrease vessel diameter and baroreceptor activity (Munch et al. 1987; Yang and Andresen 1990; Fig. 3). Hormones or drugs that relax vascular muscle and increase vessel diameter such as nitroglycerin may increase baroreceptor activity (Munch et al. 1987; Yang and Andresen 1990).

When nerve activity is compared at an equivalent vessel diameter, activity is greater when vascular tone is increased and less when tone is decreased (Munch et al. 1987; Yang and Andresen 1990; Fig. 3). These changes in nerve activity may be attributed to an "in series" coupling of baroreceptors to vascular smooth muscle.

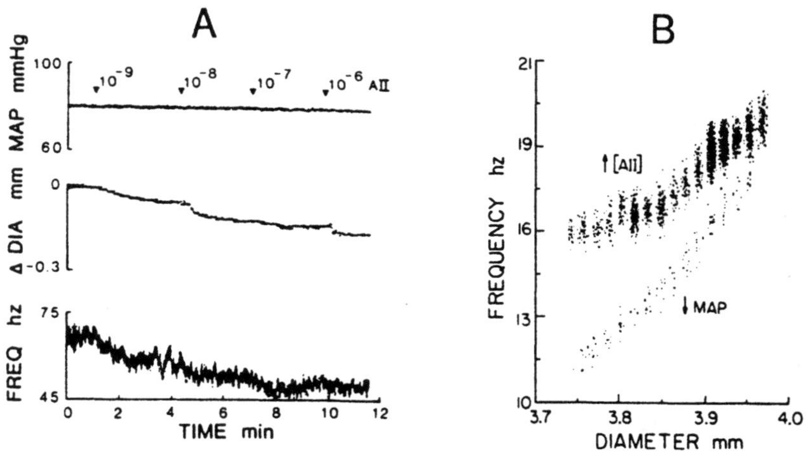

Fig. 3. Influence of vasoconstriction of the aortic arch on aortic baroreceptor activity. *A* Exposure of the isolated aortic arch to increasing concentrations of angiotensin II (*AII*) at constant pressure (*MAP*) decreases both the diameter (Δ*DIA*) of the aorta and baroreceptor activity recorded from a single fiber (*FREQ*). *B* When aortic diameter is decreased to the same degree by active vasoconstriction (↑*AII*) at constant pressure and by decreasing arterial pressure (↓*MAP*), baroreceptor activity remains higher when vascular tone is elevated (*AII*). (Munch et al. 1987)

Contraction and shortening of smooth muscle may increase tension on baroreceptors and increase nerve activity. Relaxation of smooth muscle may reduce tension on the endings and decrease activity.

3.4.2 Sensitization of Baroreceptors by the Sympathetic Nervous System and Vasopressin

The carotid sinus and aortic arch are innervated by sympathetic efferent nerves (Reis and Fuxe 1968; Kirchheim 1976). Norepinephrine released from sympathetic nerves may alter baroreceptor discharge indirectly through its vasoconstrictor action (see Sect. 3.4.1), but also may increase baroreceptor activity through a direct effect on the nerve endings (Koizumi and Sato 1969; Tomomatsu and Nishi 1981; Goldman and Saum 1984; Kunze et al. 1984; Munch and Brown 1987; Munch et al. 1987; Fig. 4). Epinephrine also increases baroreceptor sensitivity (Seagard et al.

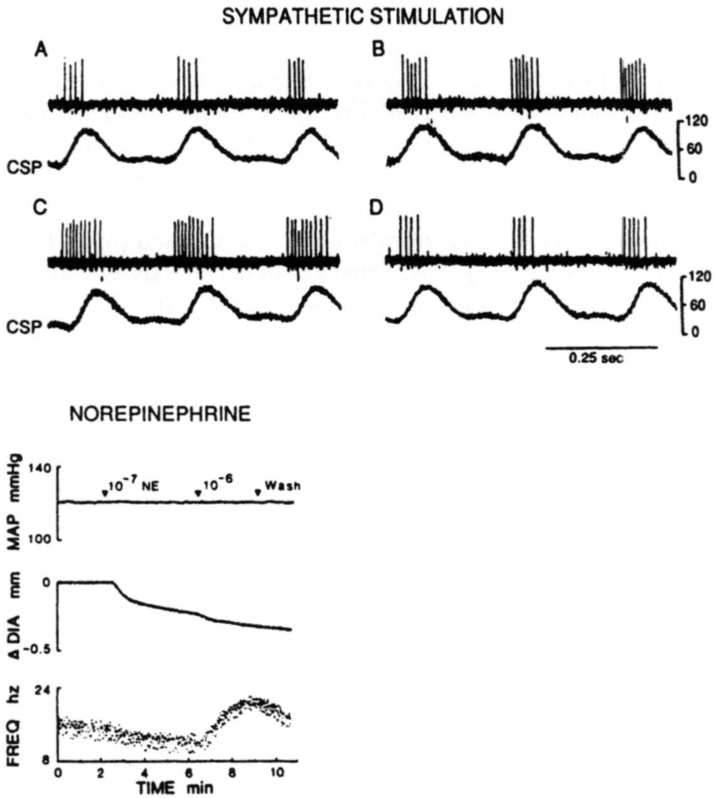

Fig. 4. Above Increased single-fiber activity recorded from the carotid sinus nerve of the rabbit during electrical stimulation of cervical sympathetic nerves. *A* control; *B* after 3 s of stimulation; *C* after 10 s of stimulation; *D* recovery. *CSP* carotid sinus pressure. *Below* Influence of norepinephrine (*NE*) on aortic baroreceptor activity in the rat. The lower concentration of NE (10^{-7} M) decreased aortic diameter (*ΔDIA*) and baroreceptor activity (*FREQ*). The higher concentration (10^{-6} M) increased nerve activity despite a further decline in diameter. (After Tomomatsu and Nishi 1981; Munch et al. 1987)

1987). The excitatory influence of norepinephrine and epinephrine is mediated through alpha-adrenergic receptors on the baroreceptor endings (Goldman and Saum 1984; Kunze et al. 1984; Munch et al. 1987; Seagard et al. 1987).

The modulation of baroreceptor activity by the sympathetic nervous system is likely to be important physiologically. A change in *endogenous* sympathetic activity to the carotid sinus exerts a significant influence on carotid baroreceptor activity (Koizumi and Sato 1969; Felder et al. 1983), and on baroreflex control of sympathetic activity to peripheral organs such as the kidney (Seagard et al. 1987).

Another circulating hormone that may sensitize baroreceptor afferents is vasopressin. Vasopressin is released from the posterior pituitary gland in response to various types of stress and in pathologic states such as heart failure (Share 1988; Johnston et al. 1988). Although a potent vasoconstrictor in vitro, systemic infusions of vasopressin in vivo often cause only minimal increases in arterial pressure, explained in part by an enhancement of baroreflex function that buffers the rise in pressure (Cowley et al. 1974). The increase in baroreflex sensitivity results from an action of vasopressin within the central nervous system and from sensitization of baroreceptors (Undesser et al. 1985; Abboud et al. 1986; Guo et al. 1986). A direct action on baroreceptors, rather than an indirect action secondary to vasoconstriction, is suggested by the augmentation of the carotid baroreflex with exposure of the isolated sinuses to vasopressin. This is not observed with exposure to another vasoconstrictor agent angiotensin II (Schmid et al. 1985). The influence of vasopressin on rat aortic baroreceptors has been attributed to vasoconstriction (Yang and Andresen 1990).

3.5 Paracrine Influence of Endothelial Factors on Baroreceptor Activity

3.5.1 Prostacyclin: An "Excitatory Factor" Contributing to Baroreceptor Activation

In addition to increasing baroreceptor activity, vascular stretch can trigger release of prostanoids such as prostacyclin (PGI_2) as well as other factors from endothelium (Frangos et al. 1985; Karwatowska-Prokopczuk 1989). Prostanoids stimulate or sensitize other visceral sensory neurons (Coleridge et al. 1976; Staszewska-Barczak 1983). Therefore, we hypothesized that prostanoids released from endothelium may sensitize baroreceptors, thereby contributing to their activation during increases in arterial pressure.

Exposure of the isolated carotid sinus in anesthetized rabbits to exogenous PGI_2 increases baroreceptor activity (Fig. 5); inhibition of prostanoid formation with indomethacin or aspirin decreases activity (Chen et al. 1990). These effects are not accompanied by changes in vessel diameter or compliance. The sensitization of baroreceptors by PGI_2 augments the carotid baroreflex control of lumbar sympathetic nerve activity (McDowell et al. 1989). An increase in the endogenous formation of PGI_2 with intrasinus administration of arachidonic acid also increases baroreceptor activity and enhances the baroreflex (McDowell et al. 1989).

Vascular endothelium is the main source of prostanoid formation within the vessel wall, and PGI_2 is the predominant metabolite formed (Moncada et al. 1977).

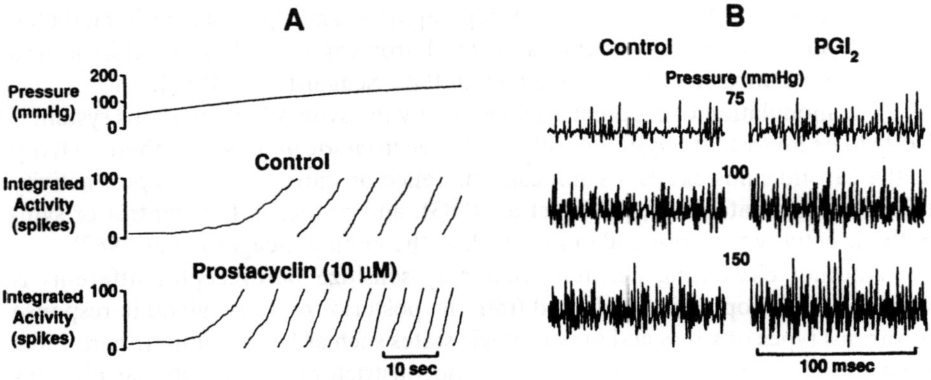

Fig. 5. Effect of prostacyclin (*PGI*$_2$) on baroreceptor activity recorded from an isolated rabbit carotid sinus. *A* Shown are the increases in nerve activity (denoted by the slope of the trace) during a slow ramp increase in carotid sinus pressure before and during exposure to PGI_2 (10 μM). *B* The neurograms recorded in this experiment at 75, 100, and 150 mmHg before and during PGI_2 are shown at a faster paper speed. (Xie et al. 1990)

Removal of, or damage to endothelial cells with a balloon catheter or a stream of a 95% O_2-5% CO_2 gas mixture, decreases baroreceptor activity and addition of exogenous PGI_2 to the denuded sinus restores activity (Chen et al. 1990). The results suggest that endothelium is an important source of prostanoids that sensitize baroreceptors.

Implication of Endothelial Dysfunction in Pathologic States. The production of prostanoids from dysfunctional endothelial cells may be impaired in pathologic states such as chronic hypertension and atherosclerosis (Gryglewski et al. 1978; Pipili and Poyser 1982; Falardeau and Martineau 1983; Rush et al. 1988). In contrast to normal rabbits, exposure of the isolated carotid sinus to indomethacin does not suppress baroreceptor activity in rabbits with renal hypertension (Fig. 6) or with hypercholesterolemia and atherosclerosis (Xie et al. 1988a, 1990). A decreased capacity for PGI_2 formation, rather than decreased responsiveness to PGI_2, appears to account for the lack of a response to indomethacin. Loss of prostanoid-dependent activation of baroreceptors in hypertension and atherosclerosis may contribute to chronic baroreceptor resetting and decreased baroreceptor sensitivity in these diseases (see Sect. 5.3).

3.5.2 Endothelial "Inhibitory Factor" Suppresses Baroreceptor Activity

In addition to producing PGI_2, the endothelium may produce a factor that suppresses baroreceptor activity (Chapleau et al. 1988a). Endothelial cell cultures grown on microcarrier beads were placed in the carotid sinus of anesthetized dogs in which the native endothelium had been removed. The endothelial cells were activated with either the calcium ionophore A23187 or bradykinin. Baroreceptor responses were compared with those obtained after placement of beads *without* endothelium combined with the same activating agent in the carotid sinus. The activated endothelial cells increase the pressure threshold and suppress the activity

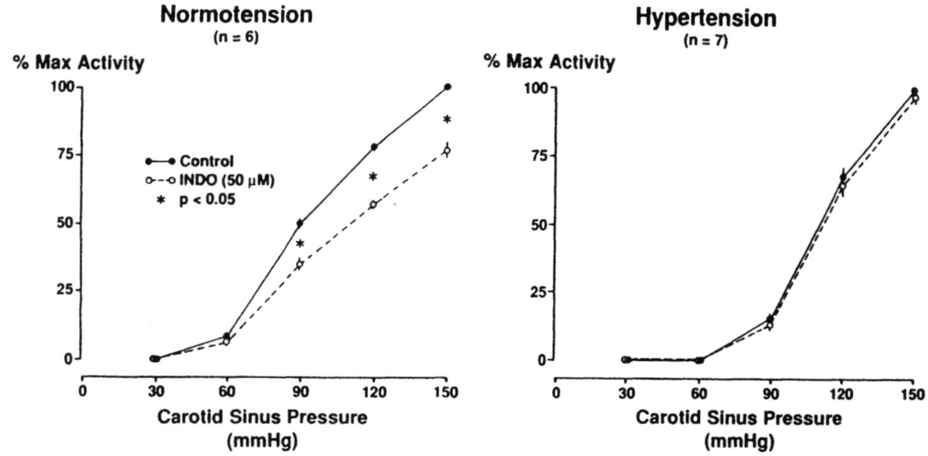

Fig. 6. Effect of indomethacin (*INDO*, 50 μM) on carotid sinus baroreceptor activity in normotensive (*left*) and chronically hypertensive (*right*) rabbits. Indomethacin in the isolated carotid sinus decreased nerve activity significantly in normotensive but not hypertensive rabbits. (After Xie et al. 1990)

recorded from single baroreceptor fibers (Fig. 7). The inhibitory influence on baroreceptors is reversible and cannot be explained by changes in vascular tone. The identity and physiological significance of the inhibitory factor released from cultured endothelium are unknown. The recently discovered peptide endothelin that is produced by endothelial cells (Yanagisawa et al. 1988) and suppresses baroreceptor activity is a candidate (Chapleau et al. 1989b).

Thus, factors released from endothelial cells such as PGI_2 and endothelin may modulate baroreceptor activity in a paracrine manner, similar to the modulation of vascular smooth muscle tone by endothelium-derived relaxing and contracting factors (Vanhoutte 1989). The main determinants of baroreceptor activity discussed so far are summarized in Fig. 8.

4 Influences of Pulsatile Pressure and Flow on Baroreceptor Discharge

4.1 Static vs Pulsatile Pressure

In many studies of baroreceptor function, a static nonpulsatile pressure has been used to activate the receptors. Normally, baroreceptors are exposed to a *pulsatile* pressure that varies in both amplitude and frequency. While some studies have downplayed the importance of pulsatility in baroreflex control maintaining that mean pressure is the important parameter (Kumada et al. 1970; Arndt et al. 1977; Yamazaki et al. 1988), others have demonstrated marked differences in the responses of baroreceptors to static and pulsatile pressure (Ead et al. 1952; Angell-James

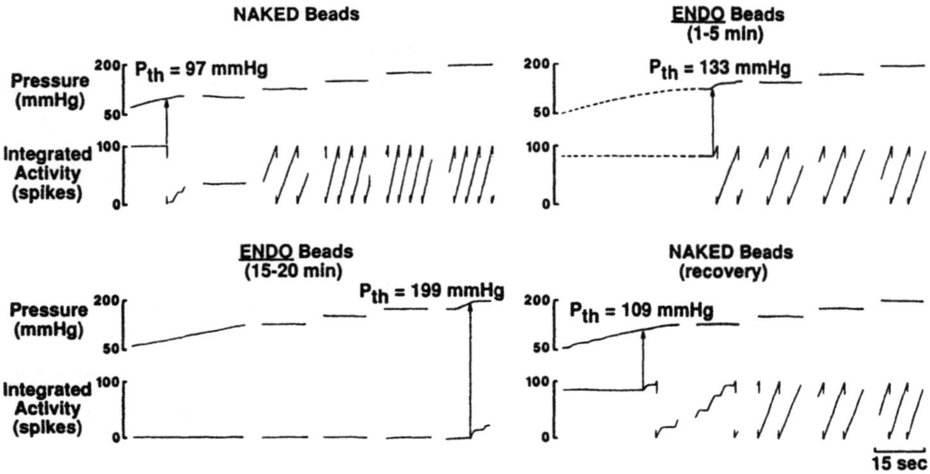

Fig. 7. Placement of endothelium cultured on microcarrier beads (*ENDO beads*) into the isolated dog carotid sinus increased the baroreceptor pressure threshold (*Pth*) and suppressed nerve activity recorded from single fibers in a time-dependent manner. Replacement of endothelialized beads with beads without endothelium (*NAKED* beads) restored baroreceptor sensitivity. Calcium ionophore (A23187, 2 μM) was added to the endothelialized beads to stimulate the endothelium and to the naked beads as an internal control. The native endothelium was removed with a balloon catheter prior to the experiment. (Chapleau et al. 1988a)

and Daly 1970, 1971; Kirchheim 1976; Chapleau and Abboud 1987, 1989; Chapleau et al. 1987, 1988b, 1989a, 1990a).

The relationship between mean pressure and baroreceptor activity measured from the whole carotid sinus or aortic nerve is sigmoid during static pressure, with the linear range occurring within a narrow range of pressure (Kirchheim 1976; Chapleau and Abboud 1987). In contrast, during pulsatile pressure the mean pressure-activity curve is linear over a much wider range, and the maximum slope or gain of the relation is much less than with static pressure (Chapleau and Abboud 1987). The extended linear range of the pressure-activity curve suggests that baroreceptors may detect changes in arterial pressure over a wider pressure range than one would expect, based on studies using static pressure.

There are two reasons for the change in the shape of the pressure-activity curve with pulsatile vs static pressure. (1) At low levels of pressure, a shift from static to pulsatile pressure of the same mean level increases baroreceptor activity (Angell-James 1971b; Chapleau and Abboud 1987). The increased activity results primarily from recruitment of fibers previously inactive during static pressure. Single baroreceptor neurons in the carotid sinus are activated during pulsatile pressure when peak systolic pressure is much lower (approximately 30 mmHg) than the static pressure required for activation (Chapleau and Abboud 1987), which reflects the sensitivity of the receptors to the rate of change of pressure (dP/dt) (Landgren 1952b). (2) At moderate to high pressures (100–150 mmHg) a shift from static to pulsatile pressure *decreases* baroreceptor activity recorded from the whole carotid sinus

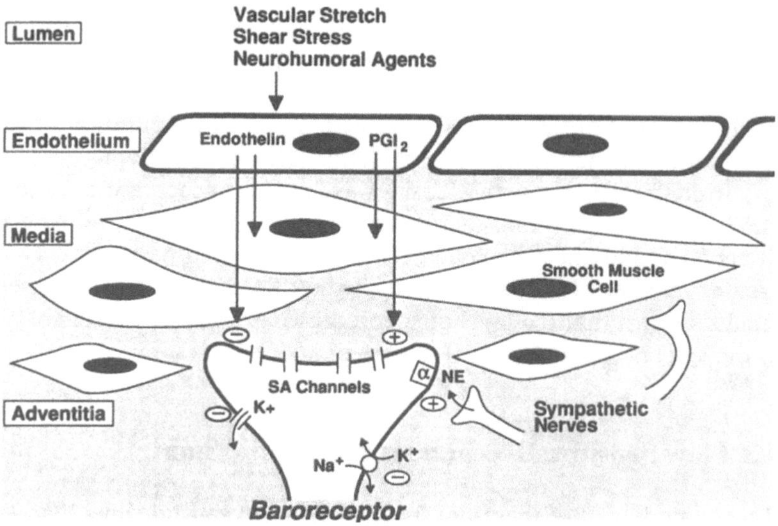

Fig. 8. Schematic summarizing mechanisms involved in activation of baroreceptors and modulation of nerve activity. Vascular stretch and deformation of baroreceptor endings most likely triggers depolarization of the receptor ending by opening stretch-activated (SA) ion channels. Factors released from endothelial cells in response to mechanical or chemical stimuli may decrease (endothelin) or increase (prostacyclin, PGI_2) the excitability of baroreceptors. Evidence suggests that activation of an electrogenic Na^+, K^+ pump and K^+ channels exert inhibitory (–) influences on baroreceptor activity while norepinephrine (*NE*) released from sympathetic nerves exerts an excitatory (+) influence mediated through alpha (α) adrenergic receptors on the baroreceptor ending. Changes in vascular tone may influence baroreceptor activity indirectly by altering deformation of the ending

nerve and from single units (Chapleau and Abboud 1987). The decrease in activity per second is caused by periods of neural silence in diastole ("diastolic silence") that occur despite pressure remaining far above the static pressure threshold.

4.2 Do Changes in Heart Rate Influence Baroreceptor Activity Independent of Changes in Mean Pressure?

At normal arterial pressures, baroreceptor activity occurs predominantly in the systolic phase of the pressure pulse. Therefore, one might expect that increases in the *frequency* of pulsation (heart rate) would increase the amount of nerve activity per second even when mean pressure does not increase. Most studies that examined the influence of pulse frequency on baroreceptor activity used *sinusoidal* pressure waves in isolated carotid sinus or aortic arch preparations. An increase in frequency of sinewave pulses increases the activity of nonmyelinated afferents (Brown et al. 1978) but does *not* alter the level of activity recorded from myelinated fibers; the greater frequency of bursts of activity is offset precisely by a decreased number of action potentials in each burst (Arndt et al. 1975; Brown et al. 1978; Abboud and

Chapleau 1988). The decreased activity per burst reflects the marked decrease in duration of systole that occurs with increased frequency of sinewaves.

In vivo, the arterial pulsations are not sinusoidal. Pressure increases rapidly in systole and decreases at a slower rate in diastole. An increase in the frequency of natural arterial pulses is associated with a decreased duration of diastole, with a minimal change in the duration of systole and positive dP/dt.

In contrast to that seen with sinewave pulses, an increase in frequency of *natural pulses* in the isolated sinus increases the activity of single myelinated baroreceptor fibers (Abboud and Chapleau 1988). The increased nerve activity is caused by a greater number of bursts of activity, with little change in activity per burst. These results suggest that the level of baroreceptor activity in vivo is not only a function of the level of pressure, but also a function of heart rate.

4.3 Flow-Sensitive Receptors in the Carotid Sinus

Recent studies have identified receptors in the carotid sinus that are sensitive to changes in flow at constant pressure and vascular strain (Hajduczok et al. 1988; Chapleau et al. 1989c). Initiation of flow of a physiologic saline solution through the isolated carotid sinus, without a change in nonpulsatile pressure or diameter increases carotid sinus nerve activity (Fig. 9). The increase in nerve activity is proportional to the magnitude of flow and the calculated increase in wall shear stress. The presence of flow decreases the pressure threshold of single baroreceptor units.

The carotid sinuses would seem to be an ideal location for a flow sensor, since the sinuses and internal carotid arteries are major conduit vessels that supply blood

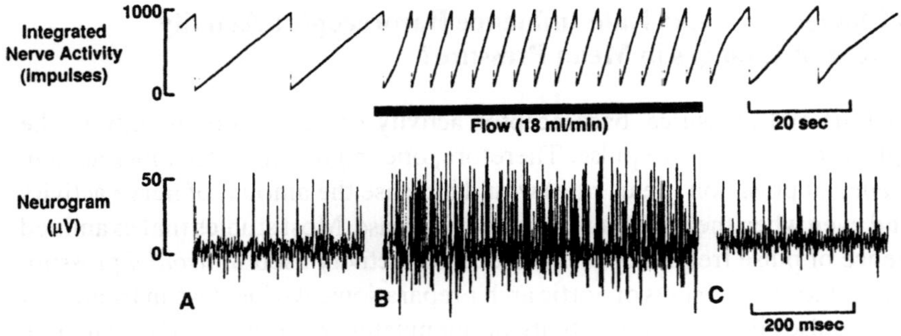

Fig. 9. Introduction of flow through the vascularly-isolated carotid sinus of the dog without altering pressure increases carotid sinus nerve activity. The neurograms recorded at a faster speed are shown in the *lower* trace (*A* no flow; *B* flow; *C* no flow). (Hajduczok et al. 1988).

flow to the brain. A decrease in arterial pressure, accompanied by a decrease in blood flow through the carotid sinuses, may cause a greater decrease in carotid sinus nerve activity and trigger a more pronounced reflex sympathoexcitatory response than the same fall in pressure without a decrease in flow. The greater sympathoexcitatory response may raise arterial pressure and help restore blood flow to the brain.

5 Adaptation and Resetting of Baroreceptors

5.1 Adaptation During Sustained Elevation of Pressure

Baroreceptor activity increases with a rise in arterial pressure, but then declines or adapts as the elevated pressure is maintained (Bronk and Stella 1935; Landgren 1952b; Munch et al. 1983; Chapleau et al. 1990a). During a sustained increase in static, nonpulsatile pressure in the isolated carotid sinus or aortic arch, baroreceptor activity declines rapidly over the first few seconds after the rise in pressure and then falls more slowly for up to 5–15 min. The adaptation of baroreceptor activity that occurs over a period of minutes is not complete, i.e., activity remains significantly above the control level present before the rise in pressure.

This property of *adaptation* is also present in many other sensory neurons, including various types of mechanoreceptors (Nakajima and Onodera 1969a,b; Husmark and Ottoson 1971a,b; Ottoson and Swerup 1982; Edwards 1983).

5.2 Postexcitatory Depression of Baroreceptors

When pressure is returned to a control level after a brief period of elevated pressure, baroreceptor activity is suppressed with absence of activity for a short period and a gradual return of activity to control levels over a period of seconds (Landgren 1952b; Saum et al. 1976). This postexcitatory depression of activity is related to the level and duration of elevated pressure. The baroreceptors are not completely refractory during the period of neural silence, since a sufficient increase in pressure can still activate the endings.

Postexcitatory depression of baroreceptors is also evident after electrical stimulation of baroreceptor neurons distal to the receptors without a change in pressure (Saum et al. 1976). The magnitude of post-excitatory depression is proportional to the number of spikes that antidromically invade the receptor ending.

5.3 Resetting of the Baroreceptor Pressure-Activity Relation

The decreased responsiveness of baroreceptors during and after a sustained rise in pressure results in an increase in the pressure threshold necessary for baroreceptor activation and a rightward shift of the pressure-activity relation (McCubbin et al.

1956; Krieger 1982, 1986; Chapleau et al. 1987, 1988b, 1989a; Andresen and Yang 1989a; Korner 1989) (Fig. 10). This resetting occurs rapidly during acute changes in pressure. A form of resetting is evident even within one cardiac cycle, when nerve activity is much less or absent in diastole, compared with systole at equivalent levels of pressure and vascular strain (Bronk and Stella 1932; Chapleau and Abboud 1987; Abboud and Chapleau 1988).

The magnitude of *rapid baroreceptor resetting* is related to the level and duration of the elevated pressure, reaching a maximum after 5–20 min of hypertension. Rapid resetting is partial and not complete, since the magnitude of the increase in pressure threshold is only 15–40% of the change in the "conditioning" pressure and the baseline level of baroreceptor activity remains above control (Fig. 10). Although adaptation and resetting of baroreceptors may be disadvantageous in that they limit the ability of baroreceptors to buffer a sustained increase in pressure, the advantage is that the decline in activity may return the set-point or prevailing level of pressure to the steep portion of the pressure-activity curve (Fig. 10). As a result, the baroreceptors maintain the ability to detect and effectively buffer rapid fluctuations in pressure around the new higher set-point level of pressure.

Long-term elevation in arterial pressure (chronic hypertension) for a period of weeks to years exacerbates baroreceptor resetting. Pressure threshold increases further, and there may be a decrease in slope or "gain" of the pressure-activity curve as well as the maximum level of activity at high pressure (McCubbin et al. 1956;

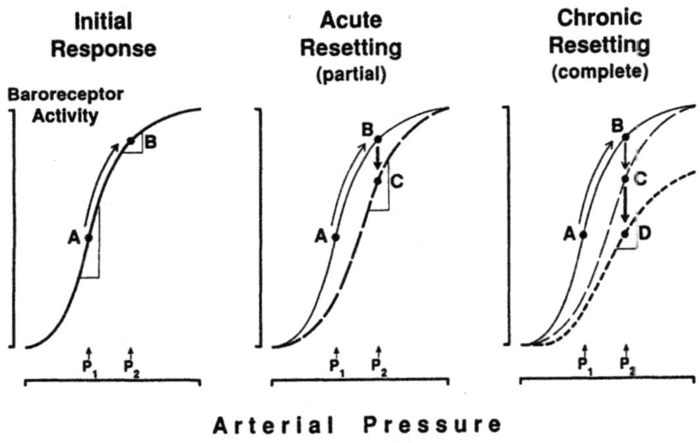

Fig. 10. Resetting of baroreceptors during increases in arterial pressure. *Left* An increase in arterial pressure from P_1 to P_2 initially increases baroreceptor activity from *A* to *B*. The increased afferent activity buffers the rise in pressure but there is less sensitivity to further changes in pressure because of the decreased slope of the pressure-activity curve at high pressure. *Middle* As the elevated pressure (P_2) is maintained baroreceptors rapidly or acutely reset with a decrease in activity from *B* to *C* that impairs buffering of the hypertension but restores sensitivity to additional fluctuations in pressure. *Right* Additional resetting of baroreceptors is evident in chronic hypertension. Nerve activity decreases from *C* to *D* and there is a decrease in maximum baroreceptor activity and slope of the pressure-activity curve. The *brackets* indicate fluctuations in pressure and activity around the setpoint levels [initial response (——), acute resetting (– –), chronic resetting (----)]. (Chapleau et al. 1989a)

Angell-James 1973; Andresen et al. 1978, 1980; Brown 1980; Andresen and Yang 1989a) (Fig. 10). This *chronic baroreceptor resetting* is complete in that the increase in the pressure threshold is about equal to the increase in arterial pressure, and the absolute baseline level of baroreceptor activity is similar to that in normotensive animals despite the elevated pressure.

Chronic resetting of baroreceptors also occurs with growth and development; baroreceptor activity remains relatively constant as the diameter of the aortic arch and carotid sinuses increase (Andresen et al. 1978, 1980; Brown 1980).

5.4 Mechanisms of Adaptation and Resetting of Baroreceptors

5.4.1 "Mechanical" Mechanisms

Two general mechanisms are thought to cause adaptation and resetting of mechanoreceptors: (1) *mechanical* mechanisms that interfere with transmission of stimulus to nerve ending and reduce deformation of the receptor (Catton and Petoe 1966; Nakajima and Onodera 1969b; Coleridge et al. 1984; Krieger 1987;), and (2) *ionic* or neural mechanisms that reduce nerve activity despite sustained deformation of the receptor (Nakajima and Onodera 1969a,b; Husmark and Ottoson 1971a,b; Ottoson and Swerup 1982; Edwards 1983).

Adaptation and resetting of baroreceptors has been attributed to relaxation of viscoelastic elements coupled in series with the receptor (Coleridge et al. 1984; Krieger 1987). Increased arterial pressure increases vascular strain, which is effectively transmitted to the nerve ending, causing deformation. As the elevated pressure is maintained, the viscoelastic elements gradually lengthen, and deformation of the receptor is reduced paradoxically despite *greater* overall vessel strain. The lengthening of viscoelastic elements and increased vessel strain is referred to as vascular "creep" and may contribute to decreased nerve activity at a constant level of elevated pressure (adaptation) as well as baroreceptor resetting (Coleridge et al. 1984; Krieger 1987).

Data suggest that vascular "creep" is not entirely responsible for rapid baroreceptor resetting. Resetting may occur in the absence of "creep" as measured by changes in diameter (Munch and Brown 1985; Andresen and Yang 1989b) and may be totally reversed by lowering pressure at a time when "creep" is still evident (Heesch et al. 1984b). Adaptation or rapid resetting of mechanoreceptors in vena cava (Mifflin and Kunze 1982) and in muscle spindles (Husmark and Ottoson 1971b) also has been dissociated from mechanical changes. Rapid resetting of aortic baroreceptors is evident after removal of endothelium and smooth muscle (Kunze 1985) suggesting that resetting may occur at the level of baroreceptors located in the adventitia.

A second "mechanical" mechanism that may cause *chronic* resetting of baroreceptors is decreased vascular compliance or distensibility. The decreased distensibility may result from chronic structural changes in the vessel wall that occur in pathologic states such as hypertension and atherosclerosis and with age (Angell-James 1973, 1974; Brown 1980; Andresen and Yang 1989a). The decreased vascular distensibility prevents normal deformation of the receptors and may decrease the slope of the pressure-activity curve (Fig. 10). Degeneration of baroreceptors also may contribute to decreased baroreceptor activity in chronic hypertension (Angell-James 1973).

Evidence suggests that other mechanisms in addition to decreased vascular distensibility contribute to chronic baroreceptor resetting. (1) The baroreceptor pressure threshold is relatively normal in animals with hypercholesterolemia and atherosclerosis, despite a marked decrease in vascular distensibility and slope of the pressure-activity curve (Angell-James 1974; Xie et al. 1988b). (2) In animals with chronic hypertension, the pressure threshold can be decreased and the pressure-activity relation shifted to lower pressures rapidly within minutes by lowering arterial pressure (Andresen and Yang 1989b; Xie et al. 1991). The rapid restoration of threshold to near normal levels indicates that chronic resetting cannot be explained entirely by structural changes in the carotid sinus or aortic arch. The ability of baroreceptors in chronic hypertension to reset rapidly may facilitate normalization of pressure during pharmacologic treatment of hypertension (Incalzi et al. 1989). (3) Resetting of baroreceptors may be evident *prior* to development of hypertension in Dahl salt-sensitive rats and when vascular distensibility is normal in spontaneously hypertensive rats (SHR) and Dahl salt-sensitive rats (Andresen et al. 1978; Gordon and Mark 1984). The results suggest a genetic alteration in receptor function unrelated to structural changes in blood vessels. Thus, resetting of baroreceptors may actually precede development of hypertension and contribute to the rise in pressure.

5.4.2 "Ionic" Mechanisms

A decrease responsiveness of baroreceptor neurons to mechanical deformation may also contribute to adaptation and resetting of baroreceptors. Inhibition of the Na^+, K^+ pump prevents or attenuates postexcitatory depression of baroreceptors (Fig. 11) and rapid resetting of baroreceptor pressure threshold (see Sects. 3.3.2, 5.2, and 5.3) (Saum et al. 1976; Heesch et al. 1984a). It is interesting that inhibition of the pump with ouabain can restore the sensitivity of baroreceptors in animals with heart failure, suggesting that an overactive Na^+, K^+ pump may contribute to decreased responsiveness of the sensory endings in this disease (Wang et al. 1990). Inhibition of the Na^+, K^+ pump does not influence rapid adaptation of baroreceptors that occurs within the first few seconds of a rise in pressure (Saum et al. 1976) (Fig. 11) or rapid resetting during the diastolic phase of the cardiac cycle (Abboud and Chapleau 1986), indicating that other mechanisms are responsible.

Preliminary results suggest a possible role of K^+ channels in baroreceptor adaptation and resetting. Exposure of the isolated carotid sinus of dogs to 4-aminopyridine (a blocker of the A current, see Sect. 3.3.3) reduces baroreceptor adaptation during a sustained increase in pressure (Chapleau et al. 1990b) and preferentially increases activity in the diastolic phase of the cardiac cycle.

As discussed in Section 3.5.1, endothelial dysfunction also may contribute to resetting of baroreceptors.

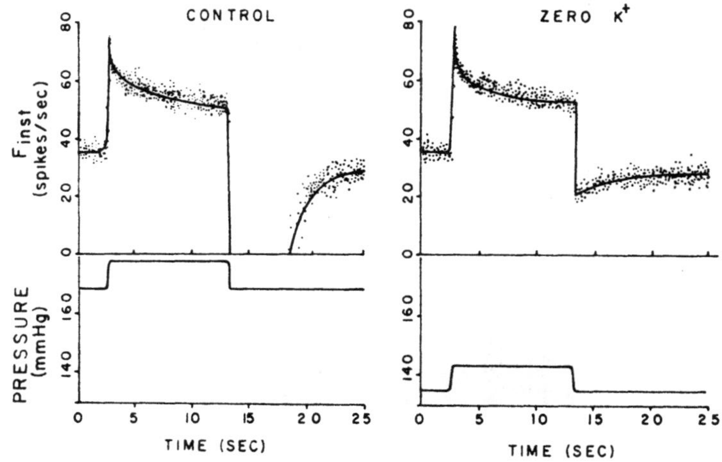

Fig. 11. Role of Na^+, K^+ pump in adaptation and postexcitatory depression of baroreceptors. *Left* Activity of a single baroreceptor fiber adapts following a step increase in pressure in the isolated aortic arch and activity is suppressed upon lowering of pressure (postexcitatory depression). *Right* After inhibition of the Na^+, K^+ pump with removal of K^+ from the extracellular fluid, postexcitatory depression is largely prevented, whereas adaptation is not altered. Similar results are seen after inhibition of the Na^+ pump with ouabain. In this experiment less pressure was required to cause similar levels of nerve activity after removal of K^+. (Brown 1980)

5.5 Influence of Pulsatile Pressure on Adaptation and Resetting of Baroreceptors

In contrast with that seen during elevated static pressure, the responsiveness of baroreceptors is relatively well maintained during acute increases in pulsatile pressure, i.e., there is much less adaptation (Chapleau et al. 1990a; Fig. 12). The adaptation that does occur results from a decrease in activity in diastole with no change in activity in systole. The lack of adaptation of systolic activity may reflect a different ionic mechanism of activation of the receptors during dynamic stretch in systole, compared with static stretch or release of stretch in diastole. Exposure of baroreceptors to procaine (Chapleau and Abboud 1986) or a low concentration of Na^+ (Thoren et al. 1982) decreases baroreceptor activity during static stretch, with a minimal effect on activity during dynamic stretch. Interventions that alter ionic mechanisms in other mechanoreceptors also dissociate responses to static vs dynamic stretch (Katz 1950; Husmark and Ottoson 1971a; Ottoson and Swerup 1982; Edwards 1983; Ito et al. 1985; Fujitsuka et al. 1987).

Pulsatility not only maintains the responsiveness of baroreceptors during dynamic stretch, but may cause a "sensitization" of baroreceptors that develops gradually over a period of minutes and outlasts the period of pulsing by up to several minutes (Chapleau and Abboud 1989). Baroreceptor activity is increased after, compared with before a period of pulsing at equivalent levels of static pressure and diameter (Fig. 13). The magnitude of sensitization is related to amplitude of the pressure pulse, pulsatile deformation of the vessel wall, and frequency of pulsation.

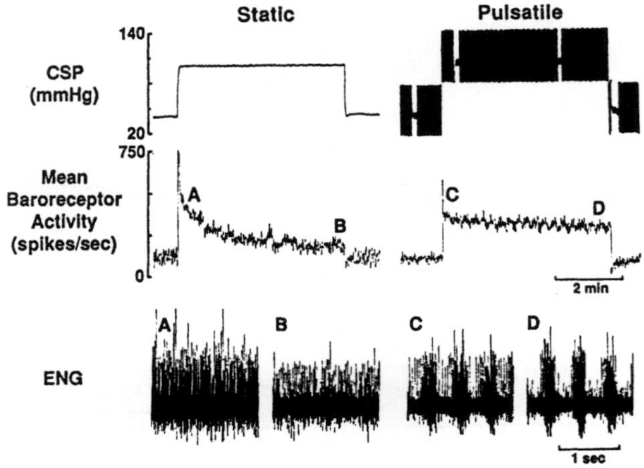

Fig. 12. A step increase in static carotid sinus pressure (*CSP*) increases baroreceptor activity and activity declines or adapts as the elevated pressure is maintained (*left*). During an equivalent increase in the mean level of pulsatile pressure (*right*) the adaptation is much less and the increased activity is sustained for a longer period compared with static pressure. The neurograms (*ENG*) recorded at a faster speed early (within 5 s) and late (at 5 min) following the step rise in pressure are shown in the *lower* traces. (Chapleau et al. 1991)

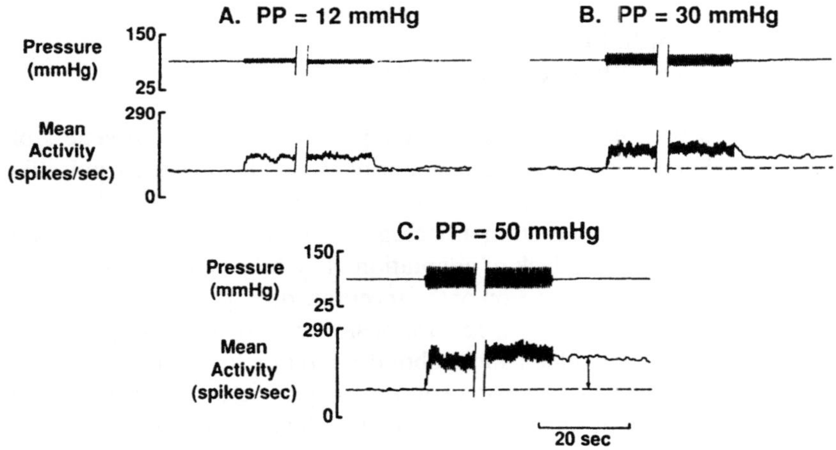

Fig. 13. Baroreceptor activity recorded from the vascularly-isolated carotid sinus of the dog is increased at the same level of static pressure after compared with before 5 min of exposure to pulsatile pressure. The increased activity after pulsing is related to the amplitude of the pressure pulse (*PP*) and is also evident when activity is not increased during the pulsing period (not shown). (Chapleau and Abboud 1989)

The increased nerve activity following a period of increased pulsatility contrasts with the decreased activity or postexcitatory depression that follows a period of increased mean pressure.

The magnitude of rapid resetting is often considered to be purely a function of changes in *mean* arterial pressure (Andresen and Yang 1989a; Korner 1989). Recent findings including those discussed above suggest that the degree of pulsatility may also be an important determinant of rapid resetting. The pressure threshold of single baroreceptor units determined with a ramp increase in *nonpulsatile* pressure is increased after a 10-min exposure to elevated *static* pressure in the isolated carotid sinus. Following equivalent exposure to elevated *pulsatile* pressure, the pressure threshold is not increased and may actually be less than control (Chapleau et al. 1987; Fig. 14). Thus, sensitization of baroreceptors by increased pulsatility may offset or prevent the resetting that usually occurs after acute increases in mean pressure.

Attenuation of adaptation and resetting of baroreceptors during pulsatile vs static pressure contributes to more sustained reflex responses and attenuation of baroreflex resetting when pressure is pulsatile (Mendelowitz and Scher 1988, 1990; Chapleau et al. 1989a,d; O'Leary et al. 1989). In addition, *phasic bursts* of afferent baroreceptor activity during pulsatile pressure are more effective than *continuous* afferent discharge during static pressure in triggering a sustained reflex response (Chapleau et al. 1989a,d; Richter et al. 1970), i.e., the phasic afferent signal minimizes adaptation that occurs within the central nervous system.

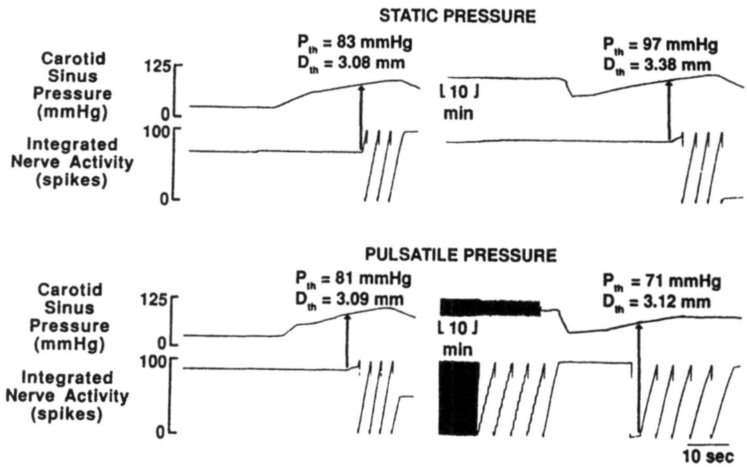

Fig. 14. The threshold carotid sinus pressure (*Pth*) and diameter (*Dth*) required to activate a single baroreceptor fiber was determined with a slow ramp of pressure both before and after exposure to 10 min of elevated static (*top*) and elevated pulsatile pressure (*bottom*). Thresholds were increased (reset) after exposure to elevated static pressure. In contrast, there was no resetting when the elevated pressure was pulsatile; Pth actually decreased and there was minimal change in Dth. (Chapleau et al. 1987)

6 Summary

Arterial baroreceptors are sensory nerve endings located in the carotid sinuses and aortic arch that are activated by mechanical deformation during increases in arterial pressure. The main function of baroreceptors is to regulate the level of arterial blood pressure.

The basic mechanism of activation of baroreceptors appears to involve SA ion channels. Baroreceptor activity is modulated by changes in the ionic environment of the receptors, activation of an electrogenic Na^+ pump and K^+ channels (A-current), neurohumoral factors such as sympathetic nerve activity, and factors released from vascular endothelium that may increase (PGI_2) or decrease (endothelin) excitability of baroreceptors.

Baroreceptors are sensitive not only to pressure but also to flow. Increased flow at constant pressure and vascular strain increases carotid sinus nerve activity in proportion to the increase in shear stress acting on the intraluminal surface of the vessel wall.

Baroreceptor activity adapts during a sustained increase in arterial pressure and is decreased (postexcitatory depression) after pressure is lowered. There is an increase in pressure threshold and a shift to the right of the pressure-activity relation after the period of hypertension. Adaptation and resetting of baroreceptors may involve both mechanical and ionic mechanisms, including viscoelastic "creep", decreased vascular distensibility, and activation of the Na^+ pump and K^+ channels.

The characteristics of baroreceptor discharge and the responsiveness of the endings are strikingly different during stimulation with pulsatile pressure, compared with static, nonpulsatile pressure. Nerve activity is markedly phasic during pulsatile pressure, with neural silence in diastole at pressures much higher than the pressure needed for activation during static pressure. The relation between mean pressure and nerve activity is linear over a much wider range and the maximum slope of the relation is less during pulsatile pressure. The responsiveness of baroreceptor endings is better maintained when pressure is pulsatile than when it is static during and following periods of acute hypertension.

Additional work in these areas should clarify the role of the arterial baroreflex in the short- and longer-term regulation of arterial pressure and circulation.

Acknowledgments. I would like to acknowledge my colleagues, Drs. Francois M. Abboud, George Hajduczok, Hsing I. Chen, Peili Xie, and Thomas S. McDowell for their contributions to the studies performed in our laboratory. I would also like to thank Drs. McDowell and Hajduczok for critical review of this chapter, Nancy Stamp for typing the manuscript, and Rodrick Faccio, Carolyn Wagner, and Deb Schiek for preparation of the figures.

References

Abboud FM, Chapleau MW (1986) "Diastolic silence" of carotid baroreceptors at pulsatile pressures above threshold – role of ionic mechanisms (abstract). Fed Proc 45(3):296

Abboud FM, Chapleau MW (1988) Effects of pulse frequency on single-unit baroreceptor activity during sine-wave and natural pulses in dogs. J Physiol (Lond) 401:295–308

Abboud FM, Aylward PE, Floras JS, Gupta BN (1986) Sensitization of aortic and cardiac baroreceptors by arginine vasopressin in mammals. J Physiol (Lond) 377:251–265

Andresen MC, Kunze DL (1987) Ionic sensitivity of baroreceptors. Circ Res 61 (Suppl I):I-66-I-71

Andresen MC, Yang M (1989a) Arterial baroreceptor resetting: contributions of chronic and acute processes. Clin Exp Pharmacol Physiol (Suppl) 15:19–30

Andresen MC, Yang M (1989b) Rapid baroreceptor resetting is unaltered by chronic hypertension in rats. Am J Physiol 256 (Heart Circ Physiol 25):H1228-H1235

Andresen MC, Krauhs JM, Brown AM (1978) Relationship of aortic wall and baroreceptor properties during development in normotensive and spontaneously hypertensive rats. Circ Res 43:728–738

Andresen MC, Kuraoka S, Brown AM (1980) Baroreceptor function and changes in strain sensitivity in normotensive and spontaneously hypertensive rats. Circ Res 47:821–828

Angell-James JE (1971a) The effects of changes of extramural, "intrathoracic," pressure on aortic arch baroreceptors. J Physiol (Lond) 214:89–103

Angell-James JE (1971b) The effects of altering mean pressure, pulse pressure and pulse frequency on the impulse activity in baroreceptor fibres from the aortic arch and right subclavian artery in the rabbit. J Physiol (Lond) 214:65–88

Angell-James JE (1973) Characteristics of single aortic and right subclavian baroreceptor fiber activity in rabbits with chronic renal hypertension. Circ Res 32:149–161

Angell-James JE (1974) Arterial baroreceptor activity in rabbits with experimental atherosclerosis. Circ Res 34:27–39

Angell-James JE, Daly M de B (1970) Comparison of the reflex vasomotor responses to separate and combined stimulation of the carotid sinus and aortic arch baroreceptors by pulsatile and non-pulsatile pressures in the dog. J Physiol (Lond) 209:257–293

Angell-James JE, Daly M de B (1971) Effects of graded pulsatile pressure on the reflex vasomotor responses elicited by changes of mean pressure in the perfused carotid sinus-aortic arch regions of the dog. J Physiol (Lond) 214:51–64

Arndt JO, Dorrenhaus A, Wiecken H (1975) The aortic arch baroreceptor response to static and dynamic stretches in an isolated aorta-depressor nerve preparation of cats in vitro. J Physiol (Lond) 252:59–78

Arndt JO, Morgenstern J, Samodelov L (1977) The physiologically relevant information regarding systemic blood pressure encoded in the carotid sinus baroreceptor discharge pattern. J Physiol (Lond) 268:775–791

Bishop VS, Malliani A, Thoren P (1983) Cardiac mechanoreceptors. In: Shepherd JT, Abboud FM (eds) Handbook of physiology Sect 2 The cardiovascular system Vol. III Peripheral circulation and organ blood flow Pt. 2. Am Physiol Soc, Bethesda, MD, pp 497–555

Bronk DW, Stella G (1932) Afferent impulses in the carotid sinus nerve. I. The relation of the discharge from single end organs to arterial blood pressure. J Cell Comp Physiol 1:113–130

Bronk DW, Stella G (1935) The response to steady pressures of single end organs in the isolated carotid sinus. Am J Physiol 110:708–714

Brown AM (1980) Receptors under pressure. An update on baroreceptors. Circ Res 46:1–10

Brown AM, Saum WR, Yasui S (1978) Baroreceptor dynamics and their relationship to afferent fiber type and hypertension. Circ Res 42:694–702

Catton WT, Petoe N (1966) A visco-elastic theory of mechanoreceptor adaptation. J Physiol (Lond) 187:35–49

Chapleau MW, Abboud FM (1986) Mechanisms of carotid baroreceptor stimulation with static and dynamic stretch (abstract). Proc Int Union Physiol Sci 16:43

Chapleau MW, Abboud FM (1987) Contrasting effects of static and pulsatile pressure on carotid baroreceptor activity in dogs. Circ Res 61:648–658

Chapleau MW, Abboud FM (1989) Determinants of sensitization of carotid baroreceptors by pulsatile pressure in dogs. Circ Res 65:566–577

Chapleau MW, Heesch CM, Abboud FM (1987) Prevention or attenuation of baroreceptor resetting by pulsatile pressure during elevated pressure. Hypertension 9 (Suppl III):137–141
Chapleau MW, Hajduczok G, Shasby DM, Abboud FM (1988a) Activated endothelial cells in culture suppress baroreceptors in the carotid sinus of dog. Hypertension 11(6) Pt 2:586–590
Chapleau MW, Hajduczok G, Abboud FM (1988b) Mechanisms of resetting of arterial baroreceptors: an overview. Am J Med Sci 31(4):327–334
Chapleau MW, Hajduczok G, Abboud FM (1989a) Peripheral and central mechanisms of baroreflex resetting. Clin Exp Pharmacol Physiol (Suppl) 15:31–43
Chapleau MW, Hajduczok G, Abboud FM (1989b) Endothelin suppresses baroreceptor activity: a new mechanism contributing to hypertension? Hypertension (Abstr) 14:336–337
Chapleau MW, Johnson SL, Hajduczok G, Abboud FM (1989c) "Rheoreceptors" in the carotid sinus are activated by flow at constant pressure: interactions between flow and pulsatility. Clin Res (Abstr) 37:874A
Chapleau MW, Hajduczok G, Abboud FM (1989d) Pulsatile activation of baroreceptors causes central facilitation of baroreflex. Am J Physiol 256 (Heart Circ Physiol 25):H1735-H1741
Chapleau MW, Hajduczok G, Abboud FM (1990a) Adaptation of baroreceptors is absent or attenuated during pulsatile compared with static pressure. Am J Physiol (Heart Circ Physiol) (submitted)
Chapleau MW, Hajduczok G, Abboud FM (1990b) Mechanism of adaptation of baroreceptors in acute hypertension: role of A-current. Hypertension (Abstr) 16:348
Chapleau MW, Hajduczok G, Abboud FM (1991) Resetting of the arterial baroreflex: peripheral and central mechanisms. In: Zucker I, Gilmore J (eds) Reflex control of the circulation. CRC Press, Boca Raton, Florida, pp 165-194
Chen HI, Chapleau MW, McDowell TS, Abboud FM (1990) Prostaglandins contribute to activation of baroreceptors in rabbits. Possible paracrine influence of endothelium. Circ Res 67:1394–1404
Coleridge HM, Coleridge JCG (1980) Cardiovascular afferents involved in regulation of peripheral vessels. Annu Rev Physiol 42:413–427
Coleridge HM, Coleridge JCG, Ginzel KH, Baker DG, Banzett RB, Morrison MA (1976) Stimulation of "irritant" receptors and afferent C-fibres in the lungs by prostaglandins. Lancet 264:451–453
Coleridge HM, Coleridge JCG, Poore ER, Roberts AM, Schultz HD (1984) Aortic wall properties and baroreceptor behaviour at normal arterial pressure and in acute hypertensive resetting in dogs. J Physiol (Lond) 350:309–326
Cowley AW Jr, Monos E, Guyton AC (1974) Interaction of vasopressin and the baroreceptor reflex system in the regulation of arterial blood pressure in the dog. Circ Res 34:505–514
Ead HW, Green JH, Neil E (1952) A comparison of the effects of pulsatile and non-pulsatile blood flow through the carotid sinus on the reflexogenic activity of the sinus baroceptors in the cat. J Physiol (Lond) 118:509–519
Edwards C (1983) The ionic mechanisms underlying the receptor potential in mechanoreceptors. In: Grinnell AD, Moody WJ Jr (eds) Neurology and neurobiology, vol 5. The physiology of excitable cells. Liss, New York, pp 497–503
Eyzaguirre C, Kuffler SW (1955) Processes of excitation in the dendrites and in the soma of single isolated sensory nerve cells of the lobster and crayfish. J Gen Physiol 39:87–120
Falardeau P, Martineau A (1983) In vivo production of prostaglandin I_2 in Dahl salt-sensitive and salt-resistant rats. Hypertension 5:701–705
Felder RB, Heesch CM, Thames MC (1983) Reflex modulation of carotid sinus baroreceptor activity in the dog. Am J Physiol 244 (Heart Circ Physiol 13):H437-H443
Frangos JA, Eskin SG, McIntire LV, Ives CL (1985) Flow effects on prostacyclin production by cultured human endothelial cells. Science 227:1477–1479
Fujitsuka N, Hama K, Ito F, Sokabe M (1987) Intramembrane particles and terminal responses following denervation of frog muscle spindles. J Neurocytol 16:185–194
Gastrelius S, Grampp W (1983) Kinetics of the TEA and 4-AP sensitive K^+ current in the slowly adapting lobster stretch receptor neurone. Acta Physiol Scand 118:125–134
Goldman WF, Saum WR (1984) A direct excitatory action of catecholamines on rat aortic baroreceptors in vitro. Circ Res 55:18–30
Gordon FJ, Mark AL (1984) Mechanism of impaired baroreflex control in prehypertensive Dahl salt-sensitive rats. Cicr Res 54:378–387

Grigg P (1986) Biophysical studies of mechanoreceptors. J Appl Physiol 60:1107–1115
Gryglewski RJ, Dembinska-Kiec A, Zmuda A, Gryglewska T (1978) Prostacyclin and thromboxane A_2 biosynthesis capacities of heart, arteries and platelets at various stages of experimental atherosclerosis in rabbits. Atherosclerosis 31:385–394
Guharay F, Sachs F (1984) Stretch-activated single ion channel currents in tissue-cultured embryonic chick skeletal muscle. J Physiol (Lond) 352:685–701
Guo GB, Schmid PG, Abboud FM (1986) Sites at which vasopressin facilitates baroreflex inhibition of lumbar sympathetic nerve activity. Am J Physiol 251 (Heart Circ Physiol 20):H644-H655
Gustin MC, Zhou X-L, Martinac B, Kung C (1988) A mechanosensitive ion channel in the yeast plasma membrane. Science 242:762–765
Hajduczok G, Chapleau MW, Abboud FM (1988) Rheoreceptors in the carotid sinus of dog. Proc Natl Acad Sci USA 85:7399–7403
Hajduczok G, Ferlic RJ, Chapleau MW, Abboud FM (1990) Mechanism of mechanotransduction of arterial baroreceptors (Abstr). FASEB J 4(3):A285
Hauss WH, Kreuziger H, Asteroth H (1949) Über die Reizung der Pressorezeptoren im Sinus Caroticus beim Hund. Z Kreislaufforsch 38:28–33
Heesch CM, Miller BM, Thames MD, Abboud FM (1983) Effects of calcium channel blockers on isolated carotid baroreceptors and baroreflex. Am J Physiol 245 (Heart Circ Physiol 14):H653-H661
Heesch CM, Abboud FM, Thames MD (1984a) Acute resetting of carotid sinus baroreceptors. II. Possible involvement of electrogenic Na^+ pump. Am J Physiol 247 (Heart Circ Physiol 16):H833-H839
Heesch CM, Thames MD, Abboud FM (1984b) Acute resetting of carotid sinus baroreceptors. I. Dissociation between discharge and wall changes. Am J Physiol 247 (Heart Circ Physiol 16):H824-H832
Hille B, Woodhull AM, Shapiro BI (1975) Negative surface charge near sodium channels of nerve: divalent ions, monovalent ions, and pH. Philos Trans R Soc Lond (Biol) 270:301–318
Hunt CC, Wilkinson RS, Fukami Y (1978) Ionic basis of the receptor potential in primary endings of mammalian muscle spindles. J Gen Physiol 71:683–698
Husmark I, Ottoson D (1971a) Is the adaptation of the muscle spindle of ionic origin? Acta Physiol Scand 81:138–140
Husmark I, Ottoson D (1971b) The contribution of mechanical factors to the early adaptation of the spindle response. J Physiol (Lond) 212:577–592
Incalzi RA, Gemma A, Cocchi A, Carbonin P (1989) Baroreceptor reset with nitroprusside and drug-resistant hypertension. Lancet 8661 (Vol. II):504–505
Ito F, Komatsu Y, Kaneko N, Katsuta N (1981) Control of the variability of the afferent discharge rate in frog muscle spindle by potassium blockers. Brain Res 216:199–202
Ito F, Komatsu Y, Fujitsuka N (1982) $G_{K(Ca)}$-dependent cyclic potential changes in the sensory nerve terminal of frog muscle spindle. Brain Res 252:39–50
Ito F, Fujitsuka N, Fan XL (1985) Reversal of the static component of spindle potential by imposed depolarizing current in the frog muscle spindle. Brain Res 326:107–116
Johnston CL, McGrath BP, Phillips P, Abrahams JM, Hodsman GP, Arnolda LF (1988) Vasopressin in congestive heart failure: clinical and experimental studies. In: Cowley AW Jr, Liard J-F, Ausiello DA (eds) Vasopressin: cellular and integrative functions. Raven, New York, pp 481–485
Karwatowska-Prokopczuk E, Ciabattoni G, Wennmalm A (1989) Effects of hydrodynamic forces on coronary production of prostacyclin and purines. Am J Physiol 256 (Heart Circ Physiol 25):H1532–H1538
Katz B (1950) Depolarization of sensory terminals and the initiation of impulses in the muscle spindle. J Physiol (Lond) 111:261–282
Kirchheim HR (1976) Systemic arterial baroreceptor reflexes. Physiol Rev 56:100–176
Koizumi K, Sato A (1969) Influence of sympathetic innervation on carotid sinus baroreceptor activity. Am J Physiol 216:321–329
Korner PI (1989) Baroreceptor resetting and other determinants of baroreflex properties in hypertension. Clin Exp Pharmacol Physiol (Suppl) 15:45–64
Krieger EM (1986) Neurogenic mechanisms in hypertension: resetting of the baroreceptors. State of the art lecture. Hypertension 8 (Suppl I):7–14
Krieger EM (1987) Aortic diastolic caliber changes as a determinant for complete aortic baroreceptor resetting. Fed Proc 46:41–46

Krieger EM, Salgado HC, Michelini LC (1982) Resetting of the baroreceptors. In: Guyton AC, Hall JE (eds) International review of physiology, vol 26. Univ Park Press, Baltimore, Maryland, pp 119–146
Kumada M, Schmidt RM, Sagawa K, Tan KS (1970) Carotid sinus reflex in response to hermorrhage. Am J Physiol 219:1373–1379
Kunze DL (1985) Role of baroreceptor resetting in cardiovascular regulation: acute resetting. Fed Proc 44:2408–2411
Kunze DL, Brown AM (1978) Sodium sensitivity of baroreceptors. Reflex effects on blood pressure and fluid volume in the cat. Circ Res 42:714–720
Kunze DL, Krauhs JM, Orlea CJ (1984) Direct action of norepinephrine on aortic baroreceptors of rat adventitia. Am J Physiol 247 (Heart Circ Physiol 16):H811-H816
Kunze DL, Andresen MC, Torres LA (1986) Do calcium antagonists act directly on calcium channels to alter baroreceptor function? J Pharmacol Exp Ther 239:303–310
Landgren S (1952a) Baroceptor activity in the carotid sinus nerve and the distensibility of the sinus wall. Acta Physiol Scand 26:35–56
Landgren S (1952b) On the excitation mechanism of the carotid baroceptors. Acta Physiol Scand 26:1–34
Lansman JB, Hallam TJ, Rink TJ (1987) Single stretch-activated ion channels in vascular endothelial cells as mechanotransducers? Nature (Lond) 325:811–813
Loewenstein WR (1971) Mechano-electric transduction in the pacinian corpuscle. Initiation of sensory impulses in mechanoreceptors. In: Loewenstein WR (ed) Handbook of sensory physiology vol I. Principles of receptor physiology. Springer, Berlin Heidelberg New York, pp 269–290
McCubbin JW, Green JH, Page IH (1956) Baroceptor function in chronic renal hypertension. Circ Res 4:205–210
McDowell TS, Axtelle TS, Chapleau MW, Abboud FM (1989) Prostaglandins in carotid sinus enhance baroreflex in rabbits. Am J Physiol 257 (Regulatory Integrative Comp Physiol 26):R445-R450
Mendelowitz D, Scher AM (1988) Pulsatile sinus pressure changes evoke sustained baroreflex responses in awake dogs. Am J Physiol 255 (Heart Circ Physiol 24):H673–H678
Mendelowitz D, Scher AM (1990) Pulsatile pressure can prevent rapid baroreflex resetting. Am J Physiol 258 (Heart Circ Physiol 27):H92-H100
Mifflin SW, Kunze DL (1982) Rapid resetting of low pressure vagal receptors in the superior vena cava of the rat. Circ Res 51:241–249
Moncada S, Herman AG, Higgs EA, Vane JR (1977) Differential formation of prostacyclin (PGX or PGI_2) by layers of the arterial wall. An explanation for the anti-thrombotic properties of vascular endothelium. Thrombosis Res 11:323–344
Munch PA, Brown AM (1985) Role of vessel wall in acute resetting of aortic baroreceptors. Am J Physiol 248 (Heart Circ Physiol 17):H843-H852
Munch PA, Brown AM (1987) Sympathetic modulation of rabbit aortic baroreceptors in vitro. Am J Physiol 253 (Heart Circ Physiol 22):H1106-H1111
Munch PA, Andresen MC, Brown AM (1983) Rapid resetting of aortic baroreceptors in vitro. Am J Physiol 244 (Heart Circ Physiol 13):H672-H680
Munch PA, Thoren PN, Brown AM (1987) Dual effects of norepinephrine and mechanisms of baroreceptor stimulation. Circ Res 61:409–419
Nakajima S, Onodera K (1969a) Membrane properties of the stretch receptor neurons of crayfish with particular reference to mechanisms of sensory adaptation. J Physiol (Lond) 200:161–185
Nakajima S, Onodera K (1969b) Adaptation of the generator potential in the crayfish stretch receptors under constant length and constant tension. J Physiol (Lond) 200:187–204
Nakajima S, Takahashi K (1966) Post-tetanic hyperpolarization and electrogenic Na pump in stretch receptor neurone of crayfish. J Physiol (Lond) 187:105–127
O'Leary DS, Scher AM, Bassett JE (1989) Effects of steps in cardiac output and arterial pressure in awake dogs with AV block. Am J Physiol 256 (Heart Circ Physiol 25):H361-H367
Ottoson D, Swerup C (1982) Studies on the role of calcium in adaptation of the crustacean stretch receptor. Effects of intracellular injection of calcium, EGTA and TEA. Brain Res 244:337–341
Paintal AS (1973) Cardiovascular receptors. In: Neil E (ed) Handbook of sensory physiology vol III. Springer, Berlin Heidelberg New York, pp 1–45

Pipili E, Poyser NL (1982) Release of prostaglandins I_2 and E_2 from the perfused mesenteric arterial bed of normotensive and hypertensive rats. Effects of sympathetic nerve stimulation and norepinephrine administration. Prostaglandins 23:543–549
Quest JA, Gillis RA (1974) Effect of digitalis on carotid sinus baroreceptor activity. Circ Res 35:247–255
Reis DJ, Fuxe K (1968) Adrenergic innervation of the carotid sinus. Am J Physiol 215:1054–1057
Richter DW, Keck W, Seller H (1970) The course of inhibition of sympathetic activity during various patterns of carotid sinus nerve stimulation. Pflügers Arch Eur J Physiol 317:110–123
Rogawski MA (1985) The A-current: how ubiquitous a feature of excitable cells is it? Trends Neurosci 8:214–219
Rudy B (1988) Diversity and ubiquity of K channels. Neuroscience 25:729–749
Rush DS, Kerstein MD, Bellan JA, Knoop SM, Mayeux PR, Hyman AL, Kadowitz PJ, McNamara DB (1988) Prostacyclin, thromboxane A_2, and prostaglandin E_2 formation in atherosclerotic human carotid artery. Arteriosclerosis 8:73–78
Sachs F (1987) Baroreceptor mechanisms at the cellular level. Fed Proc 46:12–16
Sagawa K (1983) Baroreflex control of systemic arterial pressure and vascular bed. In: Shepherd JT, Abboud FM (eds) Handbook of physiology Sect 2 The cardiovascular system, vol III Peripheral circulation and organ blood flow Pt. 2. Am Physiol Soc, Bethesda, MD, pp 453–496
Saum WR, Brown AM, Tuley FH (1976) An electrogenic sodium pump and baroreceptor function in normotensive and spontaneously hypertensive rats. Circ Res 39:497–505
Saum WR, Ayachi S, Brown AM (1977) Actions of sodium and potassium ions on baroreceptors of normotensive and spontaneously hypertensive rats. Circ Res 41:768–774
Schmid PG, Guo GB, Abboud FM (1985) Different effects of vasopressin and angiotensin II on baroreflexes. Fed Proc 44:2388–2392
Seagard JL, Hopp FA, Kampine JP (1987) Effect of sympathetic sensitization of baroreceptors on renal nerve activity. Am J Physiol 252 (Regulatory Integrative Comp Physiol 21):R328-R335
Seagard JL, van Brederode JFM, Dean C, Hopp FA, Gallenberg LA, Kampine JP (1990) Firing characteristics of single-fiber carotid sinus baroreceptors. Circ Res 66:1499–1509
Share L (1988) Role of vasopressin in cardiovascular regulation. Physiol Rev 68:1248–1284
Sokolove PG, Cooke IM (1971) Inhibition of impulse activity in a sensory neuron by an electrogenic pump. J Gen Physiol 57:125–163
Spyer KM (1981) Neural organization and control of the baroreceptor reflex. In: Baker PF, Grunike H, Habermann E et al. Reviews of Physiology Biochemistry and Pharmacology, vol 88. Springer, Berlin Heidelberg New York Tokyo, pp 23–124
Staszewska-Barczak J (1983) Prostanoids and cardiac reflexes of sympathetic and vagal origin. Am J Cardiol 52:36A-45A
Thoren P, Andresen MC, Brown AM (1982) Effects of changes in extracellular ionic concentrations on aortic baroreceptors with nonmyelinated afferent fibers. Circ Res 50:413–418
Tomomatsu E, Nishi K (1981) Increased activity of carotid sinus baroreceptors by sympathetic stimulation and norepinephrine. Am J Physiol 240 (Heart Circ Physiol 9):H650-H658
Undesser KP, Hasser EM, Haywood JR, Johnson AK, Bishop VS (1985) Interactions of vasopressin with the area postrema in arterial baroreflex function in conscious rabbits. Circ Res 56:410–417
van Brederode JFM, Seagard JL, Dean C, Hopp FA, Kampine JP (1990) An experimental and modeling study of the excitability of carotid sinus baroreceptors. Circ Res 66:1510–1525
Vanhoutte PM (1989) Endothelium and control of vascular function. State of the art lecture. Hypertension 13:658–667
Wang W, Chen J-S, Zucker IH (1990) Carotid sinus baroreceptor sensitivity in experimental heart failure. Circulation 81:1959–1966
Xie P, McDowell TS, Hajduczok G, Chapleau MW, Abboud FM (1988a) Contribution of cyclooxygenase metabolites to baroreceptor activation in hypercholesterolemic (HC) rabbits. Circulation (Abstr) 78:II-177
Xie P, Chapleau MW, McDowell TS, Hajduczok G, Abboud FM (1988b) Chronic resetting of baroreceptors differs in hypercholesterolemic vs renal hypertensive rabbits. Physiologist (Abstr) 31:A198
Xie P, Chapleau MW, McDowell TS, Hajduczok G, Abboud FM (1990) Mechanism of decreased baroreceptor activity in chronic hypertensive rabbits – role of endogenous prostanoids. J Clin Invest 86:625–630

Xie P, McDowell TS, Chapleau MW, Hajduczok G, Abboud FM (1991) Rapid baroreceptor resetting in chronic hypertension. Implications for normalization of arterial pressure. Hypertension 17:72–79

Yamazaki T, Brunner MJ, Sagawa K (1988) Pulsatile vs mean component of baroreflex compensation for posthemorrhage hypotension. Am J Physiol 254 (Heart Circ Physiol 23):H1074-H1080

Yanagisawa M, Kurihara H, Kimura S, Tomobe Y, Kobayashi M, Mitsui Y, Yazaki Y, Goto K, Masaki T (1988) A novel potent vasoconstrictor peptide produced by vascular endothelial cells. Nature (Lond) 332:411–415

Yang M, Andresen MC (1990) Peptidergic modulation of mechanotransduction in rat arterial baroreceptors. Circ Res 66:804–813

Yang X-C, Sachs F (1989) Block of stretch-activated ion channels in *Xenopus* oocytes by gadolinium and calcium ions. Science 243:1068–1071

Chapter 8

Comparative Physiology of Cutaneous Mechanoreceptors

W. Hamann[1]

Contents

1 History

Until 1910, histological investigation of the skin had revealed a variety of structurally developed end organs in the skin, all of them connected to myelinated afferent nerve fibres. The function of these receptor structures had to remain a matter of speculation until it became possible to record electrophysiologically from identified single nerve fibres. Unitary recording techniques were pioneered by Adrian and Zottermann (1926). Adrian and Umrath (1929) were among the first to combine structure and function in their work on Pacinian corpuscles. Frankenhaeuser's (1949) Lindblom's (1965) and Tapper's (1965) work perfected techniques linking structure and function.

[1]Departments of Anaesthetics and Anatomy, United Medical and Dental Schools, Guy's & Lewisham Hospitals, London SE1 9RT, UK

Advances in Comparative and Environmental Physiology, Vol. 10

Werner and Mountcastle (1965) set out to describe the mathematical interdependence between stimulus and response. However, this work was impeded by the fact that the authors did not discriminate between the now well-recognized different types of slowly adapting cutaneous mechanoreceptors. In the skin senses, it was Iggo's school where the stimulus-response characteristics were determined in detail for hair receptors (Brown and Iggo 1967), Merkel cell receptors (Iggo and Muir 1969) Ruffini endings (Chambers et al. 1972), Meissner's corpuscle (Ogawa and Iggo 1977) and vibrissa (Gottschaldt et al. 1973). All these receptors were found to be sensitive mechanoreceptors. Heinbecker et al. (1933), in their experiments using pressure block, were able to establish a connection between activity in non-myelinated nerve fibres and nociception. However, not all non-myelinated fibres are nociceptive. Using the collision technique in the rabbit, Douglas and Ritchie (1957) were able to abolish a large proportion of the electrically evoked C-compound potential by gently stroking the fur of the rabbit. In the cat, Iggo (1960) described a population of low-threshold mechanoreceptors with afferent C-fibres. Recently, Nordin (1990) was able to record unitary electrical activity in the human supra-orbital nerve after stimulation with low forces. In combination with the fact that cold fibres form a subpopulation of primary afferent C-fibres (Iggo 1959, 1964), the above findings put into question the habit of equating C-fibre input into the central nervous system with pain.

Already at the level of amphibia, cutaneous mechanoreception has developed to a level of high specification. Frogs possess slowly adapting type I and type II mechanoreceptors very similar to those in mammals (Ogawa and Iggo 1977). In terms of survival value, cutaneous mechanoreception appears to be at least as significant as nociception, because of its important role in orientation and alerting. With the advent of hair, a new concept developed in cutaneous mechanoreception. Among their many other functions, hairs can be employed as levers for the amplification of mechanical force. In hairy skin, the roots of hair are a common location for mechanoreceptors. This principle has been brought to great refinement in vibrissae and sinus hairs. The original classification of cutaneous sensory receptors was anatomical, according to the structure of individual receptor types. The physiological stimulus response properties could only be speculated upon, and inaccurate assumptions were made.

In the absence of reliable information about the stimulus response properties of cutaneous sensory receptors, speculation about the organization of the skin senses oscillated between theories emphasizing either a specific role of cutaneous sensory receptors permitting a stimulus specific input from one population of sensory receptors into the central nervous system (Iggo 1982), or a more diffuse organization where different response patterns in the same receptor population would signal the presence of specific stimuli (Sinclair 1981).

Between 1960 and 1980 the argument about the peripheral organization of the skin senses was firmly resolved in favour of the specific mode. There are good reviews covering this period (Iggo and Findlater 1984; Munger and Ide 1988; Johannson and Vallbo 1983; Akoev and Adrianov 1989). The present chapter attempts to be an update rather than providing comprehensive coverage. The main thrust of work during the following years went into two directions, first the role of the known receptor types in man using microneurography (Johannson and Vallbo 1983) and the cellular, trophic and efferent functions of component structures of cutaneous sensory receptors (Iggo and Findlater 1984). Merkel cell receptors were a common topic of research, because conclusive proof is still lacking that these sensory receptors are of the secondary sensory type. With reference to the type of response maintained, stimulation sensory receptors are classified as slowly or rapidly adapting. In the following section, the presently known cutaneous mechanoreceptors will be described as they fit into this classification.

2 Types of Receptors

2.1 Slowly Adapting Cutaneous Mechanoreceptors

Three types of slowly adapting cutaneous mechanoreceptors have been described in a large variety of vertebrates. The type I receptor (s.a.I) has an irregular and the type II (s.a.II) a regular discharge pattern (Iggo and Muir 1969; Chambers et al. 1972). Both have afferent fibres in the A beta-range. The third type is the C-fibre mechanoreceptor (Iggo 1960; Bessou et al. 1971).

There is some confusion about the nomenclature. Type I cutaneous mechanoreceptors receptors consist of clusters of primary afferent endings in close association with secondary sensory cells at the epidermo-dermal border or the base of the epithelium in hair bulbs. There is agreement that in mammals these secondary cells are called Merkel cells, and Winkelmann (1977) restricts the term Merkel cell to mammals only. Others (Andres 1966; Diamond et al. 1988) extend the use of the term Merkel cell to other vertebrates. In this chapter the term Merkel cell is used in its extended term. Furthermore, in the species *Xenopus laevis*, Mearow and Diamond (1988) describe Merkel cell receptors as rapidly adapting.

Type-II slowly adapting cutaneous mechanoreceptors appear to consist of Ruffini type endings in the dermis (Chambers et al. 1972).

2.1.1 Merkel-Cell Receptor

Figure 1 (top) shows a typical electrophysiological response from a slowly adapting type-I cutaneous mechanoreceptor in the hairy skin of the cat. The recordings were made from a small strand dissected from the saphenous nerve. During the initial ramp phase of displacement, the receptor responds with a high rate of discharge. It is dynamic stimuli like brushing the skin with an artists brush or gently drawing a glass rod over the skin surface which are the most effective ways of exciting type I receptors. Following the ramp phase, the force of stimulation was maintained at 20 mN, resulting in a slowly adapting response. In this case, the record displayed covers only 2 s. However, with maintained displacement, a slowly adapting response may be recorded over many minutes. Up to four time constants may describe the various phases of adaptation (Iggo and Muir 1969).

In the absence of mechanical stimulation, type-I cutaneous slowly adapting mechanoreceptors are silent. There is no ongoing activity (Brown and Iggo 1967;

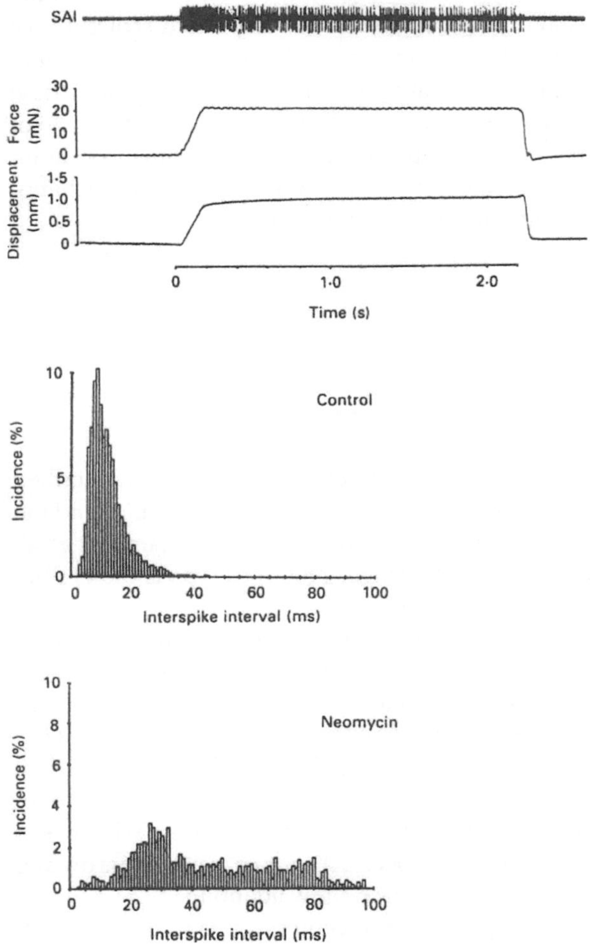

Fig. 1. Force, displacement, nervous response and interspike intervals from s.a.I receptors in the cat. *Top* Response from a single unit during force stimulation rising over 200 ms to a plateau force of 20 mN, where it was kept for 2 s. Note, there is no on going discharge in the absence of stimulation. During the period of stimulation with constant force there is a very gradual increase in the indentation, due to the plastic properties of the skin. *Middle* Interspike intervals, averaged from three s.a. I receptors. The measurements were taken from 1200–2200 ms of the above stimulation. *Bottom* Same as *middle*, but at the end of a 20-min period of close arterial infusion of neomycin. Note the change from a Poisson to an almost even distribution. (Baumann et al. 1990)

Iggo and Muir 1969; Baumann et al. 1990). During the dynamic phase of stimulation, stimulus response characteristics can be described by a power function and in the adapted state the interval histogram follows a Poisson distribution (Iggo and Muir 1969).

Merkel-cell receptors consist of clusters of Merkel cells at epithelial basement membranes. They may be positioned at the epidermo-dermal junction or the epithelial bulb of vibrissae or sinus-hairs. Merkel cell receptors are present both in glabrous and hairy skin (Munger and Ide 1988). In glabrous skin they tend to be positioned at dermal ridges, whereas in hairy skin they form the characteristic domes. Depending on the species they may or may not be associated with hairs (Haar- scheibe, Pinkus 1905).

Merkel cells are egg-shaped cells with large lobulated nuclei. The cells measure around 15 μ in the long axis, which is orientated in the same plane as the basement membrane of the epithelium to which they are attached. On the epithelial side, Merkel cells are anchored to adjacent epithelial cells through desmosomes. As a further feature at the epithelial side, Merkel cells possess cytoplasmic processes of approximately 1 μ length extending into tightly fitting epithelial recesses. Cytoplasmic processes are never the site of desmosomes (Iggo and Muir 1969). The consistent absence of desmosomes from these structures indicates that these processes do not subserve a primarily structural function.

Functional Organization of Merkel-Cell Receptors. Cauna (1962) produced the first electron-micrographic picture of a Merkel cell. This work and later work by Andres (1966) and Iggo and Muir (1969) was concerned with the whole Merkel-cell primary-afferent terminal complex. At the interface between Merkel cells and primary afferent terminal, postsynaptic thickenings were described resembling the ultrastructure of known chemical synapses. The concept of a chemical synapse was further supported by the presence of large numbers of dense-cored vesicles in the cytoplasm of Merkel cells, with a preferred location close to the nerve terminal. However, these vesicles do not seem to contain catecholamines, because it is not possible to deplete them by chronic treatment with reserpine (Iggo and Muir 1969).

The question of presynaptic vesicular release is still a matter of debate. Type I cutaneous mechanoreceptors do not produce any nervous discharge in the absence of specific mechanical stimulation. There is no ongoing nervous activity. For this reason only very limited spontaneous release of transmitter substance can be expected in the absence of specific stimuli at the time of fixation. To date, only Chen et al. (1973) have described fusion of dense cored vesicles with the assumed presynaptic membrane in Merkel cells. Although there is good ultrastructural support for the concept of chemical synaptic transmission, matters become less clear when pharmacological support is sought. Mildly excitatory action of histamine, 5-hydroxy tryptamine and acetylcholine has been described by Fjaellbrandt and Iggo (1961). Smith and Creech (1967) showed that nicotine had some excitatory action. However, none of these substances has been identified as transmitter substance. More recent results about opiates and peptides are equally inconclusive from the functional aspect. Hartschuh and Weihe (1988) found substance P and calcitonin gene-related peptide but no opioids in Merkel cells from fetal pigs, and vasoactive intestinal peptide as well was found in adult pig (Hartschuh and Weihe 1988). Hartschuh et al. (1979) demonstrated met-encephalin-like immunofluorescence in rodents, but not in the cat, dog, pig or man.

Neither Gottschaldt and Vahle-Hinz (1982) in the cat, nor Handwerker in the rat (unpubl.) were able to modulate naturally evoked responses from mechanoreceptors with Naloxone. In contrast, Pacitti (1988) observed enhanced responsiveness in a proportion of her experimental sample of Merkel cell receptors.

At the moment it is difficult to suggest a comprehensive pattern of action from the above pharmacological actions. From the physiological point of view, there is evidence in favour as well as against a chemical synaptic arrangement between Merkel cells and primary afferent fibres. Results against the synaptic model came from a variety of experiments:

1. The ability of Merkel cell receptors to follow vibration at frequencies in excess of 1500 Hz without flutter was thought by Gottschaldt and Vahle-Hinz (1981) not to be compatible with chemical synaptic transmission because of the short synaptic delay necessary for the generation of such high frequencies.
2. In other experiments, Gottschaldt and Vahle-Hinz (1981) compared the latencies of responses evoked by topical electrical and specific mechanical stimulation in s.a.I receptors. The time difference between both modes of stimulation was found to be around 0.1 ms and therefore thought to be too short to allow chemical synaptic transmission. However, Llinas (1977) has described chemical synapses transmitting with a synaptic delay of 0.2 ms, which is of the same order of magnitude as Gottschaldt and Vahle-Hinzes' results.
3. More recently, Diamond et al. (1988) showed that the dye quinacrine is taken up by Merkel cells and that illumination with ultraviolet frequencies producing fluorescence causes selective destruction of Merkel cells. Using a prodder in a semiquantitative way, it was still possible to elicit responses from Merkel cell receptors after they had been treated in this way. The authors took this finding as evidence that Merkel cells were not essential for the normal function of type-I slowly adapting cutaneous mechanoreceptors, and by implication that there was no chemical synapse between sensory nerve terminal and Merkel cell. These results could, however, be interpreted in a different way (Baumann et al. 1990). It is known that peripheral nerve fibres may themselves become sensitive to innocuous mechanical stimulation following injury (Brown and Iggo 1963). The connection between Merkel cells and primary afferent terminal is tight. Injury of one component is likely to produce some damage of the opposite structure resulting in sensitization to mechanical stimulation. In order to make this experiment conclusive, it would be necessary to record before and after the quinacrine treatment and determine the stimulus response properties of individual receptors under investigation in a quantitative fashion. This is not possible, as it takes many hours for the degeneration of Merkel cells to take place.
4. In a promising preliminary experiment, Diamond et al. (1988) recorded the membrane potential from isolated amphibian Merkel cells with patch-clamp electrodes. They were unable to record any changes of the membrane potential in response to mechanical stimulation. No information is available about the condition of the Merkel cells under investigation, in particular about presence or absence of cytoplasmic processes. It is difficult to discuss these experiments in their preliminary state. However, the concept of these experiments, looking at the function of cellular components, is likely in the end to provide a conclusive answer about the organization of the Merkel cell receptors.

Experimental work in support of chemical synaptic linkage between Merkel cells and primary afferent terminal is based on a variety of experimental approaches. Initially mathematical analysis of patterns of nervous discharge alone was the main tool (Iggo and Muir 1969; Horch et al. 1974). More recently, Anand et al. (1979) and Iggo and Findlater (1984) were looking at the interrelationship between excitability, hypoxia and presence or absence of dense cored vesicles in Merkel cells of s.a.I receptors in the cat. Experiments were performed involving recirculation of venous blood, as well as the provision of an oxygen-free environment for the limb under investigation and fast fixation of tissue for electronmicrography.

The presence of dense-cored vesicles was dependent on an adequate supply of oxygen. It was interesting to note that provision of oxygen through the skin was sufficient to maintain the vesicles in the dense-cored state. Furthermore, s.a. I receptors were only excitable by mechanical stimuli as long as dense-cored vesicles were present in the Merkel cells. In the absence of dense-cored vesicles, it was still possible to elicit afferent responses by transcutaneous electrical stimulation, indicating that there was no conduction block in afferent nerve fibres.

The above experiments provide good circumstantial evidence in favour of a chemical synaptic link between Merkel cells and afferent terminal. However, the extensive hypoxia used in these experiments would also have affected the sensory terminals making contact with Merkel cells. Like other sensory endings, these terminals contain large numbers of mitochondria. If one assumed that the transducer mechanism was located in the primary afferent terminal, there would be fast depletion of energy-rich substances by the extensive hypoxia, presumably resulting in failure of the receptor. In this context it is relevant to note that in the rabbit a proportion of the vesicles in the Merkel cells are clear under normoxic circumstances (Anderson et al. 1979). However, it is not known whether s.a. I receptors in the rabbit are more easily fatigued than in the cat.

Chemical synaptic transmission involves the influx of Ca^{2+} preceding the release of transmitter substance (Llinas 1977). Block of relevant ionic Ca^{2+} channels could therefore be expected to interrupt chemical synaptic transmission. The possible role of Ca^{2+} in the transduction process of s.a.I receptors has been discussed in detail by Iggo and Findlater (1984). Three research groups have performed experiments applying organic as well as inorganic Ca^{2+} antagonists to Merkel cell receptors, and strong, often reversible blocking action could be demonstrated independently by each group. Cooksey et al. (1984) and Pacitti and Findlater (1988) used Verapamil and Ca^{2+} in close arterial application to s.a.I receptors in the cat, causing profound reduction in the responsiveness to physiological mechanical stimuli.

Yamashita et al. (1986) applied Verapamil and a variety of heavy metal ions topically to the slowly adapting type I (Ft I) as well as and type II (FT II) in the frog. Their results showed differential sensitivity between the primary sensory endings in the Ft II receptors and the much more sensitive Ft I receptors.

Finally, Baumann et al. (1986c, 1990) took the similarity between hair cell receptors in the inner ear and Merkel cells as a lead and applied the aminoglycoside Neomycin to s.a.I receptors in the cat and rat. Acute application of aminoglycosides to hair cells causes a nonspecific decrease in permeability for cations which is counteracted by an increase in the extracellular Ca^{++} concentration (Hudspeth and Kroese 1983).

In the cat, close arterial infusion of Neomycin causes an immediate, dose-dependent drop in the responsiveness to standard mechanical stimuli with subsequent slow recovery (Baumann et al. 1986). The slow recovery is due to the circulating Neomycin. Aminoglycosides distribute primarily through the extracellular space, and Baumann et al. (1988) found that, although close arterial infusion produced much stronger suppression of responsiveness of s.a.I receptors than systemic intravenous, the time course of recovery was similar. In contrast, there is almost complete recovery within 1 h following near-total suppression when much smaller quantities of Neomycin are infused interstitially close to the receptor (Baumann et al. 1990).

The study by Baumann et al. (1990) was designed as a comparison with the s.a.II receptor, a primary sensory receptor. Stimulus evoked activity in these receptors is suppressed to a lesser degree by close arterial Neomycin than s.a.I. Furthermore, when comparing the interspike-interval histograms, the typical narrow outline covering only a small range of intervals was retained by s.a.II receptors. Neomycin merely shifted the whole complex towards longer intervals (Fig. 1, middle, below). The effect on s.a.I receptors was much more dramatic in that the typical Poisson distribution was distorted to a near random distribution, indicating a fundamentally different mode of action by the Neomycin at the s.a.I receptor (Fig. 2, middle, below). It was interesting to note that when the mechanical strength of stimulation was reduced to a level where per stimulus the same number of impulses were elicited as during, e.g. 60% suppression by close arterial Neomycin, the interspike-interval histogram still retained the Poisson distribution (Baumann et al. unpubl.).

Finally, the amounts of Neomycin infused close-arterially, did cause only partial suppression of mechanical excitability. At the neuromuscular junction, comparable doses affected only presynaptic mechanisms of transmission (Fiekers 1983).

In conclusion, the evidence presented here favours the synaptic mode of transmission between Merkel cells and nerve terminal, and by implication Merkel cells are mechanoelectric transducers. However, most results are indirect evidence, because of the physical inaccessability of Merkel cells for physiological investigations at the level of single cells. The final answer as to the functional organization of Merkel cell receptors is likely to come with advances in techniques of investigation from direct electrophysiological recordings from both Merkel cells and nerve terminals.

2.1.2 Ruffini Endings

In the early electrophysiological work on slowly adapting cutaneous mechanoreceptors, no discrimination is made between different types of receptors with afferent myelinated nerve fibres. Careful analysis of stimulus response characteristics indicated that there was a receptor different from the Merkel cell type (Brown and Iggo 1967). Chambers et al. (1972) were able to show that dermal Ruffini endings were the origin of this response.

Figure 2 (top) shows a single unit recording from an afferent fibre supplying a s.a.II receptor in the leg of the cat. In the absence of external stimulation, these units show a very regular ongoing nervous discharge, which is probably an expression of the tension inherent to the skin. The response displayed is elicited as

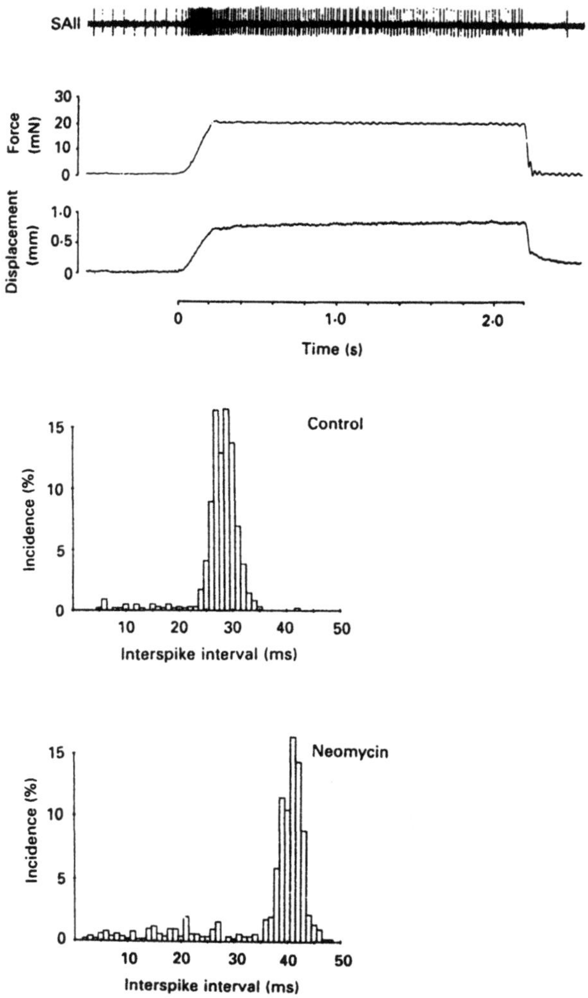

Fig. 2. Force, displacement, nervous response and interspike intervals from s.a. II receptors in the cat. *Top* Response from a single unit during force stimulation rising over 200 ms to a plateau force of 20 mN, where it was kept for 2 s. Note the regular on-going discharge in the absence of stimulation. During the period of stimulation with constant force there is a very gradual increase in the indentation, due to the plastic properties of the skin. *Middle* Interspike intervals, averaged from ten satic responses during the time interval between 1200–2200 ms of the above stimulation. *Bottom* Same as *middle,* but at the end of a 20-min period of close arterial infusion of neomycin. Note retention of the mode of distribution except for a shift to longer intervals. (Baumann et al. 1990)

above by a perpendicular ramp stimulus rising to a force of 20 mN over 300 ms followed by a plateau of 1.7-s duration. Again the dynamic component of the stimulus is most effective in exciting the receptor, eliciting a response that can best be described by a power function. In a slowly adapting manner, these responses can be maintained over minutes. The interspike interval distribution in the adapted state follows a Gaussian distribution, with most intervals falling into a very small range. A small number of shorter intervals are evenly distributed (Fig. 2, bottom). The regular rate of discharge may be explained by the fastest pacemaker always generating all nervous impulses at any point in time, with shorter intervals occurring when there is a change in pacemaker (Baumann et al. 1990).

For reasons of convenience, it is common to stimulate these receptors in experimental situations with perpendicular stimuli. However, in contrast to s.a.I receptors they may be excited by dermal stretch, and their relatively deep position in the dermis (Chambers et al. 1972) suggests that detection of dermal tension is probably their main function. This point is emphasized further by the fact that there are very few if any s.a.II receptors in the loose gluteal skin of the rat (Baumann et al. 1990), whereas they can be found more easily in the tighter skin of the leg (Lynn and Carpenter 1982).

2.1.3 C-Mechanoreceptors

Iggo (1960) reported unitary recordings from well-identified afferent C-fibres supplying sensitive mechanoreceptors in the cat. In response to maintained perpendicular low force stimulation these fibres responded with a slowly adapting discharge. Bessou et al. (1971) looked at the dynamic response properties of cutaneous C-mechanoreceptors. In comparison with mechanoreceptors in the A-fibre range, these receptors fatigue relatively easily. It is therefore difficult to determine the proportional significance of these units. In the rabbit, Douglas and Ritchie (1957) were able to reduce the size of the C-compound action potential by about 50% in an experiment using the collision technique in combination with gently stroking the fur.

Nordin (1990) was able to demonstrate C-mechanoreceptors in man in unitary microneurographic recordings from the supraorbital nerve (Fig. 3).

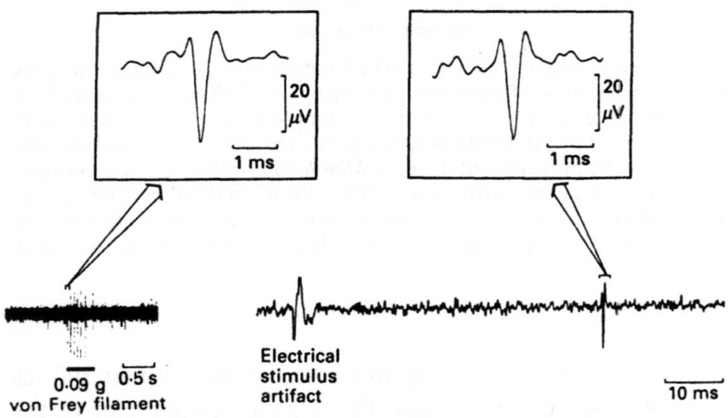

Fig. 3. Neuromicrographic recording from a cutaneous low threshold C-mechanoreceptor unit in the human supraorbital nerve. *Left* Response to stimulation with v. Frey hair. *Right* Response to electrical stimulation. (Nordin 1990)

2.2 Rapidly Adapting Receptors

In the organization of the rapidly adapting receptors, there is more diversity than in slowly adapting receptors. In mammals, different types of sensory receptors are present in glabrous and hairy skin. For the purpose of classification it is convenient to use this mammalian subdivision.

Rapidly adapting receptors are velocity detectors. They do not respond to maintained indentation of the skin. They do not show on-going activity. They are all primary sensory receptors, where the transducermechanism is located in a terminal specialization of the afferent nerve fibre.

2.2.1 Glabrous Skin

Different types of velocity detectors in the glabrous skin have optimum response frequencies to vibration. At the upper end of 200–500 Hz or more, it is Pacinian Corpuscles or Golgi-Mazzoni receptors that are most sensitive. In the lower frequency range, Meissner and Krause corpuscles are the most sensitive detectors (Iggo 1982). They tend to be located close to the epidermo-dermal border (Munger and Ide 1988), making them suitable receptors for shape discriminating tactile tasks. Pacinian corpuscles are positioned in the dermis (Munger and Ide 1988), a position that is not suitable for the detection of minute differences in shape of objects. Their receptive fields are much larger than those of Meissner corpuscles in the same area.

2.2.2 Hairy Skin

In the hairy skin, rapidly adapting mechanoreception is closely linked to the hairs acting as levers. Brown and Iggo (1967) classified hair structurally according to Straile (1960) into Tylotrich, Guard and Down hairs. Receptor structures at the hair bulbs are either lanceolate or Ruffini type endings (Andres 1966).

In the cat and rabbit, most down hairs are innervated (Brown and Iggo 1967), whereas not every individual hair possesses its own sensory apparatus in the rat (Millard and Wolff 1988). Physiologically, hair receptors are velocity receptors. Each type of unit shows different degrees of convergence, determining the typical pattern formation of receptive field.

The stimulus response characteristics can be described by a power function. No differences could be found between the exponents of the three receptor types. However, there are differences in threshold, with the down hairs being the most sensitive type (Brown and Iggo 1967). Burgess et al. (1968) confirmed this classification in principle. In addition, these authors described a type of receptor not responding to the movement of individual hairs, but only to stroking of a larger area of hairy skin, the field receptor. As a refinement, the guard hair population was subdivided into G1 and G2 on the basis of sensitivity to the slope of a stimulus (Tuckett et al. 1978). At threshold amplitude G1 hairs require a higher frequency of stimulation than G2 hairs. However, there is considerable overlap in their stimulus response characteristics between both populations (Pierce and Roberts 1981).

3 Special Features in Different Species

3.1 Mammals

There is remarkable uniformity in the organization of cutaneous mechanoreceptors in mammals. The purpose of this section is to discuss special developments and exceptions.

In the glabrous skin of the raccoon, Pubols (1983) was able to discriminate between slowly adapting and very slowly adapting receptors among the s.a.I receptors. It has not been possible to determine, whether this discrimination represents a physiological difference or whether it has its origin in topically differing mechanical properties of the skin. Another feature of the slowly adapting receptors in the raccoon glabrous skin is that there is a less clear-cut differentiation between s.a.I and s.a.II in respect of the regularity of discharge. In the raccoon glabrous skin, it is not possible to discriminate between both receptor types on the basis of the coefficient of variation.

The sensory part of sinus hairs is a composite sensory organ containing several types of the above cutaneous mechanoreceptors. The term sinus hair includes vibrissae as well as the feline carpal hairs, according to the definition of Andres (1966). Two types of slowly and two types of rapidly adapting responses have been described in response to bending of vibrissae in the cat. All these responses were found to be mediated through fast-conducting myelinated nerve fibres (Gottschaldt et al. 1973). S.a.I units showed the same adaptation and irregular discharge patterns as described for touch corpuscles (Iggo and Muir 1969), whereas the s.a.II units could be subdivided into units with either an irregular or regular pattern of nervous discharge. Directional sensitivity for bending of vibrissae into specific direction was displayed by only a few s.a.I units.

Rapidly adapting units could again be subdivided into slow and fast velocity detectors. Fast velocity detectors could follow vibration frequencies of up to 800 Hz. These responses probably originated from the lamellated Golgi-Manzoni complexes in the root bulb.

3.2 Birds

The major tactile organ in birds is the beak, and physiological investigation of cutaneous mechanoreceptors in birds have focussed on this part of the body. Gregory (1973) exclusively described rapidly adapting responses when recording unitary responses from the beak of the duck. Necker (1974) found the majority of cutaneous mechanoreceptors in the beak of the pigeon to be rapidly adapting with a minority of slowly adapting receptors. Gottschaldt and coworkers (1973, Gottschaldt et al. 1982) were able to differentiate between units predominantly responding to ramp stimuli applied to the beak, which they attributed to Grandry corpuscle and responses to high- frequency vibration apparently originating from Herbst corpuscles. Functionally, Grandry corpuscles thus appear to correspond to Meissner corpuscles or Krause end bulbs in other vertebrates. This finding is surprising. Structurally Grandry corpuscles have many similarities with Merkel cell receptors (Gottschaldt and Lausmann 1974) and could therefore expected to be slowly adapting.

Herbst corpuscles are analogous to Pacinian corpuscles. Recently, Gentle (1989) described similar stimulus response characteristics of rapidly adapting cutaneous mechanoreceptors in the beak of the domestic chicken. Slowly adapting mechanoreceptors were described in the distal part of the beak of the goose (Gottschaldt 1974), pigeon (Necker 1974) and chicken (Gentle 1989). There is agreement that slowly adapting receptors respond more vigorously during cooling.

3.3 Reptiles

The fibre spectrum of mammalian cutaneous afferent myelinated nerve fibres from mechanoreceptors are distributed bimodally with maxima in the A-beta and delta fibres. At least for alligators and caimans there is a unimodal distribution (Kenton et al. 1971). In the slowly adapting range these authors were able to discriminate between s.a.I- and s.a.II-type responses. However, not all responses could be described by a power function. Both receptor types showed increased responses with lowering of the temperature, a phenomenon that may be of increased importance in these animals, as they do not have many cutaneous thermoreceptors.

Rapidly adapting receptors covering a similar range of responsiveness to mammals have been found as well. In the blacksnake, rapidly adapting receptors have been described which followed frequencies of up to 800 Hz (Proske 1969). Functionally these receptors correspond to Pacinian corpuscles or Herbst corpuscles. This is of particular interest, because there do not appear to be microscopically developed receptor structures in the skin of the blacksnake.

4 Mechanical Properties and Method of Stimulation

The actual stimulus reaching the transducer in a mechanoreceptor is a function of the stimulus itself as well as of the force transmission properties of the surrounding tissue. Lindblom (1965) measured the time course of recovery of the skin after mechanical stimulation of the glabrous skin of the monkey. Restitution of the dimple caused by low-force mechanical stimulation was found to be a function of the duration of stimulation. Indentation with a probe measuring 0.3 mm in diameter to a depth of 0.2 mm for 25 ms had returned to prestimulation condition after 25–30 ms, whereas a stimulus lasting 150 ms needed a recovery time of 170 ms. Similarly, Pubols (1982) observed recovery intervals of up to 10 s following mechanical stimuli in the physiological range of the glabrous skin of the raccoon paw.

In terms of stimulus response characteristics, Pubols and Pubols (1983) observed that s.a.I receptors displayed a lower rate of adaptation with constant displacement as opposed to constant force stimuli. Residual indentation is of some importance to responsiveness, when repetitive constant displacement stimulation is used. In the hairy skin of the rat, Leung (1986) compared repetitive stimulation with sequences of mechanical stimuli, starting with the displacement needed to produce a force of 20 mN with ramp and hold stimuli lasting 2 s and with interstimulus time of less than a second. Subsequent stimuli were less effective in the constant displacement than constant force mode. During the first five sequential stimuli the difference was minimal. However, during later stimuli, up to 30% of the response was lost when using stimulation with constant displacement.

Experimentally, erroneous results may be the consequence, if constant displacement stimuli are used without simultaneous measurement of force, because residual indentation or local changes in vascular perfusion may cause unnoticed alterations in the quality of skin contact.

Given the above considerations, it is remarkable how consistent performance of cutaneous mechanoreceptors is under conditions of changed compliance of the skin. Lamotte and Whitehouse (1986) observed little variability in thresholds of rapidly adapting receptors between monkeys and a correspondingly small variability in detection threshold in man.

Baumann et al. (1986a,b) and investigated the stimulus response properties of s.a.I receptors in the rat during ageing and in two models of degenerative skin conditions. Rats growing up on a Vitamin A-deficient diet needed a higher stroke amplitude to produce the same force of stimulation, than in control animals. Chronic treatment with Neomycin and ageing have the opposite effect. However, with near-maximal perpendicular stimulation, there is hardly any change in the responsiveness of ageing s.a.I receptors. In contrast, there is significantly reduced responsiveness during repetitive stimulation in both degenerative conditions. It appears that it is a direct degenerative effect on the s.a.I receptors rather than mechanical changes in the compliance of the skin that are largely responsible for the reduced responsiveness.

Following the initial determination of the stimulus-response characteristics of cutaneous mechanoreceptors, there has recently been more emphasis on the integrated function of cutaneous mechanoreceptors (Darian-Smith et al. 1980; Lamotte and Whitehouse 1986; Gardner and Palmer 1990). The common finding from these investigations is that fine dynamic explorative tactile activity using receptors in the glabrous skin of the hand relies on rapidly adapting receptors. S.a.I or s.a.II receptors are not involved in the exploration of a dot on a smooth surface or a grid pattern.

Gardner and Palmer (1989, 1990) conducted an extensive study into the excitability of cutaneous mechanoreceptors in man and in the monkey using the OPTACON stimulator. Through computer control of a grid of fine probes, it is possible with this device to simulate the sensory impression of a line moving across the surface of the skin. With individual impression times of 4 ms, these authors simulated movement of a line without any response at all from slowly adapting receptors.

5 Sympathetic Control of Mechanoreceptors

Some types of cutaneous mechanoreceptors are affected by efferent control. The specific effect of modulation depends on the individual type of receptor, and often it is not certain whether the observed effects are epiphenomena.

Slowly adapting cutaneous mechanoreceptors in the frog, if affected at all, show sensitization to adequate mechanical stimuli, when conditioned by electrical stimulation of relevant sympathetic efferents. The majority of cutaneous s.a. receptors in the frog display lowering of threshold and an increase in responsiveness (Loewenstein 1956; Calof et al. 1981). Loewenstein (1956) also demonstrated facilitation caused by application of catecholamines.

The above workers did not distinguish between s.a.I and s.a.II receptors, although both receptor types exist in the non-wharty skin of the frog (Ogawa et al. 1981). The rather large range of effects under well-controlled experimental conditions may therefore be due to some extent to the fact that both groups of authors were investigating inhomogeneous groups of receptors.

Working on cats, Roberts et al. (1985) observed facilitation in the majority of a well-identified group of s.a.I receptors. In these experiments, the sympathetically induced facilitation was not dependent on maintained arterial blood pressure, because facilitation was still possible during transient occlusion of the aorta. Slowly adapting type II receptors in the cat are only facilitated by conditioning sympathetic nervous activity if their main excitatory stimulus was skin stretch perpendicular to the main axis of the limb (Pierce and Roberts 1981).

Cash and Linden (1982) investigated the effect of sympathetic stimulation on the dynamic response of periodontal mechanoreceptors. Half of their sample of 84 units was inhibited by the sympathetic conditioning stimulus, whereas the other half was not affected at all. Intravenous infusion of phentolamine abolished this effect. Pacinian corpuscles are sensitized by catecholamines (Schiff 1974). In contrast, suppression of excitability has been observed in all other rapidly adapting cutaneous mechanoreceptors, with myelinated afferents in frogs and cats investigated to date (Loewenstein 1956; Calof et al. 1981; Pierce and Roberts 1981).

Synaptic contact has not been observed between sympathetic nerve endings and cutaneous sensory endings, although several types of cutaneous mechanoreceptors seem to be closely associated with nonmyelinated nerve endings (for review see Akoev and Adrianov 1989). However, these nerve fibres could subserve paracrine release of active substances.

Acknowledgment. This work was supported by the Croucher Foundation

References

Adrian ED, Umrath K (1929) The impulse discharge from Pacinian corpuscles. J Physiol (Lond) 68:139–154

Adrian ED, Zottermann Y (1926) The impulses produced by sensory endings. Part 2. The response from a single endorgan. J Physiol (Lond) 72:151–171

Akoev GN, Adrianov GN (1989) Synaptic transmission in the mechano- and electroreceptors of the acousticolateral system. In: Autrum H, Ottoson D, Perl ER et al. Progress in Sensory Physiology, vol 9. Springer, Berlin Heidelberg New York, pp 53–95

Anand A, Iggo A, Paintal AS (1979) Lability of granular vesicles in Merkel cells of the type I slowly adapting cutaneous receptors of the cat. J Physiol (Lond) 296:19–20P

Anderson T, Bannister LH, Hamann W (1979) Comparative ultrastructure of touch corpuscles in the rabbit and the cat. J Physiol (Lond) 287:6–7P

Andres KH (1966) Ueber die Feinstruktur der Rezeptoren an Sinushaaren. Z Zellforsch Mikrosk Anat 75:339–365

Baumann KI, Hamann W, Leung (1986a) Mechanical properties of skin and responsiveness of slowly adapting type I mechanoreceptors in rats at different ages. J Physiol (Lond) 371:329–337

Baumann KI, Cheng-Chew SI, Hamann W, Leung MS (1986b) Responsiveness and ultrastructure of slowly adapting type I cutaneous mechanoreceptors in vitamin A deficient rats. J Physiol (Lond) 371:339–349

Baumann KI, Hamann W, Leung MS (1986c) Reduced responsiveness of touch (sa I) receptors in the cat following close arterial infusion of neomycin. Brain Res 377:160–162

Baumann KI, Hamann W, Leung MS (1988) Responsiveness of slowly adapting cutaneous mechanoreceptors after close arterial infusion of neomycin in cats. Prog Brain Res 74:43–49

Baumann KI, Hamann W, Leung MS (1990) Acute effects of neomycin on slowly adapting type I and type II cutaneous mechanoreceptors in the anaesthetized cat and rat. J Physiol (Lond) 425:527–544

Bessou P, Burgess PR, Perl ER, Taylor CB (1971) Dynamic properties of mechanoreceptors with unmyelinated (C) fibres. J Neurophysiol 34:116–131

Brown AG, Iggo A (1963) The structure and function of cutaneous "touch corpuscles" after nerve crush. J Physiol (Lond) 165:28–29P

Brown AG, Iggo A (1967) A quantitative study of cutaneous receptors and afferent fibres in the cat and rabbit. J Physiol (Lond) 193:70–733

Burgess PR, Petit D, Warren R (1968) Receptor types in cat hairy skin supplied by myelinated fibers. J Neurophysiol 31:833–848

Burgess R, Mei J, Tuckett RP, Horch KW, Ballinger CM, Poulos DA (1983) The neural signal for skin indentation depth. J Neurosci Res 3:1572–1585

Calof AL, Jones RB, Roberts WJ (1981) Sympathetic modulation of mechanoreceptor sensitivity in frog skin (1981) J Physiol (Lond) 310:481–499

Cash RM, Linden RWA (1982) Effects of sympathetic nerve stimulation on intra-oral mechanoreceptor activity in the cat. J Physiol (Lond) 329:451–463

Cauna N (1962) Functional significance of the submicroscopical, histochemical and microscopical organization of the cutaneous receptor organs. Anat Anz 111 (Suppl 2):181–197

Chambers MR, Andres KH, v. Duering M, Iggo A (1972) The structure and function of the slowly adapting type II mechanoreceptor in hairy skin. Q J Exp Physiol 57:417–445

Chen SY, Gerson S, Meyer J (1973) The fusion of Merkel cell granules with synapse-like structures. J Invest Dermatol 61:290–292

Cooksey EJ, Findlater GS, Iggo A (1984) The response to mechanical stimulation of s.a.I receptors of cats and rats in the presence of calcium antagonists. J Physiol (Lond) 357:30P

Darian-Smith I, Davidson I, Johnson KO (1980) Peripheral neural representation of spatial dimensions of a textured surface moving across the monkey's finger pad. J Physiol (Lond) 309:135–146

Diamond J, Mills LR, Mearow KM (1988) Evidence that the Merkel cell is not the transducer in the mechanosensory Merkel-cell-neurite complex. Prog Brain Res 74:51–56

Douglas WW, Ritchie JM (1957) On excitation of non-medullated afferent fibres in the saphenous nerve, which signal touch. J Physiol (Lond) 139:385–399

Ferrington DG, Rowe MJ (1980) Functional capacities of tactile afferent fibres in neonatal kittens. J Physiol (Lond) 307:335–353

Fiekers JF (1983) Effects of aminoglycoside antibiotics, streptomycin and neomycin on neuromuscular transmission. I. Presynaptic considerations. J Pharmacol Exp Ther 225:487–502

Fjaellbrandt N, Iggo A (1961) The effect of histamine, 5-hydroxytryptamine and acetylcholine on cutaneous afferent fibres. J Physiol (Lond) 156:570–590

Frankenhaeuser B (1949) Impulses from a cutaneous receptor with slow adaptation and low mechanical threshold. Acta Physiol Scand 18:67–74

Gardner EP, Palmer CI (1989) Simulation of motion on the skin. I. Receptive fields and temporal frequency coding by cutaneous mechanoreceptors of OPTACON pulses delivered to the hand. J Neurophysiol 62:1410–1436

Gardner EP, Palmer CI (1990) Simulation of motion on the skin. II. Mechanisms used by rapidly adapting cutaneous mechanoreceptors in the primate hand for spatiotemporal resolution and two-point discrimination. J Neurophysiol 63:841–859

Gentle MJ (1989) Cutaneous sensory afferents recorded from the nervous intermandibularis of *Gallus gallus* var. *domesticus*. J Comp Physiol A 164:763–774

Gottschaldt KM (1974) The physiological basis of tactile sensibility in the beak of geese. J Comp Physiol 95:29–47

Gottschaldt KM, Lausmann S (1974) The peripheral morphological basis of tactile sensibility in the beak of geese. Cell Tissue Res 153:477–496

Gottschaldt KM, Vahle-Hinz C (1981) Merkel cell receptors: structure and transducer function. Science 214:183–186

Gottschaldt KM, Vahle-Hinz C (1982) Evidence against transmitter function of met-enkephalin and chemosynaptic impulse generation in "Merkel cell" mechanoreceptors. Exp Brain Res 45:459–463

Gottschaldt KM, Iggo A, Young DW (1973) Functional characteristics of mechanoreceptors in sinus hair follicles of the cat. J Physiol (Lond) 235:287–315

Gottschaldt KM, Fruhstorfer H, Schmidt W, Kraft I (1982) Thermosensitivity and its possible fine-structural basis in mechanoreceptors in the beak skin of geese. J Comp Neurol 205:219–245

Gregory JE (1973) An electrophysiological investigation of the receptor apparatus of the duck's bill. J Physiol (Lond) 229:151–164

Hartschuh W, Weihe E (1988) Multiple messenger candidates and marker substances in the mammalian Merkel cell-axon complex: a light and electron microscopic immunohistochemical study. Prog Brain Res 74:181–187

Hartschuh W, Weihe E, Buechler M, Helmstaedter GE, Feuerle GE, Forssmann WG (1979) Met-enkephalin-like immunoreactivity in Merkel cells. Cell Tissue Res 201:342–348

Heinbecker P, O'Leary JL, Bishop GH (1933) Nature and source of fibres contributing to the saphenous nerve in the cat. Am J Physiol 104:23–35

Horch KW, Whitehorn D, Burgess PR (1974) Impulse generation in type I cutaneous mechanoreceptors. J Neurophysiol 37:267–281

Hudspeth AJ, Kroese ABA (1983) Voltage-dependent interaction of dihydrostreptomycin with the transduction channels in bullfrog sensory hair cells. J Physiol (Lond) 345:66P

Iggo A (1959) Cutaneous heat and cold receptors with slowly conducting C-afferent fibres. Q J Exp Physiol 44:362–370

Iggo A (1960) Cutaneous mechanoreceptors with afferent C-fibres. J Physiol (Lond) 152:337–353

Iggo A (1964) Temperature discrimination in the skin. Nature 204:481–483

Iggo A (1982) Cutaneous sensory mechanisms. In: Barlow H, Molke D (eds) The senses text in the physiological sciences. Cambridge Univ Press, Cambridge, pp 369–408

Iggo A, Findlater GS (1984) A review of Merkel cell mechanisms. In: Hamann W, Iggo A (eds) Sensory receptor mechanisms. World Scientific, Singapore, pp 117–132

Iggo A, Muir AR (1969) The structure and function of a slowly adapting touch corpuscles in hairy skin. J Physiol (Lond) 200:763–796

Johansson RS, Vallbo AB (1979) Tactile sensibility in the human hand: relative and absolute densities of four types of mechanoreceptive units in glabrous skin. J Physiol (Lond) 286:293–300

Johansson RS, Vallbo AB (1983) Tactile sensory coding in the glabrous skin of the human hand. TINS 6:27–32

Johansson RS, Vallbo AB, Westling (1980) Thresholds of mechanosensitive afferents in the human hand as measured with v. Frey hairs. Brain Res 184:343–351
Kenton B, Kruger L, Woo M (1971) Two classes of slowly adapting mechanoreceptor fibres in reptile cutaneous nerve. J Physiol (Lond) 212:21–44
LaMotte RH, Whitehouse J (1986) Tactile detection of a dot on a smooth surface: peripheral neural events. J Neurophysiol 56:1109–1128
Leung MS (1986) Responsiveness of s.a.I cutaneous mechanoreceptors during aging and in degenerative skin conditions. PhD Thesis, Chinese University of Hong Kong, Hong Kong
Lindblom U (1965) Properties of touch receptors in distal glabrous skin of the monkey. J Neurophysiol 28:966–985
Llinas RR (1977) Calcium release and transmitter release in squid synapse. In: Cowan WM, Forendelli JA (eds) Society for neuroscience symposia vol II. Society for Neurosciences, Bethesda, pp 139–160
Loewenstein R (1956) Modulation of cutaneous mechanoreceptors by sympathetic stimulation. J Physiol (Lond) 132:40–60
Loewenstein WR, Altamirano-Orrego R (1956) Enhancement of activity in a Pacinian corpuscle by sympathomimetic agents. Nature 4545:1292–1293
Lynn B, Carpenter SE (1982) Primary afferent units from the hairy skin of the rat hind limb. Brain Res 218:29–43
Mearow KM, Diamond J (1988) Merkel cells and the mechanosensitivity of normal and regenerating nerves in *Xenopus* skin. Neuroscience 26:695–708
Millard CL, Woolf CJ (1988) Sensory innervation of the hairs of the rat hindlimb: a light microscopic analysis. J Comp Neurol 277:183–194
Munger BL, Ide C (1988) The structure and function of cutaneous sensory receptors. Arch Histol Cytol 51:1–34
Necker R (1974) In: Proc Rheinisch-westphaelische Akademie der Wissenschaften, vol 53. Westdeutichev Verlag, p 134
Nordin M (1990) Low threshold mechanosensitive and nociceptive units with unmyelinated (C) fibres in the human supraorbital nerve. J Physiol (Lond) 426:229–240
Ogawa H, Iggo A (1977) Dependence of the response characteristics of glabrous rapidly adapting units in the cat on the stratum corneum. Brain Res 126:167–171
Ogawa H, Katsushi M, Yamashita Y (1981a) Physiological characteristics of low threshold mechanoreceptor afferent units innervating frog skin. Q J Exp Physiol 66:105–116
Ogawa H, Morimoto K, Yamashita Y (1981b) Physiological characteristics of low threshold mechanoreceptor afferent units innervating frog skin. J Q Exp Physiol 66:105–116
Pacitti EG (1988) Neuropharmacology of the s.a.I sensory receptor. PhD Thesis, Edinburgh University
Pacitti EG, Findlater GS (1988) Calcium channel blockers and Merkel cells. Prog Brain Res 74:37–42
Pierce JP, Roberts WJ (1981) Sympathetically induced changes in the response of guard hair and type II receptors in the cat. J Physiol (Lond) 314:411–428
Pinkus F (1905) Ueber Hautsinnesorgane neben dem menschlichen Haar (Haarscheiben) und ihre vergleichend-anatomische Bedeutung. Arch Mikrosk Anat 65:121–179
Proske U (1969) Vibration sensitive mechanoreceptors in snake skin. Exp Neurol 23:187–194
Pubols BH (1982a) Factors affecting cutaneous mechanoreceptor response. I. Constant-force versus constant displacement stimulation. J Neurophysiol 47:515–529
Pubols BH (1982b) Factors affecting cutaneous mechanoreceptor response. II. Changes in mechanical properties of skin with repeated stimulation. J Neurophysiol 47:530–542
Pubols BH, Pubols LM (1983) Tactile receptor discharge and mechanical properties of glabrous skin. Fed Proc 42:2528–2535
Roberts WJ, Elardo SM, King KA (1985) Sympathetically induced changes in the responses on slowly adapting type I receptors in cat skin. Somatosens Res 3:223–236
Schiff JD (1974) Role of sympathetic innervation of the Pacinian corpuscle. J Gen Physiol 63:601–608
Sinclair D (1981) Cutaneous sensation 1980. Anaesthesiol Intensive Care Med 9:163–173
Smith KR Jr, Creech BJ (1967) Effects of pharmacological on the physiological response of hair discs. Exp Neurol 19:477–482
Straile WE (1960) Sensory hair follicles in mammalian skin. Am J Anat 106:133–147

Tapper DN (1965) Stimulus-response relationships in the cutaneous slowly-adapting mechanoreceptor in hairy skin of the cat. Exp Neurol 13:364–385
Tuckett RP, Horch KW, Burgess PR (1978) Response of cutaneous hair and field mechanoreceptors in cat to threshold stimuli. J Neurophysiol 41:138–149
Werner G, Mountcastle VB (1965) Neural activity in mechanoreceptive cutaneous afferents: stimulus response relations, Weber function and information transmission. J Neurophysiol 28:359–397
Yamashita Y, Ogawa H, Taniguchi K (1986) Differential effects of manganese and magnesium on two types of slowly adapting cutaneous mechanoreceptor afferent units in frogs. Pfluegers Arch Eur J Physiol 406:218–224

Chapter 9

Invertebrate Auditory Receptors

H. Römer[1] and *J. Tautz*[2]

Contents

1 Introduction

If we define ears as any structure that can detect sound waves, then a review of auditory receptors in arthropods is faced with the problem of treating a great diversity of sound waves employed in this large taxon, either for social communication, or for the detection of predators or prey. Hearing may then include the detection of

[1]Lehrstuhl für Allgemeine Zoologie und Neurobiologie, Ruhr-Universität, Postfach 102148, 4630 Bochum 1, FRG
[2]Zoologisches Institut der Universität, Lehrstuhl für Tierphysiologie, Röntgenring 10, 8700 Würzburg, FRG

Advances in Comparative and Environmental Physiology, Vol. 10

sound waves in air or water, the various kinds of waves in solids, at the water/air interface etc. At the same time, there is an enormous variety of mechanoreceptors involved in the detection of sound, and some of these are not even specialized for detecting a particular kind of sound. For example, any arthropod sensillum that usually monitors stress or strain in the cuticle may in addition respond to substrate vibrations. The sensory organ in the second segment of the antenna (Johnston's organ) may function in the near-field as a displacement sound receptor in mosquitoes and *Drosophila* (Ewing 1978), as a device for autocommunicative echolocation in gyrinid beetles using water surface waves (Rudolph 1967; Tucker 1969) or as a sense organ involved in the regulation of insect flight or the control of swimming behaviour (Burkhardt and Schneider 1957; Gewecke et al. 1974; Gewecke 1980), to mention only a few. Considerations of space prevent us from reviewing the great variety of receptor types in arthropods, and we will focus here on receptors responding to airborne sound and to substrate vibrations. However, at the end of this chapter we offer the reader a list of articles dealing with aspects of hearing in arthropods that are not covered in this review.

2 Basic Structure and Physiology of Auditory Receptors

Despite the considerable diversity among the receptor organs and their accessory structures, there is a basic morphological and functional organization which is common to all auditory sensilla. Rather than describing all the structural details in the various groups of arthropods, we shall focus here on one type of receptor which has been well researched, and detail its structure and function as far as the stimulus-transduction mechanism.

2.1 Structure of Receptors

Arthropod auditory receptors are epithelial cells with a specifically modified ciliary structure in the outer dendritic segment. The axon connects the receptor directly to the central nervous system (primary sensory neuron). Each sensory neuron is associated with two or three accessory cells and cuticular structures (such as a filiform hair, see Figs. 1A, 2 and 5). This forms the functional unit, the sensillum. The accessory cells envelope the cell body of the receptor and the apical part of the dendrite (dendritic sheath cell), or form the outer receptor lymph cavity (tormogen and trichogen cell) which surrounds the outer dendritic segment. The most distal part of the dendrite is connected with the cuticular apparatus which serves to transmit the mechanical stimulus to the receptor.

There is a striking structural peculiarity in the distal part of all epidermal mechanoreceptors in arthropods: the so-called tubular body (e.g. Barth 1971; Gnatzy and Tautz 1980; Thurm 1982). This is a modality-specific complex formed of 20–2000 regularly arranged microtubules (diameter about 20 nm) of which most

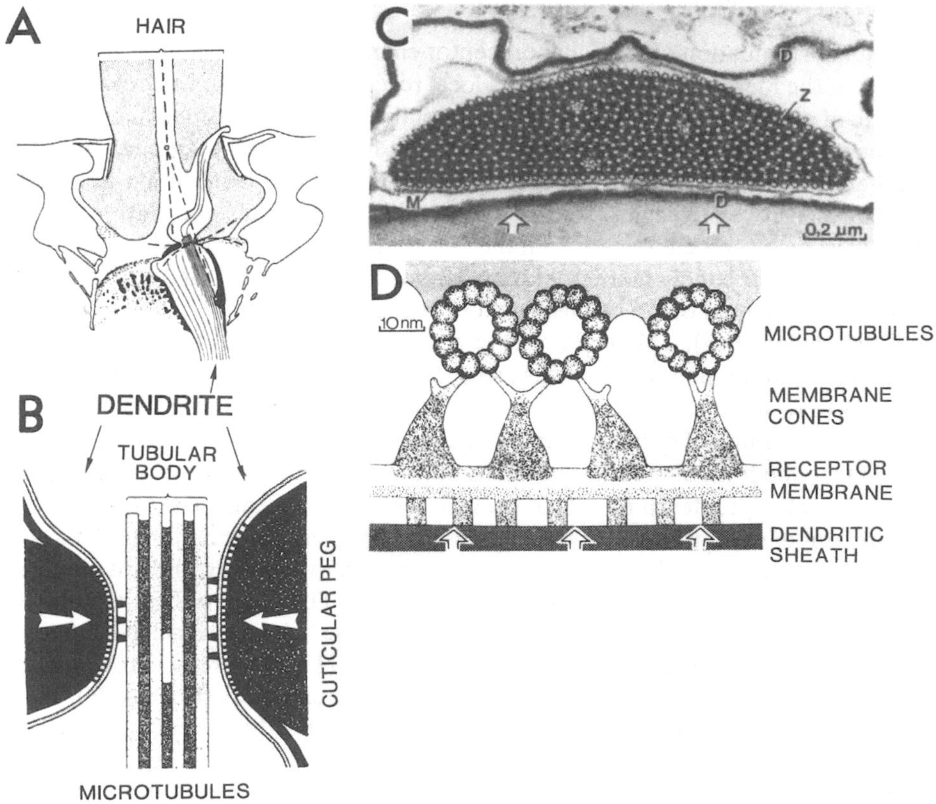

Fig. 1A. Schematic diagram of the articulation area of an insect hair sensillum (*Acheta domesticus*). *B* Stimulus transduction at the site of the tubular body. *Arrows* indicate the direction in which the forces will act onto the tubular body. *C* Cross-section of the dendrite at the site of the tubular body. *D* Dendritic sheath; *M receptor membrane; Z* electron-dense material. *D* Schematic drawing of the membrane-tubulus complex, demonstrating the relationship between membrane cones and microtubules (campaniform sensillum in *Calliphora*). (After Thurm 1982)

are embedded in a very electron-dense amorphic material (Fig. 1B, C). The outer layer of these microtubules is connected via cones to the membrane of the dendrite, as shown schematically in Fig. 1D. These tubular-membrane connections are an integral part of the dendritic membrane (very probably proteins), and they can be found only in close connection with the tubular body and nowhere else in the outer dendritic segment.

Other types of arthropod mechanoreceptors such as stretch and muscle receptors, or the muscle spindles, and Pacinian corpuscles of vertebrates (see Chap. 13, this Vol.) do contain no such tubular body or indeed any such structure that can be considered similar to this. It has been established that receptors containing a tubular body have a stimulus-accessible membrane area which is in the order of four decades smaller than the mechanoreceptors cited above (Thurm 1969).

2.2 Mechanoelectrical Transduction

Adequate stimulation of a mechanoreceptor changes the conductance of a certain region of the membrane, thus causing a local receptor current and a corresponding change of membrane potential. The latency of the increase in conductance in insect mechanoreceptors is of the order of 100 μs, indicating a tight coupling between mechanosensitive molecules and membrane channels. The localization of specific mechanosensitive molecules (so-called sensors) is still a matter of debate, although their existence can be inferred from the fact that the minimal energy necessary to elicit changes in conductance is extremely small (in the order of 10^{-18} W for the complete subgenual organ of bushcrickets or cockroaches; Autrum 1941; Autrum and Schneider 1948). This makes sense only if one postulates the presence of a relatively small number of molecules so constructed as to allow very small energy changes to result in changes of the conformation of some part of the molecule, and thus in the opening of specific membrane channels.

Due to the form of the stimulus-transmitting structures, filiform hair sensilla on the cercus of orthopteran insects or cockroaches offer a means of estimating both the amount of energy as well as the degree of mechanical deformation necessary to stimulate the dendrite. These hairs are inserted in an oval socket, and are mounted at their base on a thin flexible cuticular membrane. The plane of best mobility of the hair corresponds with the plane of symmetry of the hair base: it can be deflected most in the direction of a cuticular peg at its base. This peg projects towards the outer dendritic segment precisely at the site of the tubular body (Fig. 1A, B). Thus, when the hair is deflected by wind or low frequency sound, the cuticular peg presses the dendrite against a second cuticular formation on the opposite side (Fig. 1A, B). This is the adequate stimulus which results in a depolarizing receptor potential, whereas deflection of the hair in the opposite direction provides a pulling force at the tip of the dendrite, and this results in a hyperpolarizing potential.

Such a mechanical construction serves to concentrate most of the stimulus force onto a relatively small part of the outer dendrite which is closely associated with the tubular body, and then onto two series of membrane cones. There are no more than about 500 microtubules in each cone series (see Fig. 1C). Deflection of the hair shaft in the order of a few millidegrees from the resting position results in a displacement at the outer dendritic segment of about 0.1 nm. With these displacements, and the applied stimulus energy in the order of 10^{-19} W, the receptor potential gives rise to action potentials. It is very likely that this energy will be focussed on the minimum number of molecules driving the conductance changes. The compression of membrane cones has been calculated for small receptor responses to be in the order of 0.5 nm. These and other findings would suggest that indeed the membrane cones serve as sensors in the mechanoelectrical transduction process (for a more detailed consideration of these arguments see Thurm 1982).

2.3 Functional Organization of an Auditory Sensillum

Hill (1983a,b) and Oldfield and Hill (1986) proposed a scheme for the functional organization of an insect auditory sensillum, based on models for the insect epidermal mechanoreceptor developed by Thurm and others (e.g. Erler and Thurm 1981; Thurm 1982). On the basis of intracellular recordings from defined parts of the sensillum in the crista acustica or bushcrickets, namely the sensory neuron and the attachment cell, Hill and Oldfield observed potentials of different amplitude and direction. In the soma of the sensory neuron, large and small spikes (amplitude about 65 and 20 mV) can be recorded (Fig. 2c, d). A small spike generally appears as an inflection on the rising phase of the large spike (Fig. 2e), whereas retrograde spikes elicited by electrical stimulation of the tympanal nerve do not show this inflection. Large spikes therefore seem to be triggered by small spikes. Hyperpolarization of the soma reduces the probability of firing of the large, but not the small spikes (Fig. 2a,d), suggesting that small spikes are generated elsewhere. Further, negative going spikes in the attachment cell coincide with small spikes in the neuron (Fig. 2b).

Spike initiation in the distal part of a long sensory dendrite was first suggested by Erler and Thurm (1981). Due to the large distance between the apex and the basal part of the sensory neuron, depolarization by a graded receptor potential at the apex of the dendrite (as observed in the locust receptor, Hill 1983a) might not be sufficient to initiate spikes in the basal part. Thus, small, regenerative spikes at the apex would constitute some kind of amplification of the sensory signal, ensuring that large amplitude spikes are generated at the base of the neuron and then pro-

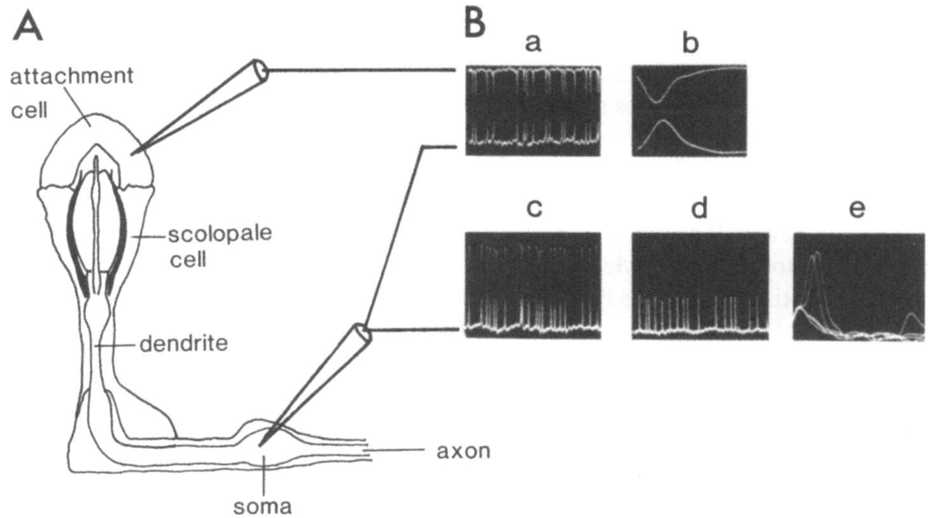

Fig. 2A. Schematic diagram of a sensillum, showing the arrangement of the cells that constitute the receptor, and the placement of the recording electrodes. *B* Spike potentials simultaneously recorded from the attachment cell (*upper traces*) and the soma of the sensory neuron (*lower traces*). Details of waveforms of neural spikes and corresponding potentials in the attachment cell are shown in *a, b*. In the soma of the sensory neuron, large and small spikes can be recorded (*c*); a small spike appears as an inflection on the rising phase of the large spike (*e*). Hyperpolarization of the soma reduces the probability of firing of the large, but not of the small spike (*d*). (After Oldfield and Hill 1986)

pagated along the soma and the axon in the tympanal nerve to the central ganglion. The amplitude of the small spikes might be explained by the fluid in contact with the apex of the dendrite, for in a specialized receptor lymph cavity, new electrochemical gradients may exist.

3 Airborne Sound Receptors

3.1 Near-Field and Far-Field Sound

Air particles near a source of disturbance are displaced and transmit the disturbance to neighbouring air particles. The disturbance propagates as a pressure variation (i.e. the variation in the packing density) through air. Thus, pressure variation is inherently correlated with particle displacement in that there can be no pressure variation without particle displacement. The opposite, however, is not always true, for under certain conditions large particle displacements may occur close to an oscillating object that do not result in a corresponding pressure variation. Such a near-field oscillation is generated because a given volume of air streams around a source rather than being compressed by the source (for details of the physics of sound sources, see Skudrzyk 1971). The efficiency of a sound source in radiating sound pressure depends very much on the amplitude of vibration and on the size of the vibrating structures relative to the wavelength of the sound. A very simple sound source like a pulsating sphere will barely radiate sound once the wavelength starting with short waves is longer than six times the length of its own radius.

A flying wasp produces sound as a by-product of its wing beat, and it can be calculated that the lower frequency limit for efficient sound pressure radiation for a source of this size will be around 3 kHz. Given that the wasp's wing-beat frequency is 150 Hz, this insect is obviously a very poor sound pressure radiator. However, the same wing beat will produce remarkably large amplitudes of air-particle displacements close to the insect, which can be detected by "sound" receptors sensitive to this type of stimulation (see below). As the amplitude of particle displacement decreases by $1/r^2$ (r = distance from source) for a monopole sound source, and by $1/r^3$ for a dipole, the range for detecting this stimulus parameter is limited compared to the propagated sound pressure variation, which decreases for all types of sound sources by only $1/r$.

3.2 Near-Field Receptors

Autrum (1936) and Pumphrey (1940) suggested that single sensory hairs on the body surface of insects should be able to respond to air particle displacement accompanied by the pressure variation of a sound source. It is not surprising that, for example, hair receptors on the cerci of crickets and cockroaches (see Fig. 1A) have been described as either sound receptors or air-flow (wind-sensitive) receptors, since there is no principle difference in responding to the oscillation of air particle displacement or to a wind puff, elicited e.g. by a predator.

Markl and Tautz (1975) have shown that no more than eight filiform hairs on the thorax of a caterpillar are sufficient to elicit an avoidance response (a stop reaction) to sound stimuli. The mechanical properties of these hairs show that within the frequency range of 100–400 Hz and with deviations not exceeding 0.2°, the hairs appear to be optimally adapted to detect medium oscillations; the hairs are as sensitive as theoretically possible (Tautz 1977). Suprathreshold responses occur at particle displacements of about 2 μm, which correspond to a sound pressure in the far field of about 90 dB (for a consideration of differing threshold values in other insects, see Tautz 1979). Compared to far-field sound receptors this would appear to be rather insensitive, but in fact the particle displacement amplitude close to a flying wasp is large enough to be detected at distances up to 70 cm, whereas the sound pressure at this distance is negligible (Tautz and Markl 1978; Tautz 1979).

The reception of medium oscillations by cercal filiform hairs is the basis for a number of escape reactions in various Orthoptera (see Bullock 1984 for review of the neuroethology of escape reactions in invertebrates). In cockroaches a wind puff as a result of the tongue-strike of a toad will elicit the insect's escape behaviour (Camhi and Tom 1978; Camhi 1980). The same type of receptors have also been reported to be involved in intraspecific communication, for when male crickets produce their calling song, they also generate low-frequency air particle oscillations by the inward and outward movements of the forewings. In the acoustic near-field, experienced within 15 cm of the calling males, these oscillations are detected by the filiform hair-to-giant-interneuron system (Kämper 1984). Thus, the same signal pattern can be picked up separately by two mechanoreceptor subsystems (tympanal receptors in the foreleg for far-field sound and filiform hairs on the cercus for near-field sound). Certain crickets that lack the stridulatory equipment to produce high-frequency far-field sound may nevertheless produce a wing-flicking signal (Dambach and Lichtenstein 1978) with air particle movements at velocities approaching 15 cm/s (Heinzel and Dambach 1987).

Another function of displacement receivers is currently being investigated for intraspecific communication in honeybees (Michelsen et al. 1986, 1987). Honeybee workers inform nestmates about both the distance and direction of food sources by means of the typical wagging dance, during which they also produce air particle oscillations ("sound") with their wings. There is increasing evidence that follower bees use acoustic cues to detect this information in the darkness of the hive: (1) the wings act as asymmetrical dipole emitters, driving air particle movements at velocities up to 1 m/s; (2) the sound field around a signaller is strongly modulated as a result of the bee's side-to-side wagging; (3) the follower bees place their antennae in the zone of the acoustic field where air particle movement is maximal. These observations suggest that the follower bees could perceive information from a dancer by means of the acoustic near field without being masked by the "noise" of other nearby dancing and sound producing individuals.

The receptor organ for this kind of behaviour is likely to be the Johnston's organ, a chordotonal organ consisting of some 30 000 receptor cells in the second segment of the antenna (Eggers 1928), which respond to movements of the antennal flagellum. In male mosquitoes, the antennae are covered with a dense layer of fibrillae that possibly may increase the surface area, resulting in a higher sensitivity to medium oscillations (Clements 1963). The Johnston's organ of male mosquitoes is sharply tuned to the wing-beat frequency of the female (350–400 Hz), which is also the resonance frequency of the antenna, whereas the male's own wing-beat frequency is 150–200 Hz higher (Tischner and Schief 1954). Males localize females during flight and behavioural observations indicate that they use a process of triangulation based on the planes of vibration of the flagella for localization (Keppler 1958; Belton 1974). It is not surprising to find that receptors of the Johnston's organ of other insects are also involved in a variety of other functions, including courtship or regulation of flight parameters (Burkhardt 1960; Gewecke et al. 1974). As in the case of the filiform hair receptor responding to the wing beat of a predator, it is instructive to note the advantages and disadvantages of a displacement receptor system such as the Johnston's organ in the near- and far-field. During courtship of the fruit fly *Drosophila melanogaster* the male produces a flicking motion of the wing at 160 Hz (wavelength about 2 m) which is perceived by the Johnston's organ in the antenna of the female. At greater distances from the male, the sound pressure follows the inverse square law, but at a distance d of about 30 cm (1/6 λ) the sound pressure is only $\lambda/2$ d greater whereas the particle velocity is $(\lambda/2\ d)^2$ greater. Bennet-Clark (1971) calculated the sound pressure at a distance of 2.5 cm from the male's wing to be 51 dB, while the particle velocity at the same distance is more than 100 times greater (95 dB), highlighting the advantage of displacement reception which rapidly diminishes with distance.

Since particle oscillation is a vector quantity, displacement receivers are inherently directional, i.e. the response depends on their orientation in the sound field. Also, the directionality of mechanoreceptive hairs may be caused by a pronounced structural asymmetry of cuticular elements at the base of the hair, allowing movements in only one plane of direction. In the case of the filiform hairs on the cercus of the cricket, the hair can be deflected furthest from the resting position in the direction of a cuticular peg at the hair base, which projects towards the lumen of the hair (Fig. 1A, B). In all cases studied, this corresponds with the flat side of the tubular body within the outer dendritic segment and with the direction in which the receptor is depolarized (Gnatzy and Tautz 1980; Gaffal et al. 1975). Finally, the mobility of the hair may be influenced by the shape of the hair itself (Tautz 1977).

3.3 Far-Field Receptors

3.3.1 Frequency Selectivity

The discovery that interneurons in the CNS of grasshoppers are tuned to different sound frequencies (Horridge 1960) provided the impetus for a number of investigations into the frequency discrimination in various auditory organs. It was tempting to believe that the frequency characteristic of an auditory receptor could be derived simply from the mechanical drive to which it is subjected, and there is much evidence that such mechanisms do exist in a number of insect ears.

In the acridid ear, the scolopidia of about 80–100 receptors are attached in four groups to different sclerites on the tympanum. The cell bodies are clumped in the Müller's organ, suspended by the length of the scolopidia from the tympanum (Gray 1960; see Fig. 3B). By recording the nervous activity of single units within the organ, Michelsen (1971) demonstrated that receptors with different sites of attachment to the tympanum exhibited a different tuning. For example, the d-cell receptors (terminology after Michelsen) with their dendrites running to the pyriform vesicle were most sensitive in the high frequency range (10–30 kHz), whereas the other receptors were tuned to much lower frequencies between 2 and 6 kHz. This situation was later confirmed by Inglis and Oldfield (1988) who recorded and intracellularly stained the somata as well as the dendrites of receptors in the organ (Fig. 3B). These findings and measurements of the vibration of the tympanum suggested that there is a "place principle" of frequency analysis, where some property of the tympanum determines the vibration amplitude of different parts of the tympanum and thus the tuning of the receptor groups. Similarly, investigations on one of the two auditory receptors of the moth *Prodenia eridania*, suggested that the tuning of this receptor depends on both the resonant properties of the scoloparium and the filtering properties of the tympanal membranes (Adams 1972).

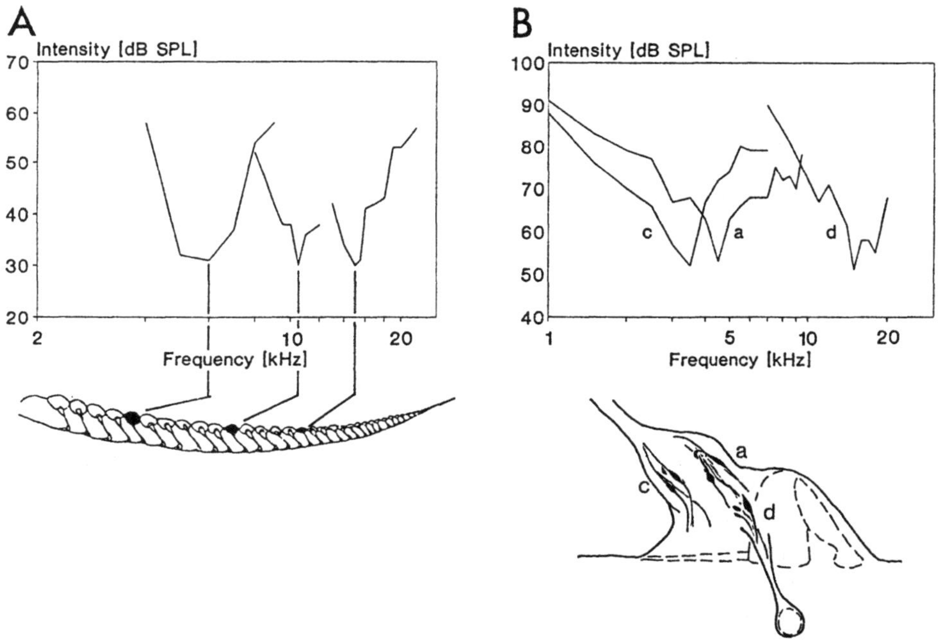

Fig. 3A. Tuning curves of three identified auditory receptors in the crista acustica of the bushcricket *Caedicia simplex*, showing the relationship between the CF of the receptors and their position in the array (after Oldfield 1982). *B* Tuning curves of three auditory receptors in the ear of the locust *Valanga irregularis*. A line drawing of some identified receptors superimposed on a standard outline of Müller's organ shows the different position of the somata and dendrites of *a*-, *c*- and *d*-receptors. (After Inglis and Oldfield 1988)

The model of Michelsen, however, was still incomplete as it did not explain the excitation of receptors by the displacement of their dendrites relative to other parts of the cells. Stephen and Bennet-Clark (1982) and Breckow and Sippel (1985) measured the relative motion between sclerites of the tympanum and cell body regions in the Müller's organ. Sites of maximum relative strains at distinct frequencies were observed to be correlated to the region of the scolopidia, and these strains seem to be the adequate stimulus for the cells. Thus it appears that pitch discrimination in the locust ear is based on the interaction of the tympanal membrane, performing spatially separated and frequency dependent modes of vibration, with the Müller's organ, which has its own resonant properties.

By contrast, in crickets and bushcrickets the vibration of the tympanal membranes does not appear to account for the frequency selectivity observed in individual auditory receptors (Dragsten et al. 1974; Larsen and Michelsen 1978). Auditory interneurons in larval instar crickets are tuned even though they lack tympanal membranes (Ball and Hill 1978). So far it is still unknown whether invertebrate auditory sensilla themselves may be tuned electrically, as shown for vertebrate hair cells (Crawford and Fettiplace 1981; See Chap. 10, this Vol.). Experiments favouring some kind of intrinsic tuning have been performed by Oldfield (1985) on the hearing organ of bushcrickets. About 20–40 receptor cells form a linear array along an auditory tracheal air sac, and the characteristic frequency of receptors increases from their proximal to their distal location (Schwabe 1906; Oldfield 1982; see Fig. 3A). Removal of both the anterior and posterior tympanal membranes does not alter the frequency selectivity of receptors. Even partial destruction of the auditory organ reduces the sensitivity of the remaining receptors without, however, affecting their characteristic frequency. Therefore, the tuning of individual receptors, particularly at frequencies of up to 60 kHz or even higher, awaits explanation.

The other mechanism for frequency analysis, namely the phase-locking of action potentials at low frequencies, has not yet been demonstrated for insect auditory receptors. However, observations on the phase-locking of whole antennal nerve responses in mosquitoes at a sound frequency of 400 Hz suggest that this principle of frequency analysis may be used as well (Ewing 1978). In the bladder cicada *Cystosoma saundersii* the carrier frequency of the male song is about 850 Hz, and the sensitivity of the whole tympanal nerve is tuned to this frequency (Young and Hill 1977). Female phonotaxis shows a remarkable frequency selectivity in a choice situation during flight, with no female choosing a song model which diverges by as little as 350 Hz from the natural carrier frequency in either direction (Doolan and Young 1989). The use of the volley principle for such a refined frequency analysis may explain the extremely large number of receptor axons (up to 2000 compared to only 1–100 in other insects) found in the tympanal nerve of cicadas.

Not all insects have evolved the potential for frequency discrimination in the ear. Moths (Roeder and treat 1957; Roeder 1972), green lacewings (Miller 1970, 1971), the praying mantis (Yager and Hoy 1986, 1987) and some beetles (Spangler 1988) all exhibit a maximum hearing sensitivity to ultrasound frequencies. Although the exact number of receptors and the tuning of individual units are not known in each case, the evidence from the ear of the noctuid moth with two receptors tuned to the same frequencies, but differing in their thresholds, indicates that these ears

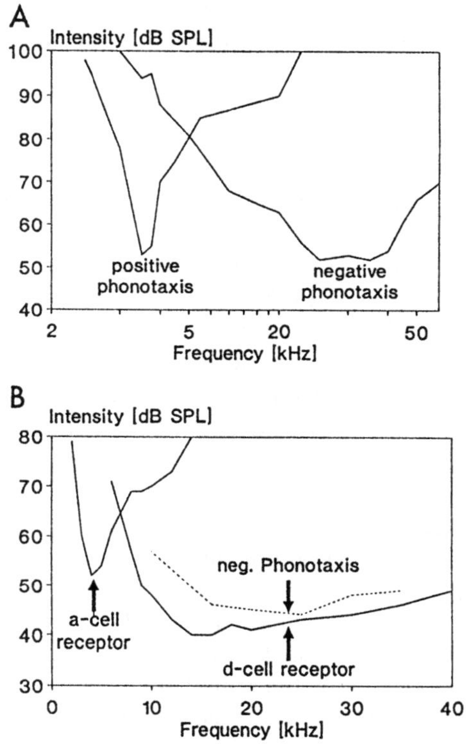

Fig. 4A. Flying crickets (*Teleogryllus oceanicus*) respond with steering movements of the abdomen and attempt to turn toward a speaker broadcasting low-frequency sound (positive phonotaxis), but they turn away when the speaker broadcasts high-frequency sound (negative phonotaxis). This demonstrates frequency discrimination at the behavioural level (Moiseff et al. 1978). *B* Flying locusts (*Locusta migratoria*) show a similar negative phonotactic behaviour to ultrasound (Robert 1989) and its tuning is similar to the tuning of d-cell receptors, but not of a-cell or other low-frequency receptors (Römer 1976)

are not designed for frequency discrimination. However, in the three orthopteran groups, the crickets, bushcrickets and acridid grasshoppers, there is sensitivity to both low and high frequency sound, either by a series of receptors each tuned to a different characteristic frequency (CF; bushcrickets) or by grouping of several receptors, each group with a different CF (acridids, crickets see Fig. 3). It is still a matter of debate as to how these animals make use of the frequency cue at the peripheral organ. In many of the ultrasound-sensitive insects studied so far, high-frequency sound is an aversive stimulus, whereas low-frequency sound elicits attraction when the temporal pattern is correct. Hoy (1989) introduced the term categorial response (following categorial perception in human psychophysics) for the observation that an insect may be sensitive to sound over a broad and continuous range of frequencies but discriminates between only certain (probably two) discrete frequencies. This is best exemplified by Moiseff et al. (1978) for the flying cricket (Fig. 4A). Similarly, the locust also exhibits negative phonotaxis during flight (Robert 1989; see Fig. 4B), and other acridid grasshoppers are known to respond by approaching an acoustic signal at low frequencies (von Helversen 1972). In the locust (and probably in the cricket), however, there is a broad, but discontinuous frequency sensitivity in the ear, based on the low-frequency sensitivity of three receptor groups (a-, b- and c-cells) and the high-frequency d-cells. The similarity of the tuning of both the d-cells and the avoidance behaviour (Fig. 4B) suggests that those receptors may have evolved in the behavioural context of predator avoidance.

It appears that categorial response in this insect is basically the property of the hearing organ, and the CNS just enhances the dual mechanism by frequency-selective excitatory and inhibitory actions onto central interneurons (Römer and Marquart 1984).

Whereas the functional significance for the discrimination of "low" and "high" frequencies is rather obvious, it is still unclear whether minor details in the spectrum of a signal may be discriminated by the insect, and the behavioural context for such discrimination could be mate choice. That the environmental filter (Michelsen and Larsen 1983; Römer and Bailey 1990) may introduce a high degree of ambiguity with respect to the frequency content of the signal favours other interpretations. Rheinlaender (1975) and Römer (1987) suggested that the high-frequency selectivity in the bushcricket ear may be used for intensity or distance discrimination, whereas Harrison et al. (1988) demonstrated that in the cricket, inhibitory frequency bands provide a mechanism to maintain the fidelity of the temporal calling pattern in the CNS of a nearby receiver.

3.3.2 Directional Sensitivity

Although a nondirectional acoustic receiver must not a priori be biologically useless (see e.g. Tautz 1989), it is convenient to assume that for most biological functions of hearing the "where" information about a sound source is essential to the animal. The directional characteristics of sound receptors depend strongly on the physical parameter of the sound wave. We have already seen how near-field receptors, sensitive to medium displacement, are inherently directional; particle displacement is a vector quantity. By contrast, far-field receptors respond to the pressure component of sound, which is a scalar quantity, and hence these receptors are not inherently directional. To overcome this problem, animals may adopt different mechanisms: the use of (1) an (inherently directional) pressure difference receiver; (2) the diffraction of sound around the head or body; or (3) the difference in the time-of-arrival of the sound at both ears.

Many insects and small vertebrates have to face the problem that their size is much smaller than the wavelength of the relevant sound. As a result, the interaural difference in sound pressure, due to the "shadow" effect of the body or head, is often negligible at low frequencies. At higher frequencies, however, much larger differences result (Römer 1976; Miller 1977; Lewis 1983); in large species of moths and for ultrasonic frequencies the intensity difference between the ears may reach 40 dB (Payne et al. 1966). Autrum (1940) realized that the observed directionality of insect ears was not consistent with a pure-pressure receiver, and he introduced the idea of a pressure-difference receiver instead, where the sound wave was able to act on the inner and outer surface of tympanal membranes.

The anatomical basis for such a pressure-difference receiver is often the complex tracheal system associated with the hearing organ, connecting the back of the tympanum with the air through an opening hole in the pleuron of the thorax (bushcrickets) or even to the ear on the opposite side of the body (crickets). In bushcrickets, this trachea has lost its respiratory function completely and serves only as a sound guide. The function of the auditory trachea in directional hearing (and the frequency selectivity of the ear) has been studied in a number of more or less closely related species of orthopteran insects using quite different methods (Lewis 1974; Zhantiev et al. 1975; Hill and Boyan 1977; Larsen and Michelsen 1978; Michelsen and Larsen 1978; Fletcher and Thwaites 1979; Oldfield 1980; Hill and Oldfield 1981; Kleindienst et al. 1981; Thorsen et al. 1982; Schmitz et al. 1982). However, the anatomical differences of the acoustic trachea between species may account for some of the controversial results and interpretations reported in the literature (Bailey 1990). Despite these differences it is clear that the amount of sound pressure guided to the back of the tympanum varies with the carrier frequency of the sound. Whereas at low frequencies the gain of the trachea may be close to one, the same trachea may amplify sound at higher frequencies by more than 20 decibels, causing the sound pressure at the back of the tympanum to be much larger than at the front. The ear changes from a pressure-difference receiver at low frequencies to almost a pure-pressure receiver at high frequencies (Hill and Oldfield 1981; Larsen 1981). This is a very similar situation to that shown for the acridid auditory system, where the body between the two tympana acts as a low pass filter, with each ear behaving like a pressure difference receiver at 4 kHz, but as a pressure receiver above 10 kHz (Michelsen 1971; Römer 1976; Miller 1977). This provides the animal with a right-left difference of 8–12 dB at 4 kHz, contrasting with only 1 dB as a result of the sound "shadow" due to the insect's body.

Unfortunately, many investigators have been interested only in the physical mechanism of directional hearing rather than the biological context in which it is used. For example, a high directionality obtained at a particular frequency may be of little relevance when the natural song of the insect contains many frequencies (broadband), and when this is used as a stimulus. Similarly, the total amount of the interaural intensity difference, often taken as a measure for the directionality of a hearing system, may not be the difference on which the localizing insect operates. Calculations based on phonotactic behaviour of crickets and bushcrickets indicate that the insect has to rely on small interaural intensity differences when localizing a conspecific (Rheinlaender 1984). Using an independent ear stimulation technique, von Helversen and Rheinlaender (1988) have shown that no more than 1–2 dB of right-left difference is sufficient for a reliable side-to-side discrimination.

As the disparity between hearing organs on each side of the insect is usually very small, it is unlikely that physical time differences in the order of 10–30 μs could be used for sound localization. However, since the latency of the response of a receptor fibre may change by about 6 ms with sound direction (Mörchen et al. 1978), a neuronal time cue may be available for the insect that exceeds the physical time difference 100–1000 times. A comparison from the input from the two ears shows that there are two neural codes (spike count and response latency), each of which can adequately specify the direction of a sound source (Boyan 1979). It has been shown that the subsequent synapses in the auditory pathway make use of both these modes of directional coding (Rheinlaender and Mörchen 1979; Römer and Rheinlaender 1983).

3.3.3 Temporal Resolution

Although behavioural evidence suggests the importance of time cues in insect auditory communication, very little is known about the physiological basis of time resolution in the ears (for review, see Michelsen 1985). The maximum amplitude-modulation frequency auditory receptors can follow, varies from 20 Hz (c-cells in the locust; Michelsen 1966) to 300–500 Hz in cicadas (Huber et al. 1980). The mechanical time resolution of the ear (the time constants of vibration of the tympanal membranes) is very short (about 60 μs in the moth ear; Schiolten et al. 1981; 50–100 μs in the locust ear; Michelsen and Larsen 1985). This does not appear to limit the temporal resolution observed either in behavioural or neurophysiological studies. For example, behavioural experiments on female grasshoppers *Chorthippus biguttulus* indicate a time resolution in the order of 1–2 ms (von Helversen 1972) and auditory receptors in both grasshoppers (Ronacher and Römer 1985) and moths (Surlykke et al. 1988) exhibit gap detection thresholds of the same magnitude.

3.3.4 "Noise" in Auditory Receptors

Invertebrate mechanoreceptors responding either to airborne sound or to substrate vibrations (see below) are, due to their developmental origin and placement in the exoskeleton, subject to activation and modulation by all kinds of body movements. For example, receptors in the tympanal organ of the locust respond to deflections of the tympanic membrane induced by respiratory movements. These deflections change the transmission properties of the hearing organ, and, as a result, responses to sound that coincide with a particular phase of respiration may be completely masked by the activity elicited by the respiratory movement (Hedwig 1988). The same is true for movements of the hind legs (Hedwig et al. 1988). During stridulation, neurons of the auditory pathway either do not respond to sound or respond only during certain phases of the leg movement (Hedwig 1986). In particular, during insect flight strong thoracic movements and forces are produced by flight muscle contractions, which may modulate the discharge of auditory receptors and decrease their activity (Hedwig 1989; for similar results in walking crickets see Schildberger et al. 1988). Thus, there is a considerable level of "internal" noise in auditory receptors of animals performing any kind of behaviour. As a consequence, hearing sensitivity or other sorts of acoustic measures should always be given with reference to the behavioural status of the animal under study.

4 Vibration Receptors

Receptors designed to detect the vibration of the substrate appear to exhibit a much greater variability with respect to their structure than receptors responding to airborne sound. In arthropods, all the sensilla that either monitor stress or strain in the cuticle or react to joint movements may as well detect substrate vibrations. Figure 5 gives an overview over the most common of these receptors.

Among the most sensitive mechanoreceptors known are those forming the subgenual organs in the proximal part of the tibia in each leg of insects. They consist of about 10–40 scolopidia (Fig. 5a), fixed to the cuticle at one point by means of their accessory cells, whereas their proximal end is supported by a trachea (Autrum 1941;

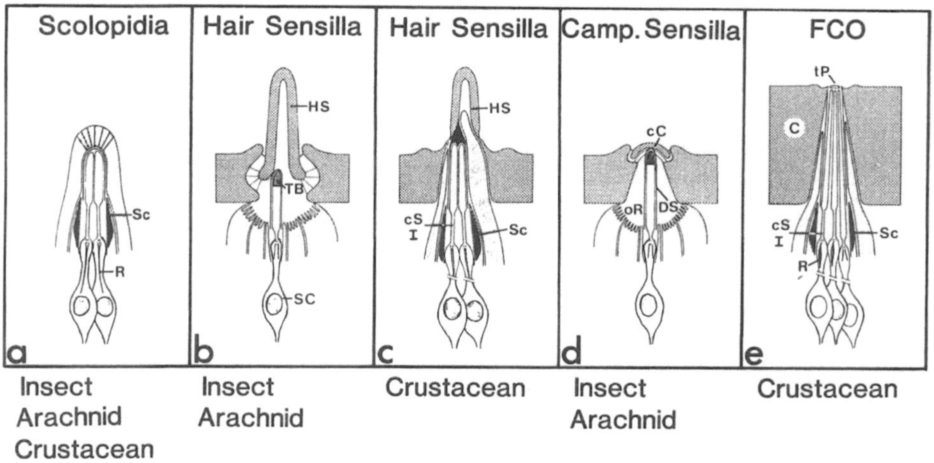

Fig. 5a-e. Schematic diagram of the structure of mechanoreceptors responding to airborne sound or substrate vibration. *cC* Cuticular cap; *cSI* ciliary segment type I; *DS* dendritic sheath; *HS* hair shaft; *oR* outer receptor-lymph cavity; *R* ciliary rootlet; *SC* sensory cell; *Sc* scolopale; *TB* tubular body. (After Schmidt and Gnatzy 1984)

Autrum and Schneider 1948; Schnorbus 1971; Michel 1974; Young and Ball 1974). The organs in the different legs may differ with respect to their absolute sensitivity and their tuning, as shown for unrestrained crickets standing in a normal position (Dambach 1972). These differences appear to result from size differences of the organs rather than the different size and mass of the legs, since additional loading of a hind leg caused no change in the sensitivity of the corresponding subgenual organ. Surprisingly, leg position does not appear to influence the tuning at all, and the absolute sensitivity only by a factor of 2 (Dambach 1972). This contrasts with findings of Schnorbus (1971) in cockroaches and Aicher et al. (1983) in fiddler crabs, where sensitivity can vary considerably with the position of the legs containing the receptors.

In campaniform sensilla, the hair is replaced by a cuticular dome, thickened along its long axis. The sensory neuron inserts with its dendrite into the thickened region of the dome (Fig. 5d). The sensitivity of the sensillum may be correlated with the size of the dome (Pringle 1938). Campaniform sensilla were also found to be associated with other types of mechanoreceptors, such as filiform or clavate hairs on the cerci of crickets, where they may monitor the deformation of the socket of the hair (Gnatzy and Schmidt 1971; Edwards and Palka 1974). Campaniform sensilla are usually arranged in groups and concentrated in those regions of the cuticle that are subject to stress or strain.

Slit sensilla in spiders or scorpions represent modified campaniform sensilla, positioned predominantly on the lateral surface of the legs. The orientation of the slits is roughly parallel to the long axis of the leg (Fig. 6A; Barth and Wadepuhl 1975). The receptors respond to minute displacements in the cuticle with threshold values close to 0.5 Å (Brownell and Farley 1979b; see Fig. 11B). Based on

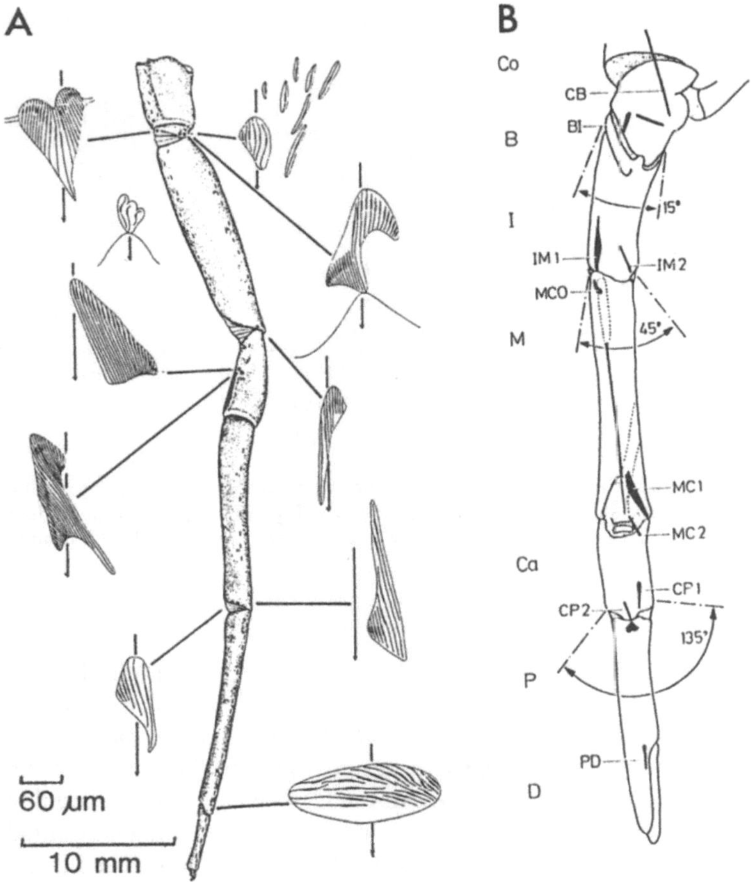

Fig. 6A. Position and orientation of slit sense organs on the walking leg of the arachnid *Cupiennius salei*, posterior view. Details of groups of slits and of lyriform organs drawn to scale are given as *insets. Arrows* run parallel to the respective leg segments (Barth and Libera 1970). *B* Ventral view of a left walking leg of *H. gammarus* with the position of the internal chordotonal organs and the extent of movement possible at each of the joints. Segments: *Co* coxopodite; *B* basipodite; *I* ischiopodite; *M* meropodite; *Ca* carpopodite; *P* propodite; *D* dactylopodite. Receptors are named by the initials of the limb segments which constitute the joint at which they are situated. An exception is the myochordotonal organ (*MCO*) which responds to movement of the *M-C* joint although the sensory cells are situated close to the *I-M* joint. (After Wales et al. 1970)

morphological and electrophysiological evidence, Barth (1972) concluded that stress in the cuticle surrounding the slit is the adequate stimulus and that deformation of the dendritic end by monoaxial compression forces leads to the mechanoelectrical transduction process. Greatest sensitivity should then be achieved when slits are compressed by forces applied perpendicular to their long axis. The two longest slits in the basitarsal compound slit sensillum of the scorpion *Paruronotus mesaensis* are oriented almost perpendicular to the plane in which the tarsus is moved by vertical substrate displacements. Since the sensillum responds to surface waves of the Rayleigh type (Table 1) in which the major displacement component is directed vertically, these sense organs appear to be ideally adapted for detecting the surface wave component of vibrations (Brownell 1977; Brownell and Farley 1979b).

In all types of receptors shown in Figure 5 the mechanoelectrical transduction is basically similar and follows the mechanism detailed in Section 2.2 (Gaffal et al. 1975; Moran and Rowley 1975; Thurm 1982). Thus it appears that the variability in the cuticular accessory structures, and therefore the way in which the mechanical force is lead onto the sensory cell, rather than any kind of intrinsic property accounts for the diversity of the mechanoreceptive function of these receptors.

4.1 Types of Substrate Waves

Waves conducted through fluid or solid media may be used for intraspecific communication or in a predator-prey interaction. The signals are perceived as vibrations of a boundary between solid/water, solid/air or air/water. Solids are excellent transmitters of mechanical disturbances. Sound propagation in rock is as high as 3600 m/s (in water, 1500 m/s; in air, 340 m/s). However, there is a considerable number of different wave types that can be transmitted. The exact type of disturbance depends on the movement performed by the single particles and this in turn depends on the forces acting between neighbouring particles. Within a solid, a particle oscillating back and forth hits its neighbour particles right and left, which then also start to oscillate etc. Thus a disturbance is propagating in which the vector of the oscillation of a single particle is in direction of the propagation of the wave (longitudinal wave, Fig. 7A). Additional strong coupling between particles allows shearing forces to develop and thus the particle movements can also be directed in other than the propagation direction of the disturbance (transversal waves, Fig. 7B). Of greater biological relevance are the disturbances on the surface of a solid where animals usually sit or walk. At the surface of a solid (as at the interface between layers of solids of different properties) surface waves can occur. The waves can be directed either parallel (surface longitudinal waves) or perpendicular to the propagation direction of the disturbance. In the latter it can be perpendicular to the surface (Rayleigh-waves, Fig. 7C) or within the surface (Love-waves, Fig. 7D). The situation becomes even more complicated if we look at structures short in one (plates) or in two dimensions (rods). Leafs and stems of plants can be approximated as plates and rods. In rods, torsion waves can occur where no component is oriented perpendicular to the surface and which are thus similar to the Love-waves. In both plates and rods, bending waves can occur as yet another type of disturbance. Furthermore, all waves listed in Table 1 differ both in the vector direction of the particle movements and in the velocity and attenuation of propagation (for quantitative formula on propagation velocity and attenuation see Markl 1983; Barth 1985; Aicher and Tautz 1990).

Local disturbances of the water surface generate concentric wave signals comprised of wave cycles of different amplitude and wavelength. They are pro-

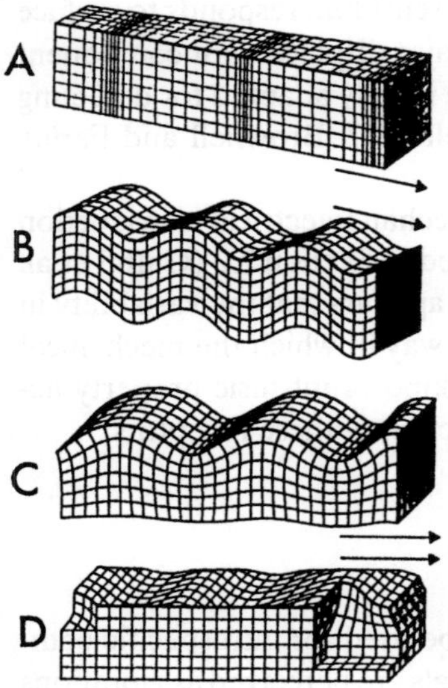

Fig. 7. Types of waves in elastic solids. Within a 3-dimensional medium: *A* compressional (P)-waves and *B* shear (S)-waves. On the boundary between two media (*C*) Raleigh waves and (*D*) love waves. Propagation direction of the disturbance is shown by *arrows*. (After Aicher and Tautz 1990)

pagated with dispersion, and thus propagation velocity is frequency dependent. Water surface waves are attenuated by several dB per centimeter propagation distance (cf. attenuation values given in Table 1 for solids). The amount of attenuation increases with increasing stimulus frequency; the water surface therefore represents a low-pass filter (for review, see Bleckmann 1988). Solid substrates may act as frequency filters as well, thus creating frequency windows for the transmission of vibratory signals (Fig. 8; Rovner and Barth 1981; Michelsen et al. 1982).

4.2 Stimulus Coding by Vibration Receptors

As explained in Section 4.1, substrate vibrations can be described and distinguished by a few parameters. As the vibrations have the properties of a vector, the directional characteristic of a receptor should give important information. The rather slow propagation velocities of the different kinds of surface waves (see Table 1) should allow the animal to determine direction and distance of a source of disturbance. Social signals may contain species-specific frequency components, and substrate vibration receptors may be tuned to certain frequency bands.

4.2.1 Directional Sensitivity

Of those receptors summarized in Fig. 5, only hair sensilla and campaniform sensilla are well investigated with respect to directionality. Hair receptors may be asymmetrically connected with the surrounding cuticle, resulting in a directionality of hair mobility. The sensory cell connected to the base of the hairs has a directional characteristic of excitability caused by the asymmetric insertion of the dendrite (see Fig. 1A). However, since substrate vibrations are directly transmitted onto the hairs the coupling to the stimulus is strong and thus the mechanical directionality is of minor importance; the hairs can be forced into each direction.

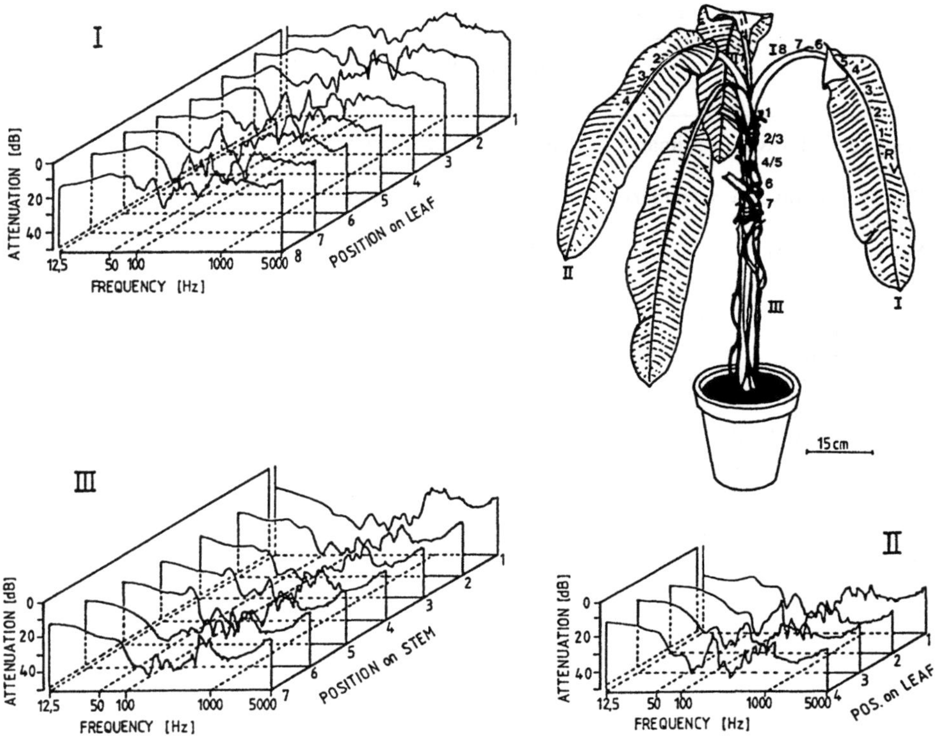

Fig. 8. Attenuation of a vibratory signal traveling through a banana plant (*Musa sapientum*) from its site of origin at leaf I (*V*) to leaf II and the pseudostem III. Graphs are based on spectral analysis of noise signals and laser-Doppler vibrometry and give attenuation values at different sites (*1–8*) on leaves and the pseudostem. 0 dB (no attenuation) corresponds to a value measured at reference point *R* close to *V* on leaf I. (Barth 1985)

Since the slit organs in arachnids are long stretched structures (see Figs. 6A, 9) a high directionality of sensitivity is the consequence. The same is true for campaniform sensilla of insects as shown by Thurm et al. (1975) for sensilla on the haltere of a fly. However, Brownell and Farley (1979c) found no behavioural or physiological evidence that individual sensilla on the scorpions leg were sensitive for vibrations from different directions in the horizontal plane around the animal. Rather, it appears that these animals make use of the unique slow conduction velocity of waves through sand, in particular of waves of the Rayleigh-type propagating at a velocity of about 50 m/s (Brownell 1977; see Table 1). This creates a sufficient delay in the order of 1 ms between stimulation of sense organs on the spatially separated legs, which appears to be used by the animal for its accurate target localization (Brownell and Farley 1979a,c).

All predatory invertebrates known to respond to water surface waves perform a more or less precise localization behaviour as well (for review, see Bleckmann 1988). Since propagation velocity of these waves is extremely low (23 cm/s at 13 Hz) time and/or phase differences can be expected to be used for target angle deter-

Table 1. Types of vibrational waves in solids

Wave type	Site of occurrence	Prop. velocity (m/s)	Attenuation (dB/cm)
Longitudinal waves (= primary waves)	Within extended solids	3600[a]	0.06[a]
Transversal waves (= secondary waves)	Within extended solids	2000[a]	0.06[a]
Surface P-waves	Surface of solids	150[b,c]	0.1–0.25[d]
Rayleigh waves	Surface of solids	40–80[b]	0.13–0.49[b]
Love waves	Surface of solids	50–60[b]	0.1–0.75[e]
Torsional waves	Rods	*	*
Bending waves	Plates, rods	5–55[f]	0.3[f]

[*]No data found.
[a]Hardin and Richard (1963).
[b]Aicher and Tautz (1990).
[c]Ramspeck and Schultze (1938).
[d]Schmidt (1954).
[e]Lee and Solomon (1975).
[f]Barth (1985).

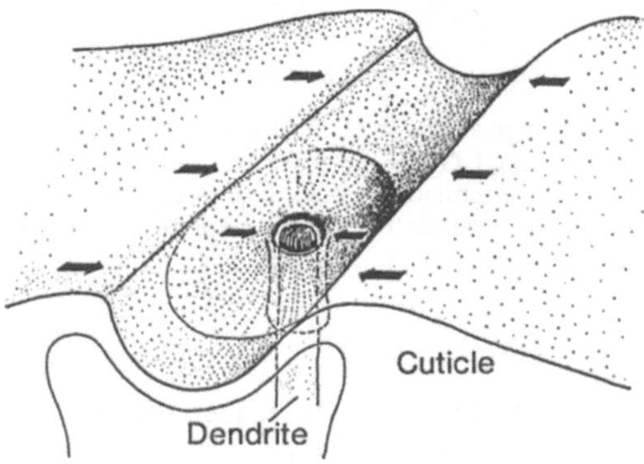

Fig. 9. Adequate deformation of a slit sensillum. *Arrows* indicate the direction of compressional forces that result in a deformation of the tip of the dendrite. (Barth 1972)

mination (Murphey 1971; Wiese 1974). A similar mechanism was proposed for the fishing spider *Dolometes triton* due to the large distance (and thus large time differences of receptor activation) between the legs (Bleckmann and Barth 1984).

4.2.2 Frequency Selectivity

The spectral threshold curves of vibration receptors can be plotted as stimulus acceleration vs frequency or as substrate displacement vs frequency, resulting in curves of much different shape. Threshold curves that show a sensitivity maximum at a certain frequency in a plot displacement vs frequency may increase steadily from low frequencies on if the acceleration is plotted instead. Thus the impression of a frequency tuning can result exclusively from the kind of data presentation. For example, the receptors in the compound slit sense organ of spiders behave as

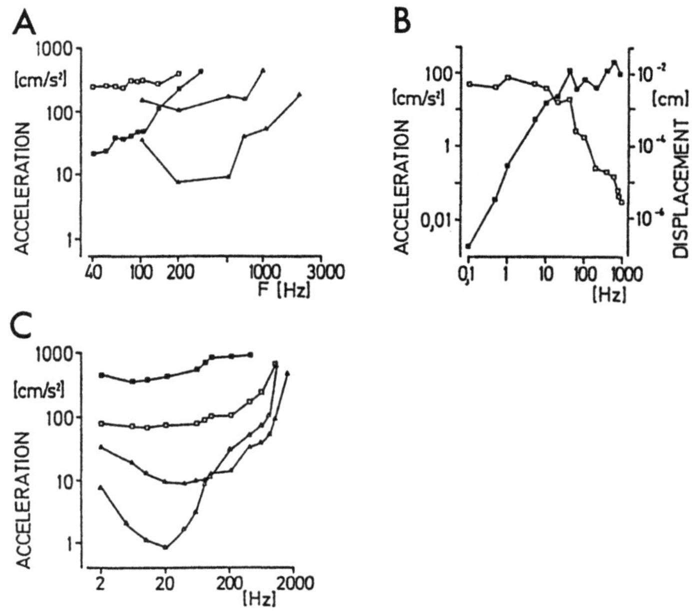

Fig. 10A. Threshold curves of four types of vibration receptors in the hindleg of the locust (*Locusta migratoria*). Stimulus duration 100 ms; repetition rate 2/s (Grosch et al. 1985). *B* Tuning curves (threshold displacement and threshold acceleration) of slit 7 in the metatarsal lyriform organ of a spider (*Cupiennius salei*) measured with vertical displacement of the tarsus and its firm coupling to the vibrator (Barth and Geethabali 1982). *C* Threshold curves for different vibration receptors in the walking legs of female fiddler crabs. (Aicher and Tautz, unpubl.)

high-pass filters as well as low-pass filters, depending on whether displacement or acceleration is plotted (Fig. 10B; Barth and Geethabali 1982). However, it may well be that the frequency range of constant threshold rather than the frequency range of highest sensitivity is of relevance. In the case of the slit sense organs, the threshold follows constant displacement up to a stimulus frequency of about 40 Hz, whereas at higher frequencies it follows constant acceleration (with constant displacement and increasing frequency the acceleration increases with the power 2). The different slit sensilla in the compound organ are similar in their frequency response, although minor differences may allow a crude frequency discrimination (Barth and Geethabali 1982). In the acceleration domain, bandpass-tuned receptors are the type-III-receptors in the subgenual organ of locusts, with an acceleration sensitivity tuned to a range of 200–500 Hz (Fig. 10A) and one unidentified vibration receptor in the legs of fiddler crabs tuned to a frequency of about 20 Hz (Fig. 10C, lowest curve; Aicher and Tautz, unpubl.).

4.3 Specificity for Particular Types of Vibrations

For animals standing on the surface of a solid, the surface can vibrate up and down, back and forth, or right and left, relative to the long axis of the animal. As animals may turn during stimulation, the only invariable remaining is the distinction between vibrations directed perpendicular to the surface and those within the surface. Basically each kind of receptor can be used to detect each type of surface wave. The specificity for the detection of a particular type of wave is given by the geometric arrangement of the single receptors on the surface of the animal. Brownell and Farley (1979a, b, c) demonstrated that the scorpion *Paruronotus mesaensis* locates prey by orienting to substrate vibrations produced by the prey in sand. Two sense organs, the basitarsal compound slit sensillum and tarsal sensory hairs, mediate this behaviour. The arrangement of the slit sensilla is such that Rayleigh-waves stimulate them maximally, while the hair sensilla protruding from the tarsus into the sand detect horizontal sand particle motions and thus Love- and surface-p-waves (Fig. 11). For the majority of vibration receptors, however, it remains to be investigated to which type(s) of wave(s) they respond best.

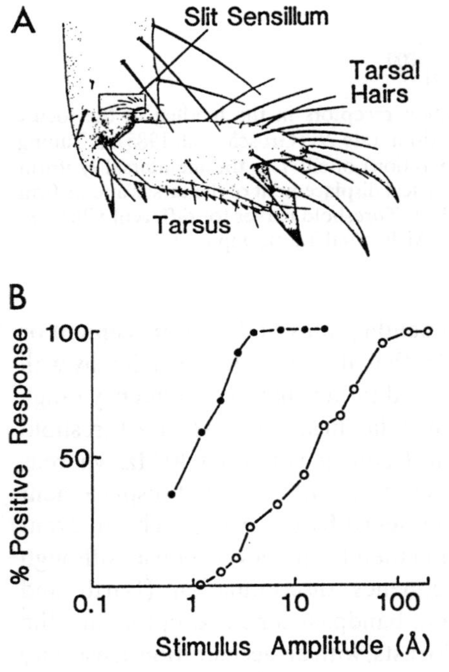

Fig. 11A. Position of tarsal hairs and the basitarsal slit sensillum on the terminal leg segments of a scorpion (*Paruroctonus mesaensis*). *B* Sensitivity of the tarsal hair (o–o) and the slit sensillum (•–•) to small displacements of the tarsus. Sensitivity represented at each amplitude as % of 40 uniform stimulations that produced action potentials characteristic of the slit sensillum and the hair in a 20-ms interval following each stimulus. (Brownell and Farley 1979b; Brownell 1984)

5 Selected Reading

An excellent introduction to the adaptive radiation of mechanoreception is given by Markl (1978), and more specifically to vibrational communication by Markl (1983). A comprehensive and still useful review on insect mechanoreception is given by Schwartzkopff (1974). The paper of Michelsen and Larsen (1985) on insect hearing, including both peripheral and central nervous processing, is highly recommended. Bleckmann (1985) reviews the vibrational perception at the water surface and compares invertebrates and vertebrates. Budelmann (1990) summarizes hearing and the corresponding receptors in nonarthropod invertebrates. Meier and Reichert (1990) establish the intra- and interspecific homologies of insect mechanoreceptors and provide a fascinating insight into the evolutionary origin of the auditory organs of Orthopterans.

Acknowledgments. We thank G.S. Boyan and W.J. Bailey for many valued discussions and for helpful criticism of the manuscript.

References

Adams WB (1972) Mechanical tuning of the acoustic receptor of Prodenia eridania (Cramer)(Noctuidae) J Exp Biol 57:297–304

Aicher B, Tautz J (1990) Vibrational communication in the fiddler crab, *Uca pugilator*. I. Signal transmission through the substratum. J Comp Physiol 166:345–353

Aicher B, Markl H, Masters WM, Kirschenlohr HL (1983) Vibration transmission through the walking legs of the fiddler crab, *Uca pugilator* (Brachyura, Ocypodidae) as measured by Laser Doppler Vibrometry. J Comp Physiol 150:483–491

Autrum H (1936) Über Lautäußerungen und Schallwahrnehmung bei Arthropoden. I. Untersuchungen an Ameisen. Eine allgemeine Theorie der Schallwahrnehmung bei Arthropoden. Z Vergl Physiol 23:332–373

Autrum H (1940) Über Lautäußerungen und Schallwahrnehmung bei Arthropoden. II. Das Richtungshören von Locusta und der Versuch einer Hörtheorie für Tympanalorgane vom Locustidentyp. Z Vergl Physiol 28:326–352

Autrum H (1941) Über Gehör und Erschütterungssinn bei Locustiden. Z Vergl Physiol 28:580–637

Autrum H, Schneider W (1948) Vergleichende Untersuchungen über den Erschütterungssinn der Insekten. Z Vergl Physiol 31:77–88

Bailey WJ (1990) The ear of the bushcricket. In: Bailey WJ, Rentz DCF (eds) The Tettigoniidae: biology, systematics and evolution. Crawford House Press, Bathhurst, pp 217–247

Ball EE, Hill KG (1978) Functional development of the auditory system of the cricket, *Teleogryllus commodus*. J Comp Physiol 127:131–138

Barth FG (1971) Der sensorische Apparat der Spaltsinnesorgane (*Cupiennius salei* Keys. Araneae). Z Zellforsch 112:212–246

Barth FG (1972) Die Physiologie der Spaltsinnesorgane. II. Funktionelle Morphologie eines Mechanoreceptors. J Comp Physiol 81:159–186

Barth FG (1985) Neuroethology of the spider vibration sense. In: Barth FG (ed) Neurobiology of arachnids. Springer, Berlin Heidelberg New York

Barth FG, Geethabali (1982) Spider vibration receptors: threshold curves of individual slits in the metatarsal lyriform organ. J Comp Physiol 148:175–185

Barth FG, Libera W (1970) Ein Atlas der Spaltsinnesorgane von *Cupiennius salei* Keys. (Chelicerata, Araneae). Z Morphol Ökol Tiere 68:343–369

Barth FG, Wadepuhl M (1975) Slit sense organs in the scorpion leg (*Androctonus australis* L., Buthidae). J Morphol 145:209–228

Belton P (1974) An analysis of direction finding in male mosquitoes. In: Barton Brown L (ed) Experimental analysis of insect behaviour. Springer, Berlin Heidelberg New York

Bennet-Clark HC (1971) Acoustics of insect song. Nature (Lond) 234:255–259

Bennet-Clark HC (1984) Insect hearing: acoustics and transduction. In: Lewis T (ed) Insect communication. Academic Press, London, pp 49–82

Bleckmann H (1988) Perception of water surface waves: how surface waves are used for prey identification, prey localization, and intraspecific communication. In: Ottosen D (ed) Progress in sensory physiology 5. Springer, Berlin Heidelberg New York Tokyo, pp 147–166

Bleckmann H, Barth FG (1984) Sensory ecology of the semiaquatic spider *Dolometes triton*. II. Release of predatory behavior by water surface waves. Behav Ecol Sociobiol 14:303–312

Boyan GS (1979) Directional responses to sound in the central nervous system of the cricket *Teleogryllus commodus* (Orthoptera: Gryllidae). J Comp Physiol 130:137–150

Breckow J, Sippel M (1985) Mechanics of the transduction of sound in the tympanal organ of adults and larvae of locusts. J Comp Physiol 157:619–629

Brownell PH (1977) Compressional and surface waves in sand: used by desert scorpions to locate prey. Science 197:479–482

Brownell PH, Farley RD (1979a) Prey-localizing behaviour of the nocturnal desert scorpion, *Paruroctonus mesaensis*: orientation to substrate vibrations. Anim Behav 27:185–193

Brownell PH, Farley RD (1979b) Detection of vibrations in sand by tarsal sense organs of the nocturnal scorpion, *Paruroctonus mesaensis*. J Comp Physiol 131:23–30

Brownell PH, Farley RD (1979c) Orientation to vibrations in sand by the nocturnal scorpion *Paruroctonus mesaensis*: mechanism of target localization. J Comp Physiol 131:31–38

Budelmann B (1990) Hearing of non-arthropod invertebrates. In: Fay RR, Popper AN, Webster DB (eds) Evolutionary biology of hearing. Springer, Berlin Heidelberg New York (in press)

Bullock TH (1984) Comparative neuroethology of startle, rapid escape, and giant fibre-mediated responses. In: Eaton RC (ed) Neural mechanisms of startle behavior. Plenum Press, New York, pp 1–11

Burkhardt D (1960) Die Eigenschaften und Funktionstypen der Sinnesorgane. Ergeb Biol 22:226–267

Burkhardt D, Schneider G (1957) Die Antennen von *Calliphora* als Anzeiger der Fluggeschwindigkeit. Z Naturforsch 12:139–143

Camhi J (1980) The escape system of the cockroach. Sci Am 243:158–172

Camhi J, Tom W (1978) The escape behaviour of the cockroach *Periplaneta americana* I. Turning response to wind puffs. J Comp Physiol 128:193–201

Clements AN (1963) The physiology of mosquitoes. Pergamon Press, New York

Crawford AC, Fettiplace R (1981) An electrical tuning mechanism in turtle cochlea hair cells. J Physiol 312:377–422

Dambach M (1972) Vibrationssinn der Grillen. I. Schwellenmessungen an Beinen freibeweglicher Tiere. II. Antworten von Neuronen im Bauchmark. J Comp Physiol 79:281–324

Dambach M, Lichtenstein L (1978) Zur Ethologie der afrikanischen Grille *Phaeophilacris spectrum* Saussure. Z Tierpsychol 46:14–29

Dragsten PR, Webb WW, Paton JA, Capranica RR (1974) Auditory membrane vibrations: measurements at sub-Angstrom levels by optical heterodyne spectroscopy. Science 185:55–57

Doolan JM, Young D (1989) Relative importance of song parameters during flight phonotaxis and courtship in the bladder cicada *Cystosoma saundersii*. J Exp Biol 141:113–131

Edwards JS, Palka J (1974) The cerci and abdominal giant fibres of the house cricket, *Acheata domesticus*. Anatomy and physiology of normal adults. Proc R Soc Lond B 185:83–103

Eggers F (1928) Die stiftführenden Sinnesorgane. Zool Bausteine 2:354

Erler G, Thurm U (1981) Dendritic impulse initiation in an epithelial sensory neuron. J Comp Physiol 142:237–249

Ewing AW (1978) The antenna of *Drosophila* as a "love song" receptor. Physiol Entomol 3:33–36

Fletcher NH, Thwaites S (1979) Acoustical analysis of the auditory system of the cricket *Teleogryllus commodus* (Walker). J Acoust Soc Am 62:350–357

Gaffal KP, Tichy H, Theiss J, Seelinger G (1975) Structural polarities in mechanosensitive sensilla and their influence on stimulus transmission (Arthropoda). Zoomorphologie 82:79–103

Gewecke M (1980) Steuerung des Schwimmverhaltens durch die Antennen beim Teichschwimmer (*Colymbetes fuscus* L., Dytiscidae, Coleoptera). Verh Dtsch Zool Ges 1980:336

Gewecke M, Heinzel H-G, Philippen J (1974) Role of antennae of the dragonfly *Orthetrum cancellarum* in flight control. Nature (Lond) 249:584–585

Gnatzy W, Schmidt K (1971) Die Feinstruktur der Sinneshaare auf den Cerci von *Gryllus bimaculatus* Deg. (Saltatoria, Gryllidae). I. Faden- und Keulenhaare. Z Zellforsch 122:190–209

Gnatzy W, Tautz J (1980) Ultrastructure and mechanical properties of an insect mechanoreceptor: stimulus-transmitting structures and sensory apparatus of the cercal filiform hairs in *Gryllus*. Cell Tissue Res 213:441–463

Gray EG (1960) The fine structure of the insect ear. Philos Trans R Soc B 243:75–94

Grosch A, Callender F, Petersen M, Cokl A, Kalmring K (1985) Vibration receptors of larvae and of imagines in locusts: location on the legs, central projections and physiology. In: Kalmring K, Elsner N (eds) Acoustic and vibrational communication in insects. Parey, Hamburg, pp 151–161

Hardin BO, Richard FE (1963) Elastic wave velocities in granular soils. J Mech Found Div, Proc Am Soc Civil Eng SM1 3407:33–65

Harrison L, Horseman G, Lewis B (1988) The coding of the courtship song by an identified auditory neurone in the cricket *Teleogryllus oceanicus* (Le guillou). J Comp Physiol 163:215–225

Hedwig B (1986) On the role of stridulation of plurisegmental interneurons of the acridid grasshopper *Omocestus viridulus* L. II. Anatomy and physiology of ascending and T-shaped interneurons. J Comp Physiol 158:429–444

Hedwig B (1988) Activation and modulation of auditory receptors in *Locusta migratoria* by respiratory movements. J Comp Physiol 162:237–246

Hedwig B (1989) Modulation of auditory information processing in tethered flying locusts. J Comp Physiol 164:409–422

Hedwig B, Land F, Elsner N (1988) The interference of sound and movement stimuli in tympanal receptors of *Locusta migratoria*. J Comp Physiol 163:243–252

Heinzel H-G, Dambach M (1987) Travelling air vortex rings as potential communication signals in a cricket. J Comp Physiol 160:79–88

Hill KG (1983a) The physiology of locust auditory receptors. I. Discrete depolarizations of receptor cells. J Comp Physiol 152:475–482

Hill KG (1983b) The physiology of locust auditory receptors. II. Membrane potentials associated with the response of the receptor cell. J Comp Physiol 152:483–493

Hill KG, Boyan GS (1977) Sensitivity to frequency and direction of sound in the auditory system of crickets (Gryllidae). J Comp Physiol 121:79–97

Hill KG, Oldfield BP (1981) Auditory function in Tettigoniidae (Orthoptera: Ensifera). J Comp Physiol 142:169–180

Horridge GA (1960) Pitch discrimination in Orthoptera (Insecta) demonstrated by responses of central auditory neurons. Nature (Lond) 185:623–624

Hoy RR (1989) Startle, categorial response, and attention in acoustic behavior in crickets. Annu Rev Neurosci 12:355–375

Huber F (1983) Neural correlates of orthopteran and cicada phonotaxis. In: Huber F, Markl H (eds) Neuroethology and behavioral physiology. Springer, Berlin Heidelberg New York, pp 108–135

Huber F, Wohlers D, Moore TE (1980) Auditory nerve and interneurone responses to natural sounds in several species of cicadas. Physiol Entomol 5:25–45

Inglis M, Oldfield BP (1988) Tonotopic organisation of the auditory organ of the locust *Valanga irregularis* (Walker). J Comp Physiol 164:49–53

Kämper G (1984) Abdominal ascending interneurons in crickets: responses to sound at the 30-Hz calling song frequency. J Comp Physiol 155:507–520

Keppler E (1958) Über das Richtungshören von Stechmücken. Z Naturforsch 13:280–284

Kleindienst HU, Koch UT, Wohlers DW (1981) Analysis of the cricket auditory system by acoustic stimulation using a closed sound field. J Comp Physiol 141:283–296

Larsen ON (1981) Mechanical time resolution in some insect ears. II. Impulse sound transmission in acoustic tracheal tubes. J Comp Physiol 143:297–304

Larsen ON, Michelsen A (1978) Biophysics of the ensiferan ear. III. The cricket ear as a four-input system. J Comp Physiol 123:217–227

Lee WB, Solomon SC (1975) Inversion schemes for surface wave attenuation and Q in the crust and the mantle. Geophys J R Astr Soc 43:47–71
Lewis B (1974) The physiology of the tettigoniid ear, I-IV. J Exp Biol 60:821–869
Lewis B (1983) Directional cues for auditory location. In: Lewis DB (ed) Bioacoustics: a comparative approach. Academic Press, London, pp 233–260
Markl H (1978) Adaptive radiation of mechanoreception. In: Ali MA (ed) Sensory ecology. Review and perspectives. Plenum Press, New York, pp 319–344
Markl H (1983) Vibrational communication. In: Huber F, Markl H (eds) Neuroethology and behavioural physiology. Springer, Berlin Heidelberg New York, pp 332–353
Markl H, Tautz J (1975) The sensitivity of hair receptors in caterpillars of *Barathra brassicae* L. (Lepidoptera, Noctuidae) to particle movement in a sound field. J Comp Physiol 99:79–87
Meier T, Reichert H (1990) Embryonic development and evolutionary origin of the orthopteran auditory organs. J Neurobiol 21:592–610
Michel K (1974) Das Tympanalorgan von *Gryllus bimaculatus* deGeer (Saltatoria, Gryllidae). Z Morphol Tiere 77:285–315
Michelsen A (1966) Pitch discrimination in the locust ear: observations on single sense cells. J Insect Physiol 12:1119–1131
Michelsen A (1971) The physiology of the locust ear. I Frequency sensitivity of single cells in the isolated ear. II Frequency discrimination based on resonances in the tympanum. III Acoustical properties of the intact ear. Z Vergl Pysiol 71:49–128
Michelsen A (1985) Time resolution in auditory systems. Springer, Berlin Heidelberg New York
Michelsen A, Larsen ON (1978) Biophysics of the ensiferan ear. I. Tympanal vibrations in bushcrickets (Tettigoniidae) studied with laser vibrometry. J Comp Physiol 123:193–203
Michelsen A, Larsen ON (1983) Strategies for acoustic communication in complex environments. In: Huber F, Markl H (eds) Neuroethology and behavioural physiology. Springer, Berlin Heidelberg New York, pp 321–332
Michelsen A, Larsen ON (1985) Hearing and sound. In: Kerkut GA, Gilbert LI (eds) Comprehensive insect physiology, biochemistry and pharmacology, vol 6. Pergamon Press, Oxford, pp 496–556
Michelsen A, Fink F, Gogala M, Traue D (1982) Plants as transmission channels for insect vibrational songs. Behav Ecol Sociobiol 11:269–281
Michelsen A, Kirchner WH, Lindauer M (1986) Sound and vibrational signals in the dance language of the honeybee, *Apis mellifera*. Behav Ecol Sociobiol 18:207–212
Michelsen A, Towne WF, Kirchner WH, Kryger P (1987) The acoustic near field of a dancing honeybee. J Comp Physiol 161:633–643
Miller LA (1970) Structure of the green lacewing tympanal organ (*Chrysopa carnea*, Neuroptera). J Morphol 131:359–382
Miller LA (1971) Physiological responses of green lacewings (*Chrysopa*, Neuroptera) to ultrasound. J Insect Physiol 17:491–506
Miller LA (1977) Directional hearing in the locust *Schistocerca gregaria* Forskal (Acrididae, Orthoptera). J Comp Physiol 119:85–98
Moiseff A, Pollack GS, Hoy RR (1978) Steering responses of flying crickets to sound and ultrasound: mate attraction and predator avoidance. Proc Natl Acad Sci USA 75:4052–4056
Mörchen A, Rheinlaender J, Schwartzkopff J (1978) Latency shift in insect auditory nerve fibers. Naturwissenschaften 65:656
Moran DT, Rowley III C (1975) High voltage and scanning electron microscopy of the site of stimulus reception of an insect mechanoreceptor. J Ultrastruct Res 50:38–46
Murphey RK (1971) Sensory aspects of the control of orientation to prey by the waterstrider, *Gerris remigis*. Z Vergl Physiol 72:168–185
Oldfield BP (1980) Accuracy of orientation in female crickets, *Teleogryllus oceanicus* (Gryllidae): dependence on song spectrum. J Comp Physiol 141:93–100
Oldfield BP (1982) Tonotopic organisation of auditory receptors in Tettigoniidae (Orthoptera: Ensifera). J Comp Physiol 147:461–469
Oldfield BP (1985) The tuning of auditory receptors in bushcrickets. Hearing Res 17:27–35
Oldfield BP, Hill KG (1986) Functional organization of insect auditory sensilla. J Comp Physiol 158:27–34
Payne RS, Roeder KD, Wallman J (1966) Directional sensitivity of the ears of noctuid moths. J Exp Biol 44:17–31

Pringle JWS (1938) Proprioreception in insects. II. The action of the campaniform sensilla on the legs. J Exp Biol 15:114–131
Pumphrey RJ (1940) Hearing in insects. Biol Rev Camb Philos Soc 15:107–132
Ramspeck A, Schultze GA (1938) Die Dispersion elastischer Wellen im Boden. Veröff Inst Dsch Forschungsges Bodenmechanik (Degebo) Techn Hochschule Berlin 6:1–28
Rheinlaender J (1975) Transmission of acoustic information at three neuronal levels in the auditory system of *Decticus verrucivorus* (Tettigoniidae: Orthoptera). J Comp Physiol 97:1–53
Rheinlaender J (1984) Das akustische Orientierungsverhalten von Heuschrecken, Grillen und Fröschen: eine vergleichende neuro- und verhaltensphysiologische Untersuchung. Habilitationsschrift, Bochum
Rheinlaender J, Mörchen A (1979) 'Time-intensity trading' in locust auditory interneurons. Nature (Lond) 281:672–674
Robert D (1989) The auditory behaviour of flying locusts. J Exp Biol 147:279–301
Roeder KD (1972) Acoustic and mechanical sensitivity of the distal lobe of the pilifer in certain Choerocampine hawkmoths. J Insect Physiol 18:1249–1264
Roeder KD, Treat AE (1957) Ultrasonic reception by the tympanic organs of noctuid moths. J Exp Zool 134:127–157
Römer H (1976) Die Informationsverarbeitung tympanaler Rezeptorelemente von *Locusta migratoria* (Acrididae, Orthoptera). J Comp Physiol 109:101–122
Römer H (1987) Representation of auditory distance within a central neuropil of the bushcricket *Mygalopsis marki*. J Comp Physiol 161:33–42
Römer H, Bailey WJ (1990) Insect hearing in the field. J Comp Biochem Physiol 97A:443–447
Römer H, Marquart V (1984) Morphology and physiology of auditory interneurons in the metathoracic ganglion of the locust. J Comp Physiol 155:249–262
Römer H, Rheinlaender J (1983) Electrical stimulation of the tympanal nerve as a tool for analysing the responses of auditory interneurons in the locust. J Comp Physiol 152:289–296
Ronacher B, Römer H (1985) Spike synchronization of tympanic receptor fibres in a grasshopper (*Chorthippus biguttulus* L., Acrididae). J Comp Physiol 157:631–642
Rovner JS, Barth FG (1981) Vibratory communication through living plants by a tropical wandering spider. Science 214:464–466
Rudolph P (1967) Zum Ortungsverfahren von *Gyrinus substriatus* Steph. (Taumelkäfer). Z Vergl Physiol 56:341–375
Schildberger K, Milde JJ, Hörner M (1988) The function of auditory neurons in cricket phonotaxis. II Modulation of auditory responses during locomotion. J Comp Physiol 163:633–640
Schiolten P, Larsen ON, Michelsen A (1981) Mechanical time resolution in some insect ears. I. Impulse responses and time constants. J Comp Physiol 143:289–295
Schmidt H (1954) Die Schallausbreitung in körnigen Substanzen. Acustica 4:639–652
Schmidt M, Gnatzy W (1984) Are the funnel-canal organs the "campaniform sensilla" of the shore crab, *Carcinus maenas* (Decapoda, Crustacea) Cell Tissue Res 237:81–93
Schmitz B, Scharstein H, Wendler G (1982) Phonotaxis in *Gryllus campestris* L. (Orthoptera, Gryllidae). I. Mechanism of acoustic orientation in intact female crickets. J Comp Physiol 148:431–444
Schnorbus H (1971) Die subgenualen Sinnesapparate von *Periplaneta americana*: Histologie und Vibrationsschwellen. Z Vergl Physiol 71:14–48
Schwabe J (1906) Beiträge zur Morphologie und Histologie der tympanalen Sinnesapparate der Orthopteren. Zoologica 20:1–154
Schwartzkopff J (1974) Mechanoreception. In: Rockstein M (ed) The physiology of Insecta. Academic Press, New York, pp 273–352
Shimozawa T, Kanou M (1984a) Varieties of filiform hairs: range fractionation by sensory afferents and cercal interneurons of a cricket. J Comp Physiol 155:485–493
Shimozawa T, Kanou M (1984b) The aerodynamics and sensory physiology of range fractionation in the cercal filiform sensilla of the cricket *Gryllus bimaculatus*. J Comp Physiol 155:495–505
Skudrzyk E (1971) The foundations of acoustics. Springer, Berlin Heidelberg New York
Spangler HG (1988) Moth hearing, defense, and communication. Ann Rev Entomol 33:59–81
Stephen RO, Bennet-Clark HC (1982) The anatomical and mechanical basis of stimulation and frequency analysis in the locust ear. J Exp Biol 99:279–314
Surlykke A, Larsen ON, Michelsen A (1988) Temporal coding in the auditory receptor of the moth ear. J Comp Physiol 162:367–374

Tautz J (1977) Reception of medium vibration by thoracal hairs of caterpillars of *Barathra brassicae* L. (Lepidoptera, Noctuidae). I. Mechanical properties of the receptor hairs. J Comp Physiol 118:13–31

Tautz J (1979) Reception of particle displacement in a medium – an unorthodox sensory capacity. Naturwissenschaften 66:452–461

Tautz J (1989) Medienbewegung in der Sinneswelt der Arthropoden. Fallstudien zu einer Sinnesökologie. In: Lindauer M (ed) Information processing in animals. Fischer, Stuttgart, pp 7–59

Tautz J, Markl H (1978) Caterpillars detect flying wasps by hairs sensitive to airborne vibration. Behav Ecol Sociobiol 4:101–110

Thorson J, Weber T, Huber F (1982) Auditory behaviour of the cricket. II. Simplicity of calling song recognition in *Gryllus*, and anomalous phonotaxis at abnormal carrier frequencies. J Comp Physiol 146:361–378

Thurm U (1969) General organization of sensory receptors. Rend Scuola Intern Fisica "E. Fermi". XLIII Corso pp 44–68

Thurm U (1982) Biophysik sensorischer Mechanismen. In: Hoppe W, Lohmann W, Markl H, Ziegler H (eds) Biophysik. Springer, Berlin Heidelberg New York Tokyo, pp 681–696

Thurm U, Stedtler A, Foelix R (1975) Reizwirksame Verformungen der Terminalstrukturen eines Mechanorezeptors. Verh Dtsch Zool Ges, Stuttgart pp 37–41

Tischner H, Schief A (1954) Fluggeräusche und Schallwahrnehmung bei *Aedes aegypti* (Culicidae). Verh Dtsch Zool Ges 1954:453–460

Tucker VA (1969) Wave-making by whirligig beetles (Gyrinidae). Science 166:897–899

von Helversen D (1972) Gesang des Männchens und Lautschema des Weibchens bei der Feldheuschrecke *Chorthippus biguttulus* (Orthoptera, Acrididae). J Comp Physiol 81:381–422

von Helversen D, Rheinlaender J (1988) Interaural intensity and time discrimination in an unrestrained grasshopper: a tentative behavioural approach. J Comp Physiol 162:333–340

Wales W, Clarac F, Dando MR, Laverack MS (1970) Innervation of the receptors present at the various joints of the pereiopods and third maxilliped of *Homarus gammarus* (L.) and other macruran decapods (Crustacea). Z Vergl Physiol 68:345–384

Wiese K (1974) The mechanoreceptive system of prey localization in *Notonecta*. II. The principle of prey localization. J Comp Physiol 92:317–325

Yager DD, Hoy RR (1986) The cyclopean ear: a new sense for the praying mantis. Science 231:727–729

Yager DD, Hoy RR (1987) The midline metathoracic ear of the preying mantis, *Mantis religiosa*. Cell Tissue Res 250:531–541

Young D, Ball E (1974) Structure and development of the auditory system in the prothoracic leg of the cricket *Teleogryllus commodus* (Walker). I. Adult structure. Z Zellforsch 147:293–312

Young D, Hill KG (1977) Structure and function of the auditory system of the cicada, *Cystosoma saundersii*. J Comp Physiol 117:23–45

Zhantiev RD, Kalinkina IN, Tshukanov VS (1975) The characteristics of the directional sensitivity of tympanal organs of *Gryllus bimaculatus* Deg. (Orthoptera, Gryllidae). Rev Ent USSR 54:249–257 (in Russian)

Chapter 10

Mechanoelectrical Transduction in Vertebrate Hair Cells

H. Ohmori[1]

Contents

1 Introduction

Various membrane-bound proteins function to exchange ions across the membrane. Some of these proteins are called a carrier since ionic transport must be coupled with the transport of another substances or it requires metabolic energy. The other transporting protein is called a channel and serves as a rapid and a large-scale ion transport across the membrane. The channel is opened and closed by the inherent gate. Ions transported are selected by its filter function. The transport is fast and many thousands of ions are carried in milliseconds, and achieves the unitary

[1]Department of Physiology, Faculty of Medicine, Kyoto University, Kyoto, 606–01, Japan

Advances in Comparative and Environmental Physiology, Vol. 10

conductance of 10 to 100 pS. The gate is the most unique function of the channel, and depending on its nature the channel becomes either voltage-dependent, agonist-dependent or achieves an energy-transducing function. The mechanoreceptor cell is therefore a cell equipped with an ion channel whose gate is mechanically opened and closed. The ionic flow regulated in this manner generates membrane- potential changes and triggers other voltage-dependent channel events. Ca channels are activated by depolarizing, transducing potentials, allow Ca ions to flow into the cell for synaptic transmission, and initiate higher order of information processing.

This chapter summarizes the fundamental characteristics of mechanoelectric transduction, based mostly on the work performed on the isolated hair cells from the inner ear of a chick. The cellular mechanism of efferent innervation by the olivocochlea bundle onto the hair cell will also be discussed.

2 Electrophysiology of Hair Cells

2.1 Voltage-Gated Ion Channels on the Hair Cell

When hair cells were voltage clamped by a patch electrode using a KCl-EGTA based intracellular medium (155 KCl, 5 EGTA buffered with 10 K-HEPES to pH 7.4, concentrations in mM), a nonlinear current vs voltage relationship was observed (Fig. 1), demonstrating a large outward-going potassium conductance at potentials more positive than –50 mV and a large inward-going potassium conductance at potentials more negative than –100 mV. The outward-going K current was carried through the Ca-activated K channel and shows the N-shaped current vs voltage relationship as having an outward current peak at about +40 mV. The inward-going K current was carried through the anomalous rectifier K channel, also demonstrating a negative conductance region at potentials more negative than –150 mV. This negative conductance was generated by the channel-blocking action of extracellular Na ions (Ohmori 1978, 1980, 1984a). A relatively high-input impedance region was observed in between these two conductance mechanisms. At about the resting potential of –60 mV (–62 ± 7 mV, n = 7), the input impedance was 3.7 ± 2.3 Gohm (n = 7; Ohmori 1984a), and some hair cells demonstrated a slight nonlinear current vs voltage relationship in this high impedance region, showing a small outward current peak at about –65 mV and approaching the zero-current level at about –50 mV. This nonlinear high-impedance region seems to be created by the crossing of the skirt of the anomalous rectifier conductance and the Ca-activated K conductance, together with a trace amount of the voltage-gated Ca conductance. This high-impedance region would facilitate changes in the membrane potential, even when a small amount of ionic current flowed into the cell through the transducer channel. In some hair cells, for example, in the outer hair cell of the guinea pig cochlea, membrane conductance is dominated by a background, nonspecific conductance; a high-impedance membrane potential region has not been observed (Ashmore and Meech 1986).

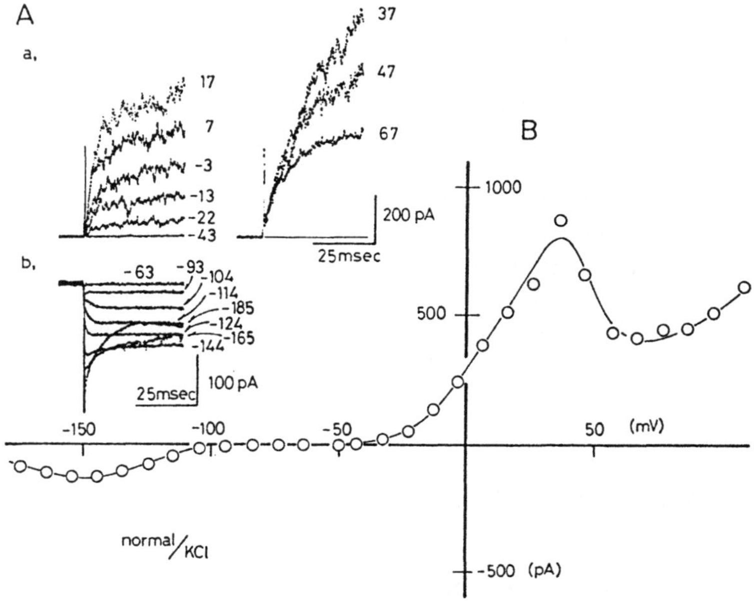

Fig. 1. Whole cell currents recorded in a normal bathing medium (155 NaCl, 5 KCl, 2.5 $CaCl_2$, 1 $MgCl_2$, 17 glucose, pH 7.4, concentrations in mM), while the patch electrode was filled with KCl, EGTA based internal medium. *A-a* Outward currents generated by depolarizing voltage steps from the holding level of –60 mV. *A-b* Inward currents through the anomalous rectifier channel. *B* Steady state amplitudes of the current in *A* were plotted against membrane potentials. (Ohmori 1984a)

When the extracellular concentration of Ca ions was increased close to the isotonic level and the hair cell was whole-cell voltage-clamped using a patch electrode filled with a CsCl-EGTA based intracellular medium (160 CsCl, 5 EGTA-Na, buffered to pH 7.4 by 10 Na HEPES, concentrations in mM), inward- going Ca currents could be generated by depolarizing voltage steps. The Ca current showed time-dependent decay and at large positive potentials the current became outward-going. This current decay and the outward flow are due to the activation of the Ca-activated K conductance. When 100 μM quinine was added to the extra- cellular medium, the same set of voltage pulses generated a series of noninactivating Ca currents (Ohmori 1984a). Quinine and quinidine are known to block the Ca-activated K channel, but its specificity is weak (Hermann and Gorman 1984).

Ca channels demonstrate voltage-dependent inactivation kinetics in some preparations (Nowyky et al. 1985), but the Ca channel of hair cells did not have inactivation kinetics. The activation phase reasonably fit the curve of the m^2 or m^3 kinetics of Hodgkin and Huxley (1952).

Ba and Sr ions permeate the Ca channel more than Ca ions. However, the amount of Ca-channel current was small in the chick hair cells (–6 to –273 pA in 100 mM Ba saline, at about 0 mV; Ohmori 1984a). If we take 1 pA as the size of the single Ca-channel current under this condition, the maximum number of Ca channels on the hair cell would be 270. Since the surface area of the cell body is about 500 μm^2 (Ohmori 1985), the density of the Ca channels is about 0.5/μm^2.

Hair cells from the chick inner ear have a capacitance of 4.5–7.4 pF (5.8 ± 1.4 pF, n = 4). The surface area of the cell body was 486 μm^2, assuming a cylindrical shape of 9.5 μm diameter and 13.6 μm length. The extra surface area added by the ciliary bundle would be 252 μm^2 if we assume 90 cilia of 0.25 μm diameter, stepwise increasing the height arrangement. Since the estimation of electrical capacitance of the cell is influenced by the speed of the recording amplifier, the capacitance reported above might have been underestimated. If we therefore take the largest capacitance recorded as a typical value, the specific capacitance of the hair cell would be close to 1 $\mu F/cm^2$. The hair cell capacitance and the input impedance determine the membrane time constant to be of the order of 10 ms (Ohmori 1985).

3 Mechanoelectrical Transduction

The nature of mechanoelectrical transduction (m-e.t.) in hair cells have long been estimated indirectly from the microphonic potentials or from the microphonic currents (Furukawa and Ishii 1976a,b; Corey and Hudspeth 1983a,b) or from the activity of afferent nerve fibers (Furukawa and Ishii 1967a,b; Furukawa 1986). The direct intracellular recording of the transduction potential was first made in 1977 by Russell and Sellick from the guinea pig cochlea. Although qualitative observations of the transducer activity of the hair cell have been made by the delicate microelectrode recordings in the epithelial preparation by Hudspeth and Corey (1977) and Corey and Hudspeth (1979), more quantitative and accurate observations have been performed after the introduction of the isolated hair-cell preparation and the patch-electrode whole-cell recording technique, as has been the case in the study of other sensory receptor cells, and in the study of ion channels in general (Lamb 1972; Neher and Sakmann 1976; Ohmori 1984b).

3.1.1 Electrophysiology of the Transduction Channel

The mechanoelectrical transduction current was first observed in the isolated hair cell preparations by Ohmori (1984b) under a whole-cell patch-electrode voltage clamp. The mechanical displacement of the hair bundle towards the taller stereocilium generated inward currents and the displacement in the opposite direction generated apparent outward currents. This latter phenomenon reflected the closure of the steady, open transduction channels. Since approximately only 10% of transducer conductance is open at the resting position of the hair bundle, the symmetrical vibration of the hair bundle could generate net inward currents and could depolarize the membrane.

3.1.2 Single-Channel Conductance of the Mechanoelectrical Transducer

Single-channel conductance (Fig. 2, 50 pS, at 30 °C) was determined in the chick hair cell from the discrete amplitude current steps detected in the transduction

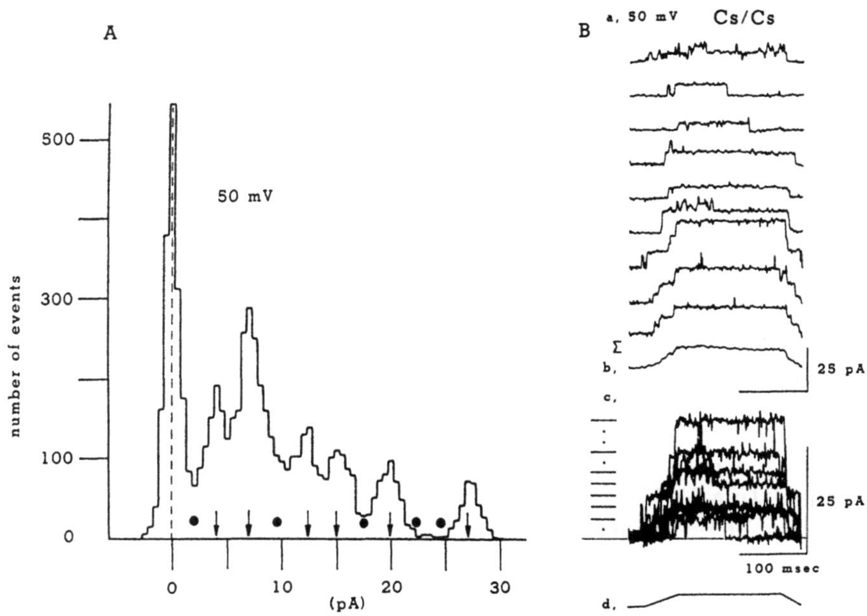

Fig. 2. M-e.t. currents generated in response to a trapezoidal hair bundle displacement in a CsCl saline. The glass rod deflected the hair bundle by 1 μm and was held at that bent position for 148 ms. *A* The probability density histogram for current amplitudes in *B-a*. Current levels which are not obvious in the histogram as peaks but likely to exist are indicated by *dots* here and in *B-c*. *B-a* Individual current traces; *B-b* the grand average of traces in *B-a; B-c* the superimposed traces of *B-a*; *B-d* the driving waveform. (After Ohmori 1985)

currents under the whole-cell voltage clamp (Ohmori 1985). The number of current steps was limited (11 steps) and the number of m-e.t. channels involved seems very small in this experiment, but this is most likely due to the hair cell transducer deterioration during isolation. By comparing the maximum whole-cell m-e.t. current conductance (2 nS) and the single-channel conductance (50 pS), the number of m-e.t. channels in a single hair cell of chick was estimated as 40. Since the transducer seems a labile structure and the small damage to the hair bundle significantly reduces the current amplitude, this estimation of the channel number would give the lower limit of the number of m-e.t. channels operating.

The absolute permeability coefficient of Cs ions can be calculated from this 50-pS single m-e.t. channel conductance (Ohmori 1985); the single channel permeability coefficient of Cs ions at –50 mV is 7.04×10^{-14} cm^3/s.

Holton and Hudspeth (1986) have estimated the single m-e.t. channel current and the conductance by the ensemble noise analysis technique, applied to a series of m-e.t. currents recorded in the bullfrog hair cell an epithelial preparation. Their estimation of 13 pS (at 10 °C) was much smaller than the above observation (50 pS); the conductance was about half of the chick's even at 37 °C when the temperature coefficient of 1.3 was assumed (25 pS). This small value might reflect different natures of the m-e.t. channels of chick (Ohmori 1985) and bullfrog (Holton and

Hudspeth 1986), but more likely reflects a common problem in the noise analysis, resulting mostly from the underestimation of current fluctuation in the higher frequency region.

3.1.3 Ion Selectivity of the Transduction Channel

Ion dependence of the transducer mechanism was first studied by Sand (1975) on the lateral line organ in the mudpuppy and was then studied systematically in the bullfrog saccular hair cell by Corey and Hudspeth (1979). Corey and Hudspeth compared the size of extracellular microphonic currents by bathing the hair cell epithelial preparation in solutions of various ionic compositions. Mean signal amplitudes were almost the same in the Li, Na, K, Rb, Cs and in NH_4 medium, while they were much smaller in the TMA and in the Ca medium. They have observed Sr to carry microphonic currents, while Ba and Mg did not.

When the hair bundle was mechanically displaced towards the taller stereocilium, depolarizing transduction potentials were generated at negative membrane potentials (Fig. 3). The amplitude of depolarization decreased with the positive shift of membrane potential and disappeared at +6 mV. With a further positive shift of the membrane potential, hyperpolarizing m-e.t. potentials emerged. The mechanoelectrical transduction of the chick hair cell has therefore a defined reversal potential (Ohmori 1985). The level of the reversal potential was determined by the composition of intracellular and extracellular medium. When using a CsCl-based intracellular medium with EGTA, the most positive reversal potential was observed in an isotonic $CaCl_2$ extracellular medium (+28.2 mV) and the most negative reversal potential was observed in an isotonic TEA-Cl (–25.3 mV) or TMA-Cl (–23.5 mV) medium. All alkaline cations and most alkaline earth cations were permeable through the transducer channel. Relative permeabilities in reference to Cs ions were determined from the reversal potential as follows: Li(1.39); Na(1.22); K(1.17); Rb(1.12); Cs(1.0); choline(0.33); TMA(0.20); TEA(0.17) for monovalent

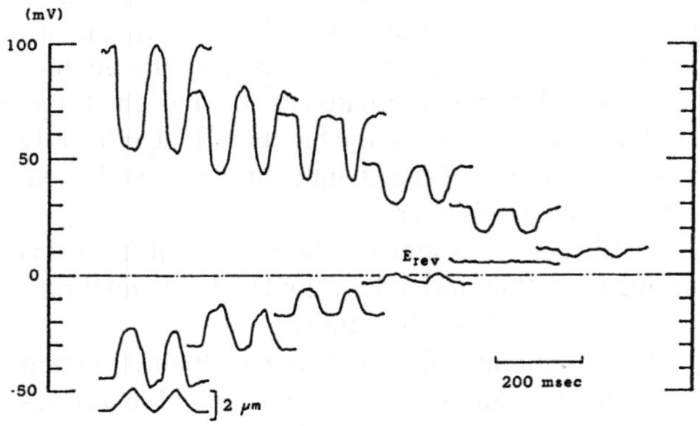

Fig. 3. M-e.t. potentials recorded in the normal saline from a cell under whole-cell, current-clamp conditions. A mechanical stimulation of the hair bundle generated membrane potential changes. Amplitudes of m-e.t. potentials were changed by the change in the holding membrane potential. The reversal potential was 6 mV. The glass rod motion was monitored optically and was plotted in the figure. (Ohmori 1985)

cations, and Ca(4.65); Sr(2.82); Ba(2.73); Mn(2.50); Mg(2.41) for divalent cations. Numbers in parenthesis are the relative permeabilities to Cs ions.

From this ion-selectivity study, the m-e.t. channel is suggested to have a high field-strength ion-selectivity site (the XI series, Eisenman 1962), which is consistent with the relatively high permeability of divalent cations through the m-e.t. channel. If we assume the diameter of the least permeable organic cations TMA, or TEA as the size of pore, the m-e.t. channel might be a pore of 6–7 Å diameter.

From the relative permeability of Ca to Cs ions, we can calculate the influx of Ca ions during m-e.transduction by assuming a 30 μM Ca concentration in the endolymph (Bosher and Warren 1968). Since the permeability of Ca ions was 4.7 times larger than Cs ions through the m-e.t. channel, a 10-ms opening of the m-e.t. channel at –50 mV will carry 3.8×10^{-22} mol Ca ions (approximately 230 Ca ions). Even such a small quantity of Ca-ion influx through the m-e.t. channel will raise the Ca concentration close to the μM level within a volume equivalent to a single stereocilium (0.25 μm^3). This calculation might still be underestimating the exact Ca influx through the channel, since the positively polarized endolymphatic potential would facilitate the Ca influx (Davis 1965). By the mechanoelectrical transduction, the intracellular Ca concentration could be raised close to the μM level at least in the vicinity of the m-e.t. channel, and this level of Ca concentration would be sufficient to trigger many Ca-dependent biological reactions.

3.1.4 Gating Speed of the m-e.t. Channel

The hair cell membrane has a time constant of about 10 ms (16 Hz), and all the m-e.t. currents must first be integrated to depolarize the membrane with this time constant in order to be transmitted to the afferent nerve via the hair cell-afferent nerve synapse. Since the basilar membrane vibration frequency is topographically arranged in a systematic way in mammalian cochlea, from the high frequency region at the base to the low frequency region at the apex (von Békésy 1960), the hair-cell frequency tuning could be made, for the most part, at the level of basilar membrane mechanical vibration in mammals (Rhode 1971; Wilson and Johnstone 1975). However, the detection of the highest auditory frequency (20 kHz for man) still requires the m-e.t. channel-gating kinetics up to that frequency level. The gating time constant was from 100–500 μs at 4 °C in the bullfrog (estimated from extracellular microphonic currents; Corey and Hudspeth 1983b), and in the turtle (calculated from whole-cell-recorded transduction current kinetics; Crawford et al. 1989).

3.1.5 Electrical Resonance of the Membrane

Membrane electrical resonance has been observed in hair cells of some species (in the turtle, Crawford and Fettiplace 1981; in the bullfrog, Lewis and Hudspeth 1983; in the goldfish, Sugihara and Furukawa 1989; in the chick, Fuchs et al. 1988). The resonance frequency has some topographical distribution in the basilar papilla of turtle (Art and Fettiplace 1987) and the density of Ca channels and the activation time constant of K channels have a close relationship with the resonance frequency of the individual hair cell. The Ca channel density was increased and the K-channel activation time constant was shortened with the increase in the resonance frequency. The resonance frequency is most likely determined by the balance of the

Ca conductance and the Ca-activated K conductance (Hudspeth and Lewis 1988). Physiological roles are considered in sharpening the hair-cell tuning frequency in these lower vertebrates.

3.1.6 Gating Sensitivity of the m-e.t. Channel

By applying hair bundle displacement quantitatively, the displacement vs response relationship was measured in various hair cells (Hudspeth and Corey 1977; Corey and Hudspeth 1983b; Russell and Sellick 1983; Crawford and Fettiplace 1985; Ohmori 1987). Figure 4A demonstrates a series of current traces generated by mechanical stimulation applied at 5 µm above the insertion of the hair bundle to the cuticular plate. The smallest displacement amplitude which could generate a transduction current was 100 Å in the chick hair cell. This displacement stimulation corresponds to 0.1° angular displacement of the hair bundle. The displacement vs response relationship shows that the amplitude of the m-e.t. current increases linearly up to the displacement angle of about 8° and the amplitude was saturated at about 10° displacement (Fig. 4B). This was exactly 100 times greater than the smallest displacement angle which has generated m-e.t. currents (0.1°), and this number is approximately the same as the number of stereocilia in a single hair bundle. This coincidence might suggest an association of one m-e.t. channel with one stereocilium. Crawford et al. (1989) have reported a similar displacement vs response relationship where the transducer current amplitude was increased up to about 5° displacement.

In other hair cells, much smaller displacement amplitudes have generated transduction current (Russel et al. 1986; Crawford and Fettiplace 1985), but the exact height of the applied displacement stimulus was not clear.

Crawford and Fettiplace (1985) reported the threshold angular displacement of the hair bundle as 0.03°. The threshold basilar membrane vibration amplitude was estimated in the guinea pig cochlea by linear extrapolation as 0.3 nm, which corresponds to the 0.04 mm/s basilar membrane velocity (Sellick et al. 1982). The isovelocity curve matches with the neural frequency tuning curve, but this 0.3-nm, isoamplitude curve systematically deviates towards the subthreshold sound pressure level at the low-frequency region of the neural frequency tuning curve. This indicates the threshold basilar membrane vibration amplitude could be much larger than this estimation. Moreover, the exact hair bundle displacement can be much larger at the threshold level by some mechanical amplification.

3.1.7 Angular Displacement of the Hair Bundle Gates the Transduction Channel

Figure 4B shows the normalized amplitude of the transduction currents plotted against the angular displacement of the hair bundle, summarizing several experiments made on hair cells having different hair bundle lengths (Ohmori 1987). When a displacement stimulus was applied to a short hair bundle shorter than 7.5 µm, the displacement was applied at (1) 5 µm above the insertion of the hair bundle to the cuticular plate; (2) at 5 µm; or (3) at 10 µm when applied to the tall hair bundle taller than 12.5 µm (see inset in Fig. 4B). The transduction current amplitude was plotted against the angular displacement after scaling, and all three types of experiments determined a single relationship (Fig. 4B). The angular displacement of the hair bundle is therefore the primary factor in the gating of transduction channels.

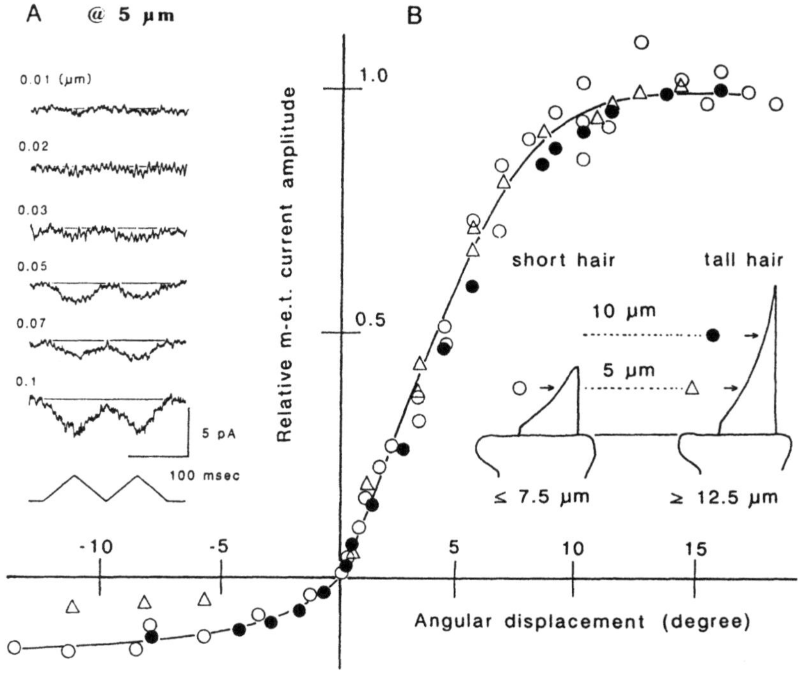

Fig. 4. Angular displacement vs m-e.t. current response relationship. *A* M-e.t. currents generated by mechanical displacements of the hair bundle applied at 5 μm above the insertion of the hair bundle to the cuticular plate. *B* M-e.t. current amplitudes were plotted against the angular displacement of the hair bundle after scaling. Displacement stimulus was applied at 5 μm above the insertion of the hair bundle to the cuticular plate for the cell whose hair bundle was shorter than 7.5 μm (*open circles*). Hair bundle displacement was applied at 5 μm (*open triangles*) and at 10 μm (*filled circles*) to the hair bundle taller than 12.5 μm (see inset). (After Ohmori 1987)

3.1.8 Adaptation of the Transduction Current

When a steady displacement stimulus was applied to the hair bundle, the amplitude of m-e.t. current decreased during stimulation; this phenomenon is called an adaptation. These current adaptation are dependent on recording conditions and have been observed in the bullfrog in extracellular microphonic currents (Corey and Hudspeth 1983a), in hair cells epithelial preparations under voltage clamp with two microelectrodes (Howard and Hudspeth 1987), in isolated bullfrog hair cells, and in turtle hair cells under the patch-electrode whole-cell clamp (Assad et al. 1989; Crawford et al. 1989). Adaptation-free m-e.t. currents have been observed in bullfrog hair cells which were whole-cell clamped with patch electrode voltage in the epithelial preparation (Holton and Hudspeth 1986), and in isolated chick hair cells (Ohmori 1985, 1987). Assad et al. (1989) have demonstrated the extracellular Ca concentration (0.1 and 4 mM) dependence of the adaptation. They have also demonstrated that the adaptation was membrane-voltage-dependent and was not observed at positive membrane potentials. Similar observations were made by Crawford et al. (1989). Since the Ca influx through the m-e.t. channel is significant, as calculated above, and the Ca influx could be minimized at positive membrane potentials, they have argued that the voltage dependence of adaptation is controlled by the Ca influx through the transducer channel and that Ca effects are on the intracellular adaptation site. Since the offset kinetics of the m-e.t. current becomes slower as the adaptation is lost, Crawford et al. (1989) further reported the possibility of the accumulation of some intracellular factor that promotes channel opening. The whole-cell patch recording condition provides almost complete replacement of the intracellular medium with the patch electrode solution, and some essential factors in the cytoplasm are likely to be washed out, leading to the running down of several membrane ion channels with time. The adaptation kinetics of the transduction channel could be one of these labile functions, and the delicate balancing of the intracellular Ca concentration could play a primary role in the hair cell transduction.

3.1.9 Blockers of the m-e.t. Channel

Streptomycin and neomycin and other aminoglycoside antibiotics are known as ototoxic agents and their effects on the m-e.t. channel were studied under voltage clamp (Ohmori 1985). Both streptomycin and neomycin reversibly blocked the m-e.t. channel in a voltage-dependent manner, suggesting that the postively charged antibiotics plug into the mechanically gated ion channel. The degree of m-e.t. channel blocking became larger with the hyperpolarization of the membrane. Blocking effects of gentamycin and dihydrostreptomycin were studied on the hair cell of the bullfrog sacculus (Kroese et al. 1989).

The voltage-dependent block of the m-e.t. channel was studied in detail for amiloride, a blocker of the epithelial Na channel (Jørgensen and Ohmori 1988). The m-e.t. current was suppressed when amiloride was present in the bathing medium, particularly at negative membrane potentials. Efflux of ions through the m-e.t. channel seems to clear the plugged channel at positive membrane potentials. The amiloride blocking of the m-e.t. channel was apparently independent of the mechanical gating of the channel and had a K_D of 50 μM and an apparent Hill coefficient of 1. Although amiloride is most likely a monovalent cation at neutral pH, the blocking kinetics had much greater dependence on the membrane voltage (equivalent valence was 1.67) than the expectation from a single monovalent cationic molecule. Jørgensen and Ohmori (1988) have explained this discrepancy by assuming one blocked state and two open states whereby both open states have affinity to amiloride molecules; hence, the association of two amiloride molecules with a single m-e.t. channel is expected. Since only the direct transition from the open state to the blocked state could be monitored electrophysiologically, association of these two amiloride molecules have most likely generated channel- blocking kinetics with an apparent Hill coefficient of 1, but with an equivalent valence greater than 1.

3.1.10 The Site of the Mechanoelectrical Transduction Channel

The site of transduction channel in hair cells is still controversial. By measuring the extracellular field potential changes while vibrating the hair bundle, Hudspeth (1982) detected small-field potential changes around the distal end of the hair bundle. These field potential changes were reversibly eliminated by applying aminoglycoside antibiotics, i.e., blockers of mechanoelectrical transduction. Hudspeth has proposed the transduction channel as being at the distal end of individual stereocilium, somewhwat associated with the tip-link structure which connects adjacent stereocilia (Pickles et al. 1984). His idea of mechanoelectrical transduction has been described in a recent review (Howard et al. 1988). As discussed above, the angular displacement of the hair bundle is the primary gating parameter; however, the angular displacement-dependent gating kinetics of the m-e.t. channel could still be explained by tip links if the stereocilia and tip links were rigid enough. The rigidity of the ciliary bundle has been measured; the stereocilium is probably rigid enough to produce a similar angular displacement of tip links as that of the stereocilium about the base of the hair bundle (Strelioff and Flock 1984; Crawford and Fettiplace 1985). However, the rigidity of tip links is not yet known.

The fluorescence imaging technique was applied to the hair cell by Ohmori (1988) to determine whether the site of mechanoelectrical transduction is the site of Ca influx during energy transduction (Fig. 5). Since there is significant Ca influx through the m-e.t. channel during energy transduction (Ohmori 1985), and the Ca-sensitive dye fura-2 (Grynkiewicz et al. 1985) can be loaded easily into the cell by the ester form of the dye, the intracellular distribution of Ca ions and changes in the Ca concentration during mechanoelectrical transduction have been imaged by combining the advantages of a fluorescence microscope and an image-processing computer system.

Fura-2 fluorescence was monitored by stimulation at two wavelengths, 340 and 380 nm (Fig. 5B,C). Fluorescence intensities increase with an increase in Ca concentration when excited at 340 nm, whereas they decrease when excited at 380 nm.

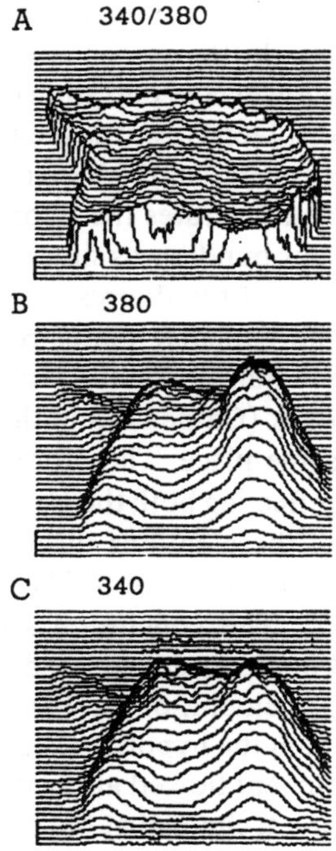

Fig. 5. Fluorescent image of the chick hair cell. The ratio intensity (*A*), and two fluorescence intensities (*B, C*) were plotted in a three-dimensional histogram form. The hair cell was loaded with Fura-2 AM and the fluorescence at 510 nm was monitored by stimulation at 340 nm (*C*) and at 380 nm (*B*) wavelengths. The ration between these two fluorescences gives the distribution profile of intracellular Ca concentration and is shown in *A*. The location of the cuticular plate and the nucleus demonstrates the lower level of intracellular Ca concentration

Taking the ratio between these two fluorescences, influences such as uneven dye distribution and cell thickness on the fluorescence intensities can be compensated, and the intracellular Ca concentration can be estimated with sufficient time resolution.

Figure 6A demonstrates the ratio intensity distribution over the hair cell, calculated from an experiment in 25 mM Ca saline. When the hair bundle was stimulated at a small amplitude with 1 kHz sinusoidal waveform, a sharp peak emerges around the insertion of the hair bundle to the cuticular plate. Figure 6C (left column) also demonstrates the fluorescence ratio intensities in planes which longitudinally cross the hair bundle. A peak is observed at the insertion of the hair bundle to the cuticular plate during mechanical stimulation. However, when a larger-amplitude mechanical stimulation was applied to the hair bundle, the ratio intensity increased over the entire cell body (Fig. 7A is the control, and Fig. 7B is during stimulation). This is probably due to the Ca influx through the Ca channel, since the depolarizing transduction potential certainly activates voltage-gated Ca channels, which are most likely located on the basolateral membrane. Under such circumstances, the diffusion of Ca ions from the cell body towards the hair bundle could not be negligible. The ion selectivity and the pharmacology of the Ca channel and the m-e.t. channel are relatively close (Ohmori 1985) and pharmacological isolations of m-e.t. channels are difficult at the present time. However, Ca channels of the hair cell were impermeable to Mn^{2+}, while Mn^{2+} is one of the most permeable cations through the m-e.t. channel (Ohmori 1985) and is known to quench the fluorescence of fura-2 (Grynkiewicz et al. 1985).

When mechanical stimulation was applied to the hair bundle in the Mn saline enriched with a small amount of Ca, the m-e.t. current carried by Mn ions was observed (Ohmori 1985). Figure 7C also demonstrates the disappearance of hair bundle fluorescence when Mn saline was puff applied against the hair bundle. When a lower concentration of Mn^{2+} (2 mM) was puff applied to the hair bundle and it was displaced towards the taller stereocilium, a beltlike area of fluorescence quenching appeared at the base of the hair bundle (Fig. 7E). Figure 6C (right column) further demonstrates the fluorescence quenching in a Mn medium (2 mM Mn), when the hair bundle was sinusoidally stimulated at the small amplitude of 1 kHz. A depression of the fluorescence ratio appeared around the insertion of the hair bundle to the cuticular plate. By superimposing the fluorescence ratio profiles in the Ca medium and in the Mn medium (center column in Fig. 6C), the locations of the ratio peak (Ca-peak) and the ratio depression (Mn-peak) are found to be exactly at the same place. Therefore, the fluorescence changes observed here are confirmed as reflecting both the Ca- and Mn-concentration changes induced directly by the influx of these cations through the m-e.t. channel. The mechanoelectrical transduction channels are most likely located at the base of the hair bundle.

This conclusion accords with the electrophysiological knowledge that gating kinetics are regulated by the angular displacement, but not by the absolute displacement, of the hair bundle (Fig. 4). However, this finding differed completely from the hypothesis of Hudspeth (1982) described above. If the extracellular m-e.t. current flowed along the hair bundle, the transduction current passing through the m-e.t. channel at the base of the hair bundle would generate extracellular potential

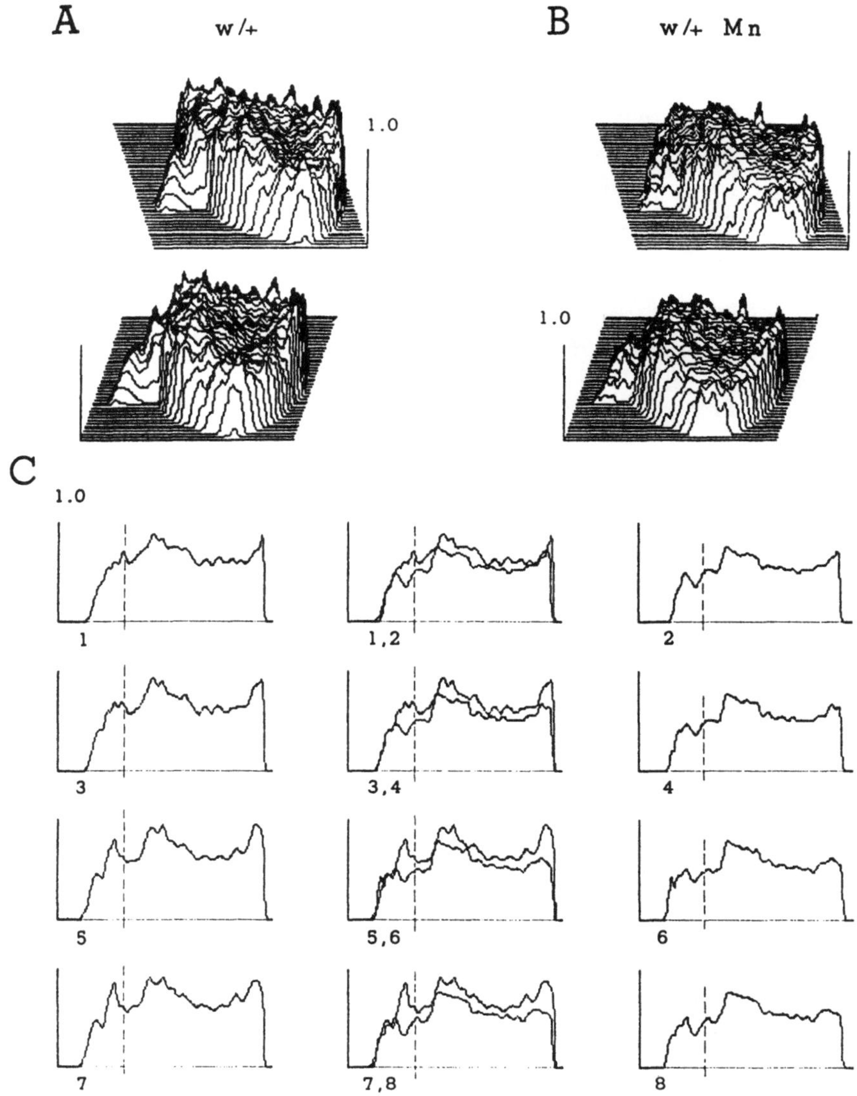

Fig. 6. Sinusoidal stimulation and Mn quenching. 1-kHz sinusoidal mechanical stimulation was given to the hair bundle first within *A* the Ca-rich bathing medium followed by the stimulus within *B* the Mn medium. *C* Fluorescence ratio profiles in the hair bundle and the cell body are demonstrated along the axis which longitudinally crosses the hair bundle and the cell body. Traces labeled with *odd numbers* are in the Ca medium corresponding to the image *A* and traces labeled with *even numbers* are in the Mn medium corresponding to the image *B*. The fluorescence peak in the Ca medium and the fluorescence depression in the Mn medium emerged exactly at the same site, as demonstrated by superimposing two profiles measured in Ca and in Mn medium in the *center* column. (Ohmori 1988)

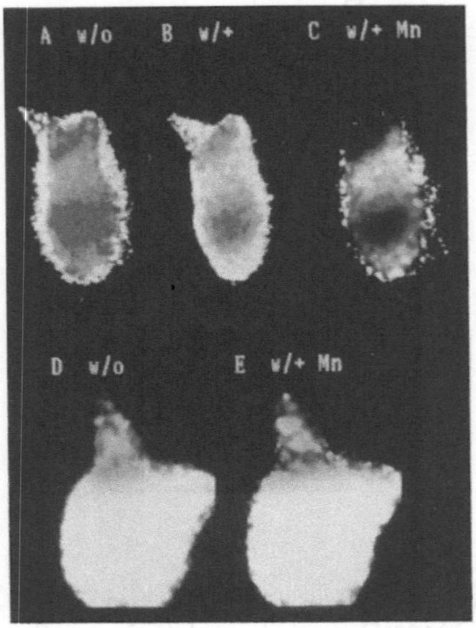

Fig. 7. Quenching by application of Mn^{2+} saline by a pressure-puff method. The hair bundle extends towards the *upper left. A* A control ratio image for the measurements in *B* and *C*. *B* A ratio image with a puff of normal saline to the hair bundle in the depolarizing direction (using pressure-puff method). *C* A ratio image with a puff of a high Mn saline on the hair bundle. Note that the hair bundle and part of the cuticular plate have disappeared from the ratio image in *C*. *D* A control image for the measurement in *E* without mechanical stimulation in 25 mM Ca saline. *E* A low Mn saline was puffed against the hair bundle from a distant location in order to stimulate the hair bundle only with the fringe of the Mn puff. A belt-shaped depression of the intensity ratio appeared around the insertion of the hair bundle to the cuticular plate. *W/o* indicates without mechanical stimulation; *w/+* with stimulation; *w/+ Mn* with stimulation in an Mn medium

fields in a displacement-dependent manner, as has been observed by Hudspeth. Alternatively, if the outer surface of the cilia was charged and the charge had an affinity to aminoglycoside antibiotics, the extracellular field potential might have been generated in a vibration- or a velocity-dependent manner at the location described by Hudspeth, and the field potential could be reduced by dihydrostreptomycin which was used to confirm the trasduction current flow. The presence of negatively charged phosphatidylinositol 4,5-diphosphate ($PI_{4,5}P_2$) and phosphatidylinositol 4-monophosphate (PI_4P) was demonstrated on the membrane of stereocilia by an antibody against $PI_{4,5}P2$ and PI_4P. The presence of neomycin (0.1–1 mM) has somewhat reduced the antibody association with the ciliary bundle in the guinea pig cochlea (Yanagisawa et al. 1988). If these phospholipids were on the surface membrane of stereocilia, they will provide large negative charges there. The presence of negative charges in the surface-coating materials has been demonstrated to exist extensively about the hair bundle by the alcian blue staining technique by Santi and Anderson (1987). The alcian-blue- positive, negatively charged, surface materials extensively coat the surface of the stereocilium, and the loose filamentous networks of the surface coat extend into the endolymphatic space for 0.5 to 1.5 μm in chinchilla. If the hair bundle of the bullfrog saccular hair cell was similarly coated by the surface materials as extensively as it was in the chinchilla cochlea hair cell, the negatively charged surface materials might have affected the measurements of the field potentials where the tip of the field-sensing electrode was positioned about "0.5 μm outside of the volume occupied by the vibrating hair bundle" (Hudspeth 1982).

3.1.11 Inferences on the Gating Mechanism of the M-e.t. Channel

Immunohistochemical evidence indicates the presence of a Ca-binding protein (Rabié et al. 1983), α-actinin, myosin (Dreckhahn et al. 1982), and fimbrin (Flock et al. 1982) in the cuticular plate. Each stereocilium in the hair bundle is packed with actin filaments and is inserted into the cuticular plate, where these two structures seem to form a well-integrated machinery. The stereocilium could be suspended by the springlike structure made of the contractile elements in the cuticular plate. With these structures the hair bundle may be more easily moved when external force is applied.

The displacement of the hair bundle would lead to the distortion of the membrane structure around the insertion of the individual stereocilia into the cuticular plate. Such a membrane stretching or relaxation would be able to gate the m-e.t. channel located there. From the ion selectivity study, the largest permeable cation through the m-e.t. channel was TMA or TEA (Ohmori 1985). If the diameter of these cations was taken as the diameter of the pore, the m-e.t. channel would have a 6–7-Å diameter. From the displacement vs response relationship, the smallest hair bundle displacement that generated the m-e.t. current was 100 Å when applied at 5 μm above the insertion of the hair bundle to the cuticular plate (Ohmori 1987). This displacement corresponds to the 2Å membrane deformation, if we assume a 0.2-μm diameter stereocilium. Since the m-e.t. channel is most likely anchored to the cytoskeleton, as are many membrane proteins, this 2-Å membrane deformation would probably be condensed on the channel and probably works as the mechanical gate directly on the 7 Å diameter channel.

4 Muscarinic Cholinergic Receptor Mechanism Might Mediate the Efferent Innervation to the Hair Cell

Efferent fibers make axosomatic contacts with outer hair cells, and mainly axodendritic contacts onto afferent nerve fibers with inner hair cells of mammals (Spoendrin 1984). Electrical stimulation of these efferent nerves generates inhibitory effects in hair cells and in most primary afferent nerve fibers (Galambos 1956; reviewed by Klinke and Galley 1974). Some afferent fibers receive excitatory efferent innervations (Highstein and Baker 1985). Acetylcholine (ACh) has been considered as the most likely neurotransmitter candidate at these terminals, however, the effect of ACh has not yet been unequivocally determined (Guth et al. 1976).

Shigemoto and Ohmori (1990) have applied ACh and ATP to the fura-2-loaded isolated hair cell or cluster of hair cells of the chick cochlear organ and have detected intracellular Ca-concentration changes. These Ca-concentration changes have activated Ca-activated K conductance and have hyperpolarized the membrane. The intracellular Ca concentration was increased by the extracellularly applied 100 μM ACh. Fluorescence signals at two excitation wavelengths and ratio intensities were measured at 3-s intervals. This Ca response was blocked by atropine (Fig. 8A, 10 μM, bath application) and by d-tubocurarine (Ca response was suppressed to 30% of the control at 1 mM, pressure-puff application), but was induced in the Ca-free extracellular medium (Fig. 8B). Calcitonin-gene-related peptide enhanced the ACh-induced intracellular Ca response. Muscarine and carbamylcholine induced intracellular Ca responses similar to ACh. Therefore, ACh released from efferent terminals seems to activate the muscarinic receptor mechanism and to induce the intracellular Ca response at least in chick hair cells. The outer hair cell of guinea pig cochlea did not show intracellular Ca concentration changes when ACh was puff applied, while ACh suppressed the Ca response generated by ATP in a competitive manner (Ashmore and Ohmori 1990). In contrast, ACh-induced suppression of the outward current was observed in the semicircular canal hair cell of the frog, although enhancement of the outward current was also observed in a small proportion of these hair cells (Houseley et al. 1990).

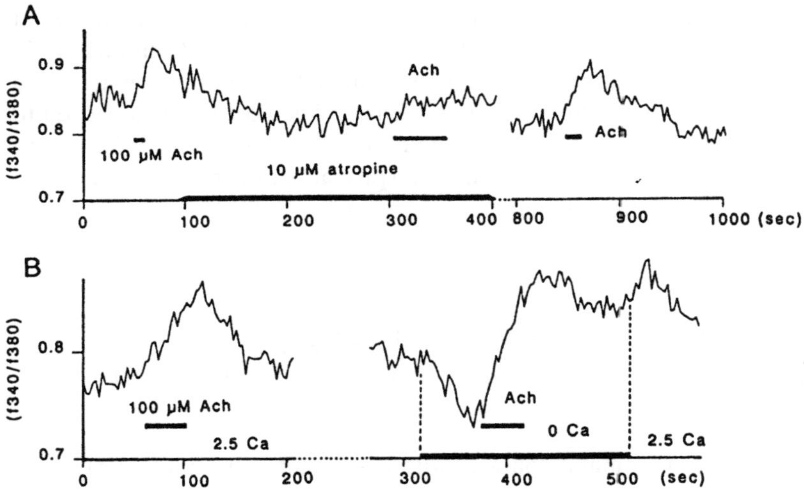

Fig. 8. Cholinergic muscarinic responses in hair cells. ACh responses are blocked by atropine and are generated in Ca-free extracellular medium. *A* A bath application of 10 μM atropine blocked most of the Ca response generated by 100 μM ACh. The application of atropine is indicated by a *bar* on the abscissa. The blocking effect was reversible. *B* Ca response was generated in Ca-free extracellular medium applied rapidly by a microperfusion method. The flow of Ca-free medium is indicated by a *bar* on the abscissa. After application of Ca-free medium the intracellular Ca concentration is decreased. A flow of 2.5 mM Ca extracellular medium introduced by suspending the microperfusion of the Ca-free solution induced a transient rise of intracellular Ca concentration. These two timings are indicated by two *vertical broken lines*. (Shigemoto and Ohmori 1990)

ATP is known to be released together with ACh at various cholinergic terminals (Burnstock 1984). ATP has generated Ca responses in chick hair cells (Shigemoto and Ohmori 1990), and in guinea pig hair cells (Ashmore and Ohmori 1990). However, the mode by which the intracellular Ca concentration increases was different from the ACh-induced Ca response. ATP induced an influx of Ca ions from the extracellular medium and also released Ca ions from the intracellular stores. ATP generates inward current, which is followed by a slowly rising outward current in the chick hair cell. An ATP-activated inward current was observed in the guinea pig outer hair cells, but the outward current was small or was not observed (Nakagawa et al. 1990; Ashmore and Ohmori 1990). The presence or absence of the outward current most likely reflects the density of Ca-activated K channels, which are abundant in chick hair cells, but are not very evident in guinea pig outer hair cells (Ohmori 1984a; Ashmore and Meech 1986).

4.1 Is Muscarinic Cholinergic Receptor a Mediator of the Efferent Synapse onto the Hair Cell?

Although it is common to induce intracellular Ca responses by ACh and ATP in the chick hair cells, there are two problems which must be explained before these two molecules or at least one of these molecules are to be considered as a neurotransmitter candidate of the efferent synapse. (1) The increase in intracellular Ca concentration would lead to the activation of the K conductance and then hyperpolarization of the membrane. This hyperpolarization might suppress the depolarizing transduction potential by increased membrane conductance and might suppress the activation of voltage-gated Ca channels. However, the increased level of intracellular Ca ions themselves has the function of releasing neurotransmitters in the other synapses (Higashida 1988). However, since the activity of the Ca-activated K channel was observed with the presence of ACh in the patch pipette, attached to the cell while the channel activity was rare when ACh was applied outside the patch membrane, the muscarinic ACh receptor is most likely localized in close association with a Ca-release mechanism and the Ca-activated K channel (Shigemoto and Ohmori, 1991). (2) The second problem is a prolonged time course of the intracellular Ca response (Fig. 8), while the hyperpolarization induced by efferent nerve activity was in the order of 100 ms (Klinke and Galley 1974). This prolonged time course of the Ca response could be due to the method of ACh application. ACh was applied by a puff method and the absolute quantity of ACh molecules could have been at a nonphysiologically high level. Neurotransmitters are released as a packet from the vicinity of postsynaptic receptors and then hydrolyzed, thus losing their activity rapidly under in vivo conditions. Such a concentrated application and the rapid breakdown of the transmitter molecule might be required to shorten the overall time course to the physiological range and could also be essential in localizing the intracellular Ca response to the particular site close to the Ca-activated K channels.

5 Length Change of the Guinea Pig Outer Hair Cell

The length of guinea pig outer hair cells could be modified by injecting the current intracellularly; applying the current extracellularly; applying ACh to the synaptic region iontophoretically; use of transmembrane potentials; and by applying ATP to the demembranated hair cells (Brownell et al. 1985; Zenner et al. 1985; Ashmore 1987; Flock et al. 1986). Only the transmembrane potential difference seems to be the active factor in controlling the length-shortening mechanism, of which detailed kinetics are yet to be discovered (Ashmore 1987). The length shortening followed the extracellular sinusoidal current flow up to 20–30 kHz (Reuter et al. 1989) and the contribution in the active sharpening of the frequency sensitivity in the inner hair cell is suggested in the guinea pig. These phenomena are not observed in the other members of the cochlear constituents of the guinea pig or in the chick hair cell.

References

Art JJ, Fettiplace R (1987) Variation of membrane properties in hair cells isolated from the turtle cochlear. J Physiol 385:207–242
Ashmore JF (1987) A fast motile response in guinea-pig outer hair cells: the cellular basis of the cochlear amplifier. J Physiol 388:323–347
Ashmore JF, Meech RW (1986) Ionic basis of membrane potential in outer hair cells of guinea pig cochlea. Nature (Lond) 322:368–371
Ashmore JF, Ohmori H (1990) Control of intracellular calcium by ATP in isolated outer hair cells of the guinea pig cochlea. J Physiol (in press)
Assad JA, Hacohen N, Corey DP (1989) Voltage dependence of adaptation and active bundle movement in bullfrog saccular hair cells. Proc Nat Acad Sci USA 86:2918–2922
Bosher SK, Warren RL (1968) Observations on the electrochemistry of the cochlear endolymph of the rat; a quantitative study of its electrical potential and ionic compositions as determined by means of flame spectrophotometry. Proc R Soc B171:227–244
Brownell WE, Bader CR, Bertrand D, Ribaupierre Y (1985) Evoked mechanical responses of isolated cochlear outer hair cells. Science 227:194–196
Burnstock G (1984) Neurotransmitters and trophic factors in the autonomic nervous system. J Physiol 313:1–35
Corey DP, Hudspeth AJ (1979) Ionic basis of the receptor potential in a vertebrate hair cell. Nature (Lond) 281:675–677
Corey D, Hudspeth AJ (1983a) Analysis of the microphonic potential of the bullfrog's sacculus. J Neurosci 3:942–961
Corey D, Hudspeth AJ (1983b) Kinetics of the receptor current in bullfrog saccular hair cells. J Neurosci 3:962–976
Crawford AC, Fettiplace R (1981) An electrical tuning mechanism in turtle cochlear hair cells. J Physiol 312:377–412
Crawford AC, Fettiplace R (1985) The mechanical properties of ciliary bundles of turtle cochlear hair cells. J Physiol 364:359–379
Crawford AC, Evans MG, Fettiplace R (1989) Activation and adaptation of transducer currents in turtle hair cells. J Physiol 419:405–434
Davis H (1965) A model for transducer action in the cochlea. Symposium on Quantal Biology, vol 30. Cold Spring Harbor, NY, USA: Cold Spring Harbor Laboratory of Quantitative Biology, pp 181–190
Dreckhahn D, Kellner J, Mannherz HG, Gröschel-Stewart V, Kendrick-Jones J, Scholey J (1982) Absence of myosin-like immunoreactivity in stereocilia of cochlear hair cells. Nature (Lond) 300:531–532

Eisenman G (1962) Cation selective glass electrodes and their mode of operation. Biophys J 2:259–323

Flock Å, Bletcher A, Weber K (1982) Immunohistochemical localization of several cytoskeletal proteins in inner ear sensory supporting cells. Hearing Res 7:75–89

Flock Å, Flock B, Uhlfendahl M (1986) Mechanisms of movement in outer hair cells and a possible structural basis. Arch Otolaryngol 243:82–90

Fuchs PA, Nagai T, Evans MG (1988) Electrical tuning in hair cells isolated from the chick cochlea. J Neurosci 8:2460–2467

Furukawa T (1986) Sound reception and synaptic transmission in goldfish hair cell. Jpn J Physiol 36:1059–1077

Furukawa T, Ishii Y (1967a) Neurophysiological studies on hearing in goldfish. J Neurophysiol 30:1377–1403

Furukawa T, Ishii Y (1967b) Effects of static bending of sensory hairs on sound reception in the goldfish. Jpn J Physiol 17:572–588

Galambos R (1956) Suppression of auditory nerve activity by stimulation of efferent fibres to cochlea. J Neurophysiol 19:424–437

Grynkiewicz G, Poenie M, Tsien RY (1985) A new generation of Ca^{2+} indicators with greatly improved fluorescence properties. J Biol Chem 260:3440–3450

Guth PS, Norris CH, Bobbin RP (1976) The pharmacology of transmission in the peripheral auditory system. Pharmacol Rev 28:95–125

Hermann A, Gorman ALF (1984) Action of quinidine on ionic currents of molluscan pacemaker neurons. J Gen Physiol 83:919–940

Higashida H (1988) Acetylcholine release by bradykinin, inositol 1,4,5-triphosphate and phorbol dibutyrate in rodent neuroblastoma cells. J Physiol 397:209–222

Highstein SM, Baker R (1985) Action of the efferent vestibular system on primary afferents in the toadfish, *Opsanus tau*. J Neurophysiol 54:370–384

Hodgekin AL, Huxley AF (1952) A quantitative description of membrane current and its application to conduction and excitation in nerve. J Physiol 117:500–544

Holton T, Hudspeth AJ (1986) The transduction channel of hair cells from the bull-frog characterized by noise analysis. J Physiol 375:195–227

Houseley GD, Norris CH, Guth PS (1990) Cholinergically-induced changes in outward currents in hair cells isolated from the semicircular canal of the frog. Hearing Res 43:121–134

Howard J, Hudspeth AJ (1987) Mechanical relaxation of the hair bundle mediates adaptation in mechanoelectrical transduction by the bullfrog's saccular hair cell. Proc Natl Acad Sci USA 84:3064–3068

Howard J, Roberts WM, Hudspeth AJ (1988) Mechanoelectrical transduction by hair cells. Annu Rev Biophys Biophys Chem 17:99–124

Hudspeth AJ (1982) Extracellular current flow and the site of transduction by vertebrate hair cells. J Neurosci 2:1–10

Hudspeth AJ, Corey DP (1977) Sensitivity, polarity and conductance change in the response of vertebrate hair cells to controlled mechanical stimuli. Proc Natl Acad Sci USA 74:2407–2411

Hudspeth AJ, Lewis RS (1988) A model for electrical resonance and frequency tuning in saccular hair cells of the bull-frog, *Rana catesbeiana*. J Physiol 400:275–297

Jørgensen F, Ohmori H (1988) Amiloride blocks the mechano-electrical transduction channel of hair cells of the chick. J Physiol 403:577–588

Klinke R, Galley N (1974) Efferent innervation of vestibular and auditory receptors. Physiol Rev 54:316–357

Kroese ABA, Das A, Hudspeth AJ (1989) Blockage of the transduction channels of hair cells in the bullfrog's sacculus by amino-glycoside antibiotics. Hearing Res 37:203–218

Lamb DMK (1972) Biosynthesis of acetylcholine in turtle photoreceptors. Proc Natl Acad Sci USA 69:1987–1991

Lewis RS, Hudspeth AJ (1983) Voltage- and ion-dependent conductances in solitary vertebrate hair cells. Nature (Lond) 304:538–541

Nakagawa T, Akaike N, Kimitsuki, Komune, Arima T (1990) ATP-induced current in isolated outer hair cells of guinea pig cochlea. J Neurophys (in press) 63:1068–1074

Neher E, Sakmann B (1976) Single channel currents recorded from membrane of denervated frog muscle fibres. Nature (Lond) 260:779–802

Nowyky MC, Fox AP, Tsien RW (1985) Three types of neuronal calcium channel with different calcium agonist sensitivity. Nature (Lond) 316:440–443
Ohmori H (1978) Inactivation kinetics and steady-state current noise in the anomalous rectifier or tunicate egg cell membranes. J Physiol 281:77–99
Ohmori H (1980) Dual effects of K ions upon the inactivation of the anomalous rectifier of the tunicate egg cell membrane. J Membr Biol 53:143–156
Ohmori H (1984a) Studies of ionic currents in the isolated vestibular hair cell of the chick. J Physiol 350:561–581
Ohmori H (1984b) Mechanoelectrical transducer has a discrete level of conductance in the vestibular hair cell of chick. Proc Natl Acad Sci USA 81:1888–1891
Ohmori H (1985) Mechano-electrical transduction currents in isolated vestibular hair cells of the chick. J Physiol 359:189–217
Ohmori H (1987) Gating properties of the mechano-electrical transducer channel in the dissociated vestibular hair cell of the chick. J Physiol 387:589–609
Ohmori H (1988) Mechanical stimulation and fura-2 fluorescence in the hair bundle of dissociated hair cells of the chick. J Physiol 399:115–137
Pickles JO, Comis SD, Osborne MP (1984) Cross-links between stereocilia in the guinea-pig organ of Corti, and their possible relation to sensory transduction. Hearing Res 15:103–112
Rabié A, Thomasset M, Legrand CH (1983) Immunocytochemical detection of calcium-binding protein in the cochlear and vestibular hair cells of the rat. Cell Tissue Res 232:691–696
Reuter G, Gitter AH, Zenner HP (1989) Electrical induced motility of outer hair cells in vitro and in situ. Proc Int Union Physiol Sci XVII, Helsinki 4429 pp
Rhode WS (1971) Observations of the vibrations of the basilar membrane in squirrel monkeys using Mössbauer technique. J Acoust Soc Am 49:1218–1231
Russell IJ, Sellick PM (1983) Low-frequency characteristics of intracellularly recorded receptor potentials in guinea-pig cochlear hair cells. J Physiol 338:179–206
Russell IJ, Richardson GP, Cody AR (1986) Mechanosensitivity of mammalian auditory hair cells in vitro. Nature (Lond) 321:517–519
Sand O (1975) Effects of different ionic environments on the mechano-sensitivity of lateral line organs in the mudpuppy. J Comp Physiol 102:27–42
Santi P, Anderson CB (1987) A newly identified surface coat on cochlear hair cells. Hearing Res 27:47–66
Sellick PM, Partuzzi R, Johnstone BM (1982) Measurement of basilar membrane motion in the guinea pig using the Mössbauer technique. J Acoust Soc Am 72:131–141
Shigemoto T, Ohmori H (1990) Muscarinic agonists and ATP increase the intracellular Ca^{2+} concentration in chick cochlear hair cells. J Physiol 420:127–148
Shigemoto T, Ohmori H (1991) Muscarinic receptor hyperpolarizes cochlear hair cells of chick by activating Ca^{2+}-activated K^+ channels, J Physiol 442:669–690
Spoendrin H (1984) Efferent innervation of the cochlea. In: Bolis L, Keynes RD, Maddrell SHP (eds) Comparative physiology of sensory systems. Cambridge Univ Press, Cambridge, UK, pp 163–188
Strelioff D, Flock Å (1984) Stiffness of sensory-cell hair bundles in the isolated guinea-pig cochlea. Hearing Res 15:19–28
Sugihara I, Furukawa T (1989) Morphological and functional aspects of two different types of hair cells in the goldfish sacculus. J Neurophysiol 62:1330–1343
von Békesy G (1960) Experiments in hearing. McGraw-Hill, New York
Wilson JP, Johnstone JR (1975) Basilar membrane and middle ear vibration in the guinea pig measured by capacitive probe. J Acoust Soc Am 57:705–723
Yanagisawa K, Horikoshi T, Yoshioka T, Sokabe M (1988) Localization of phosphatidylinositol 4,5-diphosphate on the stereocilia of cochlear hair cells. In: Hamann W, Iggo A (eds) Progress in Brain Research, vol 74. Elsevier, Amsterdam, pp 313–317
Zenner HP, Zimmermann U, Schmidt U (1985) Reversible control of isolated mammalian cochlear hair cells. Hearing Res 18:127–134

Chapter 11
Muscle Mechanoreceptors in Invertebrates

B. Rydqvist[1]

Contents

[1]Department of Physiology, Karolinska Institutet, S-10401 Stockholm, Sweden

Advances in Comparative and Environmental Physiology, Vol. 10

1 Introduction

Most mobile animals including the invertebrates control and coordinate body movement by "pattern generators" located in the central nervous system. Equally important, however, is the adjustment and coordination of movements in response to different environmental demands. This part of the coordination is dependent on different receptor structures, many of them in close relation to the muscles and the moving parts of the animal.

Invertebrates constitute the most numerous group among living organisms, ranging from unicellular organisms to highly mobile species among the arthropods. Free-moving species begin to appear among the cnidarians and ctenophorans, and starting from the platyhelmintes and annelids, proper motor control is necessary. However, in the arthropods (phyla Crustacea, Chelicerata and Uniramia) the complexity of the body necessitates an integrated control of movements, which is determined by a combination of central nervous system generators and mechanoreceptors associated with muscles and other structures. The arthropods are also the group of invertebrates containing the largest variety and number of muscle receptor organs (MRO) and other proprioceptors.

In the invertebrates, despite a "simpler" organization of the nervous system, the motor control follows the same general pattern as in the vertebrates. In analogy with the vertebrate muscle spindle, the invertebrates have developed several devices such as muscle receptor organs (MRO) that can control muscle length and tension. Examples of such MROs are the abdominal and thoracic MRO in the crustacean (Alexandrowicz 1951), the thoracic-coxal muscle receptor organ (TCMRO) as described in the leg of decapod crustaceans (Alexandrowicz and Whitear 1957) and the myochordotonal organ first described by Barth (1934).

The main reasons why invertebrate systems have attracted so much attention include (1) their functional analogy to the muscular control system in vertebrates: and (2) their large and peripherally located MROs which make it easier to apply different neurophysiological methods. In particular, the analysis of the primary processes of the mechanotransduction in many invertebrate MRO systems has made considerable progress during the last 10 years – even down to the level of stretch-activated ion channels.

The scope of the present review is thus to compare the structure and function of invertebrate muscle receptors, with special emphasis on the process of mechanotransduction in the arthropod muscle receptors. The limits are set to receptors in close contact with muscle tissue and for comprehensive reasons I have omitted many other mechanoreceptors in the skin or cuticle.

It has not been possible to cite all studies in the field; when possible, review articles have been used. A general reference to invertebrate nervous system is Bullock and Horridge (1965) and several excellent reviews focus on different aspects of proprioception in invertebrates such as Pilgrim (1960), Pringle (1961), Alexandrowicz (1967, 1972), Osborne (1970), Finlayson (1968), Bush (1981), Bush and Laverack (1982). The volume edited by Mill (1976) includes many articles relevant to this review.

2 Occurrence, Location and Gross Morphology of Muscle Receptors

Table 1 shows some important invertebrate phyla and the occurrence of different receptor organs related to muscle control.

2.1 Lower Phyla

In the Placozoan and the Poriferan phyla, the tissue is regarded to be poorly defined and regular muscle tissue, hence muscle receptors, are absent (Bullock and Horridge 1965). In the cnidarians and ctenophorans, muscle cells are present, but muscle receptors have not been identified.

In the phylum Platyhelminthes (flatworms) the first invertebrate centralized nervous system with a distinct brain occurs. As in the nemerteans and nematodes, muscle cells are present and arranged in a well-defined layer. In some species the dendrites of bipolar sensory cells project into muscle fibers. These neurons are probably sensory and sensitive to mechanical deformation (Bullock and Horridge 1965).

Table 1. Occurrence of mechanoreceptors and muscle receptors in some important invertebrate phyla[a]

Phylum	No. of species	Examples: classes, order families	Mechanoreceptors[b]
Cnidaria	9000	Jellyfish, sea anemones, corals	MeR in cnidocytes,
Platyhelminthes	12 700	Flatworms	MeR in longitud. muscle
Nematoda	10 000	Roundworms	–
Mollusca	100 000	Snails, clams, octopods, squids	Octopus substellar organ
Annelida	8 700	Segmented worms *Lumbricus terrestis* leech	Multipolar cells in longitud. muscle, leech MRO
(Arthropoda: Phyla Crustacea, Chelicerata, Uniramia)			
Crustacea	39 000	Crayfish, lobster, crab, shrimp, barnacles, wood lice	MRO, MCO, CO, TCMRO, AR
Chelicerata	63 000	Spiders, scorpions horseshoe crab	MRO, MCO, AR, SS
Uniramia	1 000 000	Millipeds, centipeds insects	MRO, MCO, CO, AR, CS

[a]Taxa and number of species according to Barnes (1984).
[b]MeR = mechanoreceptors; MRO = muscle receptor organ; MCO = myochordotonal organ; CO = chordotonal organ; TCMRO = Thoracic-coxal muscle receptor (thorax coxopodite joint); AR = apodeme receptor; CS = campaniform sensilla; SS = slit sensilla.

Phylum Annelida. Electrophysiological recordings from the body wall of centrally isolated preparations of *Nereis* and *Lumbricus* have revealed the presence of receptors which respond to an applied stretch with a train of impulses that adapt relatively slowly. However, little morphological evidence for MROs is present (see Dorsett 1976). In the leech (*Hirudo medicinalis*) Blackshaw and Thompson (1988) have described a muscle receptor organ having a peripheral cell body between the longitudinal and oblique muscle of the body wall. The two dendritic areas situated on both side of the soma are fan-shaped and associated with the longitudinal muscle of the ventral wall (Fig. 1).

2.2 Phylum Mollusca

In gastropods, little is known about muscle receptors or proprioception (Bullock and Horridge 1965). In several cephalopods a muscle receptor organ, the so-called substellar organ, is situated under the nerves radiating from the stellate ganglion. Alexandowicz (1960) first described such an organ in the mantle of *Eledone* as consisting of many multipolar neurons scattered over the muscles. Their presence

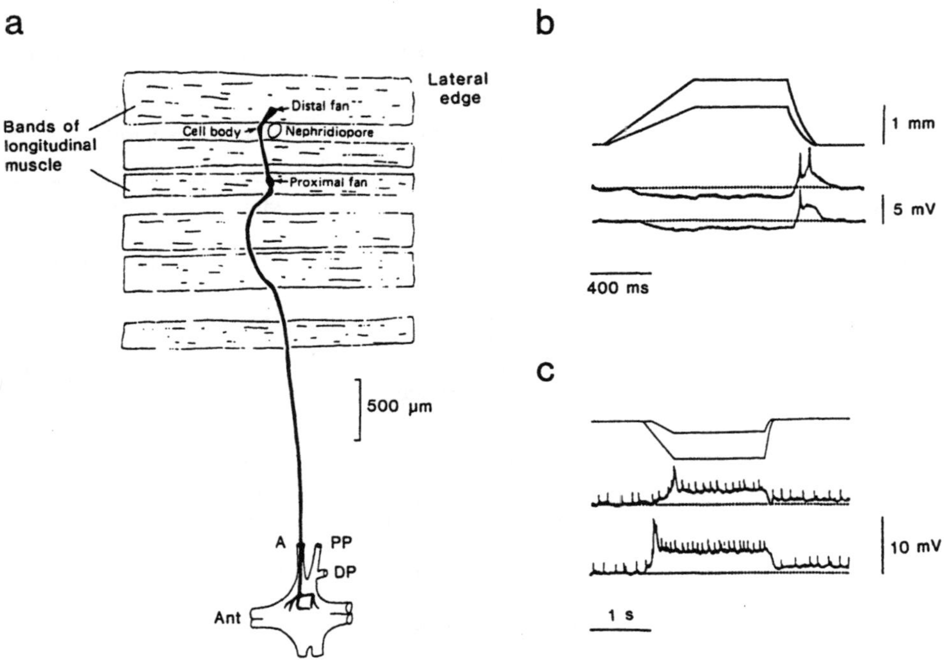

Fig. 1a. Muscle receptor neuron in the ventral longitudinal muscle of the leech (*Hirudo medicinalis*, Annelida). The cell body is located between the fan-shaped sensory terminals (50 µm) some 4 mm from the ganglion. *b* Hyperpolarizing response due to extension of the longitudinal muscle. *c* Depolarizing response due to release of stretch of the longitudinal muscle. (Blackshaw and Thompson 1988)

has also been demonstrated at a number of sites in the arms and suckers of *Octopus*. Behavioural observations do suggest that the octopus makes use of stretch information and expresses stretch reflexes (see Wells 1978). The substellar organ is similar to the so-called N-cells in thoracic muscles of some crustaceans (see below).

2.3 Phylum Crustacea

The largest amount of information on MROs is available for the decapod crustaceans and knowledge about these receptors has recently been reviewed by Bush and Laverack (1982). Three important types of muscle receptors can be found in crustaceans:

1. The dorsal abdominal and thoracic muscle receptor organ (MRO), was first described by Alexandrowicz (1951, 1952a, 1967) in the lobster *Homarus vulgaris* and *Palinurus vulgaris* and later in Paguridae (1952b), *Squilla mantis* (Stomatopoda) (1954), *Leander serratus* (1956). Florey and Florey (1955) subsequently described the same type of organ in the crayfish *Astacus*. Identical organs can be found in most decapod crustaceans, for example crabs, shrimps, etc. (see Table 1, Pilgrim 1969; Alexandrowicz 1967; Bush and Laverack 1982). In each abdominal segment there are two pairs of receptor organs (MRO1 and MRO2) located in the extensor muscles on each side (Fig. 2a). Each organ is composed of a receptor muscle to which is associated a sensory neuron. The receptor muscles is attached to intersegmental folds at each segment and the axons from the neurons join the dorsal segmental nerve to the ganglion. The MROs also receive efferent innervation: (a) one or two motor axons to the receptor muscles and (b) two or three accessory axons conveying inhibitory signals to the receptor muscle and the sensory neurons (Alexandrowicz 1951, 1967; Kuffler and Eyzaguirre 1955; Elekes and Florey 1987a).

   A similar organ as in the decapod crustaceans is also found in the order Isopoda (Moser 1976) and in the amphipodans (Alexandrowicz 1967). The general organization and structure seems to be the same in most crustaceans, but small differences exist in their location (for details see Bush and Laverack 1982).
2. The thoracic-coxal muscle receptor organ (TCMRO) is situated at the base of each leg in the thoraco-coxal joint (Fig. 3). This organ was first described in several decapod crustaceans (*Carcinus, Pagurus, Astacus* and *Homarus*) by Alexandrowicz and Whitear in 1957 and has since been the subject of extensive morphological and physiological studies (see Whitear 1965; Bush et al. 1971; Bush 1976, 1981; Bush and Laverack 1982). A receptor muscle lies parallel to the promotor muscle which moves the leg forward at the thoracocoxal joint. The receptor muscle is innervated by three sensory fibres with their cell bodies in the thoracic ganglion (See Sect. 3) and the organ is activated, i.e. stretched, by passive or active retraction of the leg and inactivated (shortened) with leg promotion. Parallel to the TCMRO is found a nonmuscular elastic strand (depressor receptor) parallel to the receptor muscle. In crabs (Brachyura)

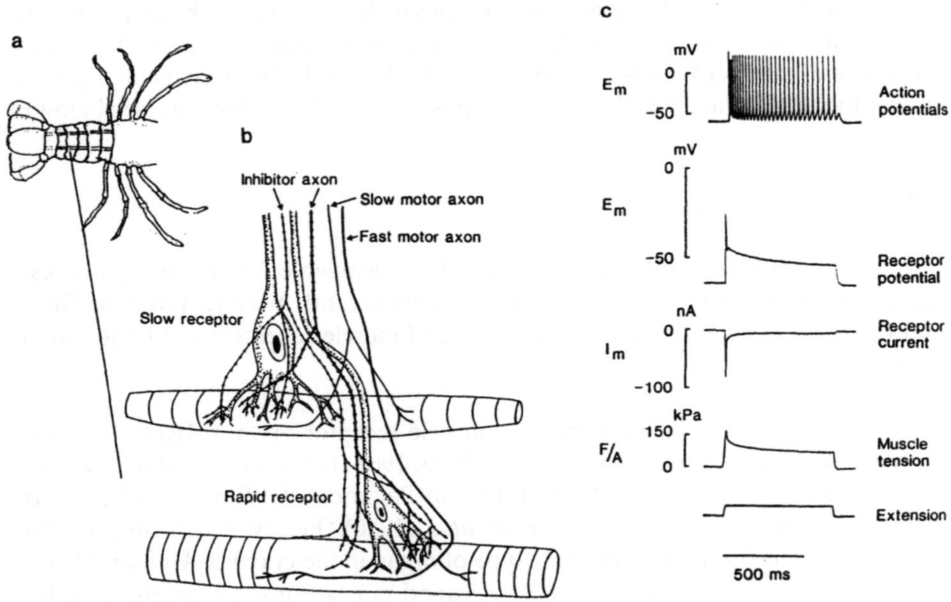

Fig. 2. Abdominal muscle receptor organ of the Crustacean. *a* Location in the abdomen. *b* Schematic drawing of the slowly adapting (MRO1) and the rapidly adapting (MRO2) receptor organs and their innervation. *c* Tension, current, potential and impulse response from a slowly adapting crayfish stretch receptor to ramp and hold extension. Receptor potential after treatment with tetrodotoxin. In many organs a third inhibitory fiber is present. *F/A* force/receptor muscle cross-sectional area; I_m membrane current; E_m membrane potential

there is also a levator receptor (Cannone and Nijland 1989). In *Astacura* a chordotonal organ is located parallel to the muscle receptor. Functionally, the receptor organ has been most thoroughly studied in the crabs (Brachyura): *Carcinus maenas* (see Bush 1981); *Callinectes sapidus* (Blight and Llinas 1980); *Scylla serrata* (Mirolli 1979a,b).

3. Another extensively studied muscle receptor in Decapoda is the myochordotonal organ (MCO) at the ischio-meropodite joint of the walking leg. First described by Barth (1934), this organ, which has been identified in *Astacura, Palinura, Anomura* and *Brachyura* with some variation (see Mill 1976), is associated with the accessory flexor muscle. The MCO is a special case of the chordotonal organ found at all joints of decapod walking legs, in antennule and antenna, chelipeds and other appendages including the mandibles (see Pringle 1961; Mill 1976; Bush and Laverack 1982). The organ and other proprioceptors in the crustacean walking leg make up an analog to the vertebrate muscle spindle and tendon organ systems (Cohen 1965). For a comparative discussion of the proprioceptors of the decapod crustacean see also Alexandrowicz (1972).

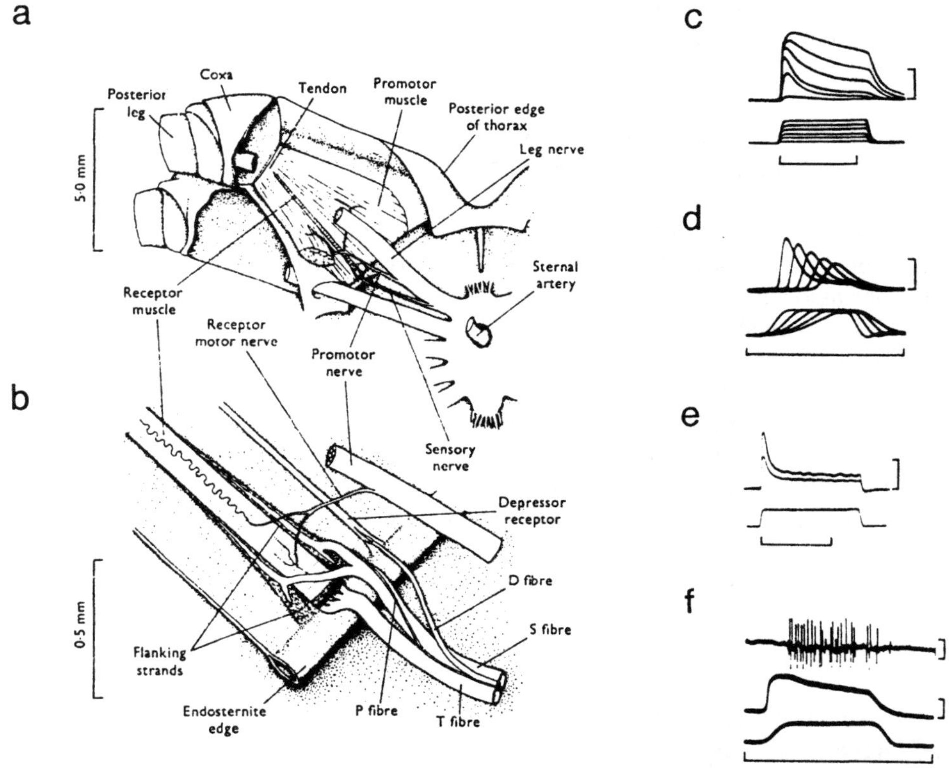

Fig. 3. Thoracic-coxal MRO (TCMRO) of the crab (*Carcinus maenas*). *a* Anatomy of TCMRO of right posterior leg in situ (dorsal view). *b* Base of TCMRO and its innervation. *c* Receptor potentials at different amplitude, D fiber. *d* Receptor potential at different ramp velocities, T fiber, *e* intracellular records from two electrodes 2 mm apart in a T fiber. *f* Promotor reflex and T fiber response to extension of receptor muscle. (*a, b, c, d, e* From Bush and Roberts 1971; *f* from Bush and Roberts 1968)

2.4 Phylum Chelicerata

Barber and Hayes (1964) described a tendon receptor organ in the trochanter of *Limulus* walking leg to which is associated two accessory muscles. The organ monitors the movement of the coxa-trochanter or the trochanter-femur joint.

Slowly and rapidly adapting MROs in the first four postabdominal segments of the scorpion *Centruroides gracilis*, similar to the crustacean MRO, have been described by Bowerman (1972). The MROs are stretched by flexion and released by extension of the posterior joint as in the crustacean abdominal MRO.

2.5 Phylum Uniramia

Subphylum Myriapoda. In centipedes (Chilipoda) segmentally organized muscle receptor organs have been described in *Lithobius, Haplophilus, Scolopendra* (for review see Finlayson 1976). The MROs are bi- or multisegmentally organized and in general up to 13 sensory neurons (bi- or multipolar) are found on each receptor muscle.

Subphylum Hexapoda (Insects). Several muscle receptors both in the abdomen and the walking legs of insects have been studied. Finlayson and Loewenstein (1958) compared several insect proprioceptors in Orthoptera, Odonata, Hymenoptera and Lepidoptera and concluded that there is a series of different proprioceptor organs, starting with neurons associated with the body wall, to neurons associated with connective tissue strands, and finally to neurons in close association with abdominal musculature. Abdominal and thoracic stretch receptors have now been identified in most orders of insects in both adult and larval stages (see Table 4.1, Finlayson 1976), but only in Lepidoptera (butterflies and moths, Fig. 4), Trichoptera (caddies-flies) and Neuroptera (ant lions) have true MROs resembling the crustacean MRO been identified (Osborn 1970).

There are relatively few MROs in the walking legs of insects, but in each leg of the honeybee there is an MRO (Markl 1965). The receptor muscle consists of four to six fibers inserted on the anterior proximal wall of the coxa and on the joint membrane at the junction of coxa and trochanter.

3 Fine Structure and Organization of Muscle Receptor Organs

3.1 Sensory Neurons with Dendrites on Ordinary Muscles

The most simple type of MRO consists of a multipolar or unipolar neurons associated with the skeletal muscle in question. This type of muscle receptors can be found in Annelida, (Bullock and Horrige 1965; Blackshaw and Thompson 1988) and on the inside of the mantle and arm muscles of *Octopus* and *Eledone* associated with a special thin layer of muscle (Alexandrovicz 1960; Wells 1978). In crustaceans, multiterminal neurons, so-called N cells, are found, with endings on thoracic muscles (Alexandrovicz 1952a; Wiersma and Pilgrim 1961). These end cells are probably degenerated muscle receptor organs in which the specialized receptor muscle (RM) has disappeared (Fig. 4c).

3.2 Types of Sensory Neurons and Their Relation to Accessory Structures

Pringle (1961) has divided mechanical receptor neurons into two main types: Type I has single sensory processes (uniterminal) and is mainly associated with cuticular components. The distal process normally contains ciliary components and is associated with auxiliary structures such as the scolopidia.

Type II are multipolar or multiterminal neurons like the abdominal MRO of the crustaceans with many fine dendritic branches, but no ciliary structure has been

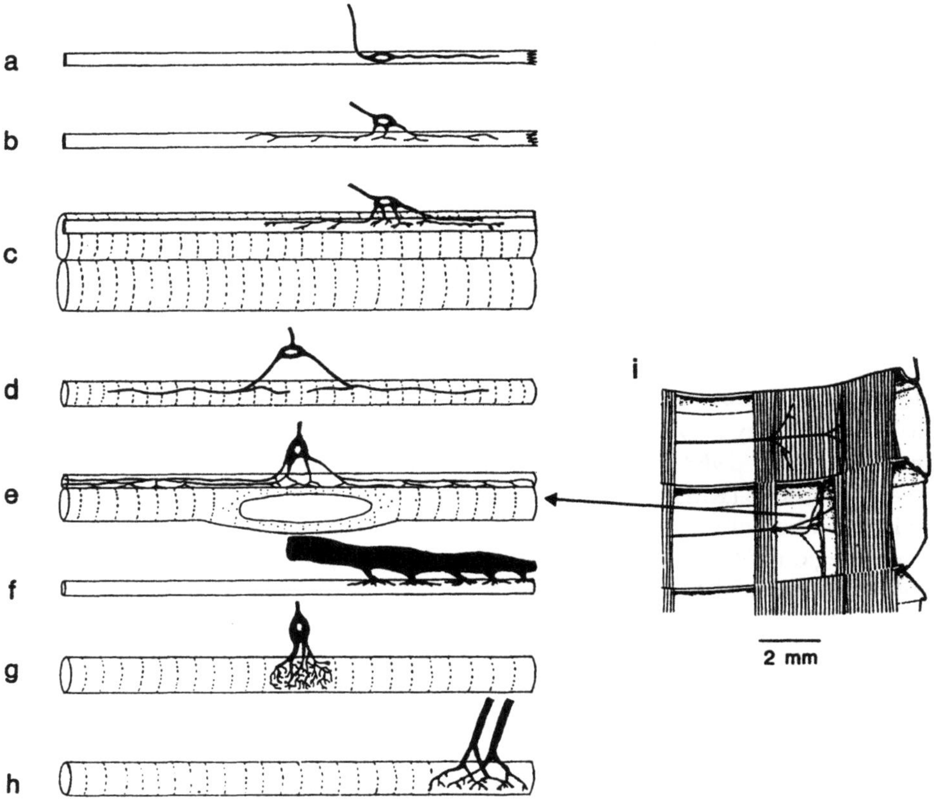

Fig. 4. Schematic representation of the relation between sensory neurons and accessory structures such as connective tissue and muscles in arthropods. *a-e* Insects; *f-h* crustacea. *a* Connective tissue strand, bipolar neuron (Plecoptera); *b* connective tissue strand, multipolar neuron (Odonata, Dictyoptera, Phasmida, Dermaptera, Hymenoptera, Diptera); *c* connective tissue on longitudinal muscle band (Orthoptera); *d* receptor muscle and multiterminal neuron, (Neuroptera); *e* connective tissue and receptor muscle with giant nucleus (Trichoptera, Lepidoptera); *f* connective tissue, neuron with cell body in CNS (Crustacea); *g* receptor muscle with central connective tissue part, "intercalated tendon" (abdominal MRO in Crustacea); *h* receptor muscle connective tissue terminal in region of dendrites, neuron in CNS (TCMRO in Crustacea); *i* location of MRO in the pupa abdomen of the moth (Lepidoptera). (*a-h* from Finlayson (1968); *i* from Finlayson and Loewenstein (1958)

found. The uniterminal neuron in the stretch receptor of *Dictyopterygella* as seen in Fig. 4a is probably a degenerated multipolar cell (Finlayson 1968).

Several nonmuscular stretch receptors made up of multiterminal neurons and associated with elastic strands of connective tissue (Fig. 4a,b,f) can be found in arthropods.

Muscle receptor organs in uniramians and crustaceans consist mainly of multipolar neurons located close to a muscle strand or a combination of elastic and muscle strand (Fig. 4d,e,g,h). One exception is the TCMRO (Fig. 4h) which has its neurons in the central nervous sytem.

The muscle receptor organ of Lepidoptera (Fig. 4e,i) resembles the crustacean MRO and consists of a multiterminal neuron covered by glial cells. The terminals ramify in a connective tissue sheet close to a modified muscle having a giant nucleus (Finlayson 1976). The muscle is innervated but lacks inhibitory efferent innervation (Osborne 1970).

3.3 Abdominal and Thoracic MROs of Crustaceans

Different aspects of the fine structure of the crustacean MROs (Figs. 2, 4g) have been studied in *Procambarus* (Bodian and Bergman 1962; Nakajima et al. 1973; Komuro 1981; Hirosawa et al. 1981; Tao-Cheng et al. 1981a,b); *Astacus* (Florey and Florey 1955; Euteneuer and Winter 1979); *Orconectes* (Peterson and Pepe 1961; Elekes and Florey 1987a,b); *Homarus* (Uchizono 1967; Alexandrowicz 1967; Nadol and DeLorenzo 1968). In *Procambarus*, both the slowly and rapidly adapting receptor muscles (RM1 and RM2) consists of a single muscle fiber which is divided by invagination of the cell membrane into numerous cytoplasmic processes in the central region of the muscle (the intercalated tendon), which is mainly made up of collagen. Some of the myofibrils insert in this region but some pass the intercalated tendon. The intermediate region is more than twice as long in RM1 (280 μm) as in the RM2 (Alexandrowicz 1951; Komuro 1981), and RM1 is shorter and thinner (30–80 μm) compared to RM2 (70–150 μm diameter). On the fibrillar level, the RM1 is characterized by thick filaments and long sacromeres (6.5 μm), whereas RM2 resembles fast arthropod muscles and have short sarcomeres (3.3 μm) (Komuro 1981). This general arrangement also holds for other species (Bodian and Bergman 1962; Alexandrowicz 1967; Nadol and DeLorenzo 1968; Nakajima and Reese 1983). In *Squilla mantis*, the two RMs are fused together in their central region (Alexandrowicz 1967). The sensory neurons (SN1, SN2) are large (30–100 μm), multiterminal cells (Type II, Pringle 1961) of mainly pyramidal to fusiform shapes, normally varying in form between different organs (see Florey and Florey 1955). They contain a nucleus (ca. 10 μm) with a clear nucleolus. The dendrites branch about four to five times and intermingle with the connective and muscle strands in the intercalated tendon. The main dendritic branches usually are about 10–20 μm in diameter. The fine terminal branches are about 2 μm long and about 0.1 μm in diameter devoid of mitochondria (Tao-Cheng et al. 1981a). The axon is in the order of 30 μm in diameter. Except for the dendritic tips, the receptor neurons of the crayfish have several layers of sheet cells that surround it.

Inhibitory Efferent Innervation. The fine structure of the inhibitory synapses has recently been investigated by Hirosawa et al. (1981), Nakajima and Reese (1983) and Elekes and Florey (1987a,b). The transmitter substance GABA (γ-amino butyric acid) was first identified by Bazemore et al. (1957) and was further studied by Kuffler and Edwards (1958), Edwards and Kuffler (1959). Recent studies (Elekes and Florey 1987a,b) of the inhibitory synapses and their interactions using serial sectioning and immunohistochemical technique have revealed a complicated array of synapses including (1) inhibitory synapses on the axon hillock and the initial segment; (2) serial synaptic terminals on the sensory neuron; (3) simultaneous synaptic contacts of the same inhibitory terminal with sensory dendrites and muscle fibres; (4) reciprocal synapses between the two types of inhibitory efferents; and (5) inhibitory synapses on the primary inhibitory axon.

3.4 Thoracic-Coxal Muscle Receptor Organ (TCMRO)

The fine structure of TCMRO has been studied in *Astacus, Carcinus,* and *Pagurus* by Whitear (1965), and in *Cancer* by Krauhs and Mirolli (1975). A recent review on the TCMRO is given by Bush and Laverack (1982) and the fine structure will be briefly summarized here (see Figs. 3, 4h). The receptor muscle originates on thoracic structures, and inserts on the anterior part of the coxa. The muscle strand is a cylindrical bundle of muscle fibers ensheathed in connective tissue (collagen) forming flanking strands dorsally and ventrally. The muscle fibers resemble slow, tonic crustacean fibers (Bush 1976). The motor innervation of the muscle is by small branches from the promotor nerve that gives off several fibers along the muscle. In *Carcinus* the sensory innervation is made up of two large diameter (50–100 μm) nerve fibers (S and T) and one or two small fibers (P and D). The cell bodies are located in the thoracic ganglion up to 10 mm from the receptor. The nerve fibers are covered by sheet cells (Bush 1976). In the crab, the T fiber inserts its dendrites in a connective tissue tendon of the proximal part of the receptor muscle. The dendritic tips are about 0.1 μm in diameter and several millimeters long (cf. abdominal MRO). The S fiber divides into two main parts which innervate the flanking strands and branch into small dendrites that run along the flanking strands. The P fiber enters the same area as the ventral S fiber. The depressor receptor is innervated by the relatively thin D fiber.

3.5 Myochordotonal Organ (MCO)

The myocordotonal organs (MCO) consists of multiple (up to 100) uniterminal or bipolar sensory neurones embedded in a connective tissue that connects to a muscle. The distal process of the neuron is in its end surrounded by one or more scolopale cells and this complex is further covered by an envelope cell. This complex make up the ovoid scolopidia, hence the name "scolopidial organ" (see Moulins 1976 for details).

4 Functional Properties of Muscle Receptor Organs

4.1 General Behaviour

One of the first studies on invertebrate MRO functions was done on the crayfish and lobster stretch receptor (Wiersma et al. 1953). In situ the receptors are activated, i.e. stretched, by flexion of the abdomen or activation of the receptor muscles (Kuffler 1954) and are involved in the control of the extensor muscles (Fields 1976). Subsequent studies (Eyzaguirre and Kuffler 1955a,b; Edwards and Ottoson 1958) on *Homarus, Panulirus, Procambarus, Astacus* and *Orconectes* showed that stretching the MRO gave rise to an impulse discharge from two components: one adapting slowly (MRO1) and one adapting rapidly (MRO2). The MRO1 can generally maintain firing for several hours. In situ there is a marked difference in threshold for activation, the slowly adapting (MRO1) having the lowest threshold. Stimulation of the motor nerve to the receptor muscles (RM1, RM2) gave rise to an increased discharge frequency from the receptor neurons (Kuffler 1954; Kuffler and Eyzaguirre 1955; Brown 1967). Intracellular recordings using tetrodotoxin to block impulse generation showed that the stretch applied resulted in a membrane depolarization, the receptor or generator potential (Fig. 2). Similar studies on the TCMRO (Ripley et al. 1968) showed that in this organ the sensory signal was conducted or electrotonically spread to the ganglion without impulses (for review see Bush 1981; Bush and Laverack 1982 and below). A similar behaviour was seen in insect MROs (Lepidoptera, Weevers 1966). A generalized muscle receptor organ can be represented by four functional blocks as illustrated in Fig. 5. The first step represents the transformation of strain to mechanical force or stress in the accessory structure. The second step is the deformation of the dendrites of the sensory neuron leading to an ionic flux through opening of stretch-activated ion channels (SA channels). As a consequence, a receptor current is generated giving rise to the receptor potential. The receptor potential is a function of the receptor current, the capacitative and resistive properties of the cell membrane, but other factors contribute to the transformation of current to membrane potential, including the geometry of the cell, as well as voltage-dependent ion channels that are present in the dendrites, soma, and axon of the sensory neuron. Transformation of current to potential would thus constitute the third step. Finally, the generation of action potentials (through Na^+ and K^+ channels) somewhere in the neuron may give rise to a frequency modulated response, transmitted to the central nervous system. However, in the thoracic-coxal muscle receptor organ (TCMRO), the sensory signal evoked by stretch is conducted by electrotonic spread to the central nervous system (Ripley et al. 1968; Bush 1976, 1981; Cannone 1987; Cannone and Nijland 1989). The decrementally conducted signal is able to elicit impulse discharges in motor neurons supplying the promotor muscle (Fig. 3f) situated in parallel with TCMRO (Bush and Roberts 1968). In the MRO of the leech, action potentials

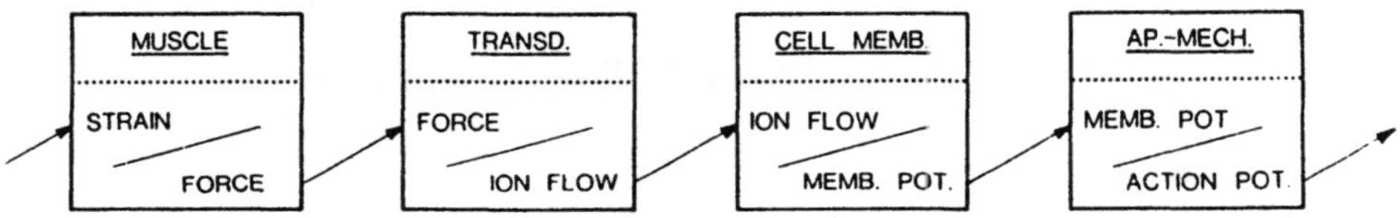

Fig. 5. Functional representation of a muscle receptor organ (MRO). *Transd.* Transducer; *Cell memb* membrane with ion channels, geometry; *Ap-mech* action potential mechanism-ion channels

generated in the soma are not actively propagated to the CNS, but decrementally spread to the ganglion (Blackshaw and Thompson 1988).

4.2 Functional Properties of Accessory Structures

4.2.1 Viscoelastic Properties

The viscoelastic properties of the crayfish stretch receptor muscle have been studied by Krnjevic and van Gelder (1961), Brown and Stein (1966), Nakajima and Onodera (1969b), Rydqvist et al. (1990). At present, no unifying description of the viscoelastic properties of the receptor muscle is available, but the dynamic properties of the slowly adapting receptor muscle (RM1) in the crayfish can be adequately described by a viscoelastic model consisting of a Voigt element in series with a nonlinear spring (Fig. 6). The power coefficient (n) for this spring is in the order of 1.5–2.0. A similar model was used to describe the dynamic properties of the

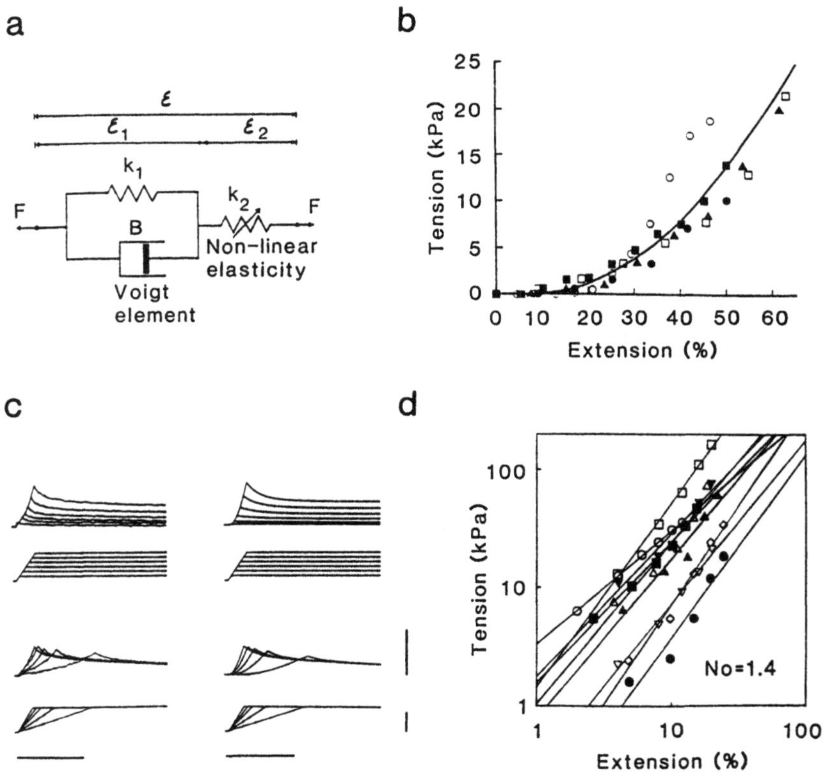

Fig. 6. Viscoelastic properties of slowly adapting abdominal MRO in *Astacus astacus*. *a* Viscoelastic model. *b* Steady state length tension relationship. *c* Comparison of recorded (*left column*) and modelled responses (*right column*) to ramp and hold extensions. *d* Log-log plot of peak tension response to ramp and hold extensions. (Rydqvist et al. 1990)

TCMRO muscle (Berger and Bush 1979). Only minor differences in viscoelastic properties between the slowly and rapidly adapting receptor muscle of the crayfish MRO have been found (Nakajima and Onodera 1969b).

4.2.2 Electrical and Contractile Properties

Kuffler (1954) and Kosaka (1969) both found excitatory and inhibitory junctional potentials (5–25 mV) in both receptor muscles as a result of motor stimulation. The excitatory synapses probably have glutamate as the transmitter and the inhibitory use GABA as the transmitter substance (Kuffler and Edwards 1958).

Upon stimulation, RM1 contracts slowly with a fusion frequency of about 5 Hz (Kuffler 1954). Trains of impulses (frequency as low as 1 Hz) in the motor nerve produce a long-lasting rise in the afferent discharge frequency from the sensory neuron which reaches a maximum directly related to stimulus frequency (Eyzaguirre and Kuffler 1955a,b; Brown 1967; Fields 1976). The RM2 contracts in a twitchlike manner to single stimuli, and complete fusion of twitches in RM2 occurs above 50 Hz (Kuffler 1954). Rapid tension development is usually more effective in causing a sensory discharge in the rapidly adapting neuron than gradual tension rise (Eyzaguirre and Kuffler 1955b). In the TCMRO repetitive stimulation of the motor nerve to the receptor muscle results in smooth tetanic tension development. Impulses elicit small postsynaptic potentials as in slow crustacean muscles with no evidence of an active membrane response. The isometric contraction of the receptor muscle depolarizes the T fiber but usually not the S fiber (Bush and Godden 1974; Bush 1976). No efferent inhibition of the TCMRO receptor muscle has been found.

4.3 Mechanotransducer Properties – Receptor Current and Potential

4.3.1 Receptor Current

The mechanoreceptor transduction is mediated by ionic channels that are mechanically activated: stretch-activated channels (SA channels). SA channels are probably abundant in the dendritic tips of the sensory neurons. These tips have been shown to deform as a result of stimulation of the receptor organs (Tao-cheng et al. 1981a,b; Krauhs and Mirolli 1975). The SA channels were first described in cultured chick muscle cells (see Chap. 00) but have since been recorded from the crayfish abdominal MRO (Erxleben 1989). Two different channel types were described on the soma and the primary dendrites: (1) an inward rectifying stretch activated (RSA) channel; and (2) a voltage-independent stretch-activated (SA) channel (see Fig. 7a,b). The open probability of the channels (P_o) can be described (Erxleben 1989) by

$$P_o = [1 + k * \exp(^{-}s * p^n)]^{-1}, \quad (1)$$

where k is a pressure-insensitive term, s the sensitivity, p the applied pressure (mmHg) in the pipette and n is 1 or 2. Voltage-clamp data of the macroscopic stretch-induced receptor current in the crayfish stretch-receptor neuron have shown that the mechanical transduction system is permeable to Na^+, K^+, and Ca^{2+} ions. The

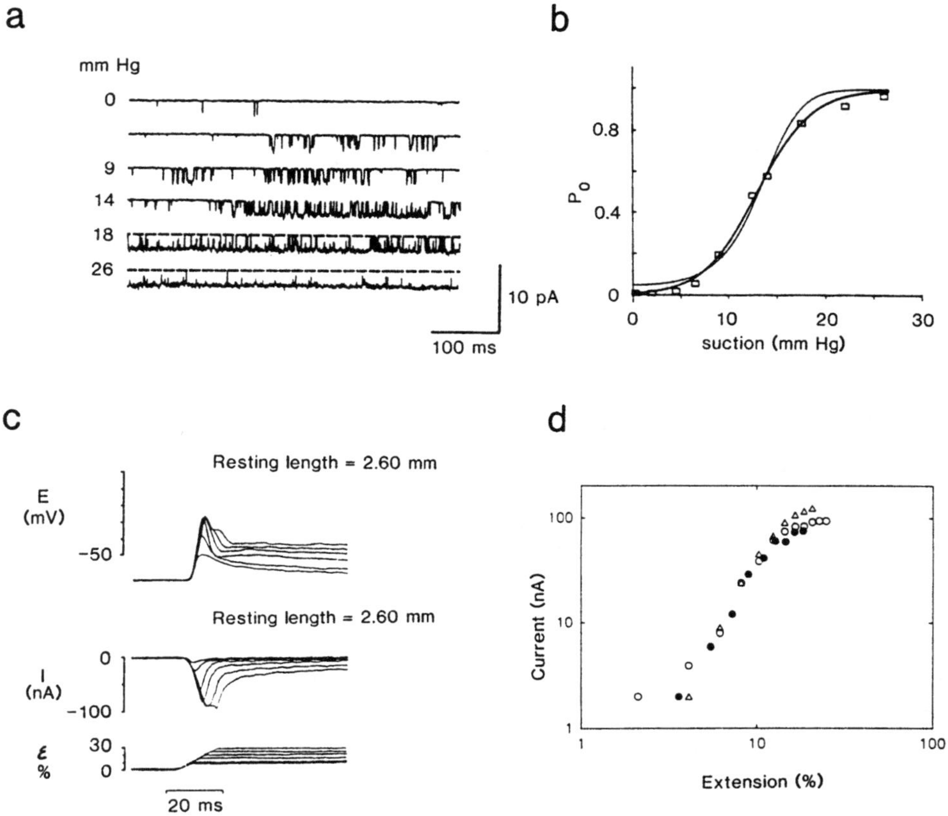

Fig. 7. Generation of receptor current and potential in slowly adapting crayfish abdominal MRO. *a* Unitary currents through SA channel at different suction levels (indicated to the *right*). *b* Open probability P_o vs suction (negative pressure in mmHg) for SA channel. *Thick line* is the fit to Boltzmann curve [Eq. (1), text]. Cell-attached patch with high Na^+ in pipette. *c* Receptor potential and current responses to ramp and hold extensions in crayfish. *d* Log-log stimulus response relation from *c* and other cells. (*a* and *b*, Erxleben 1989; *c* and *d*, Rydqvist and Swerup, unpubl.)

reversal potential was found to be between –30 and +25 mV (Terzuolo and Washizu 1962; Obara 1968; Klie and Wellhöner 1973; Chaplain 1975) and Brown et al. (1978) found a $P_{Na}/P_K = 1.6$, assuming Na^+ and K^+ as the main carrier of the receptor current and a reversal potential of +13 mV. However, the transducer system is also permeable to Ca^{2+} ions (Edwards et al. 1981) and analysis of the RSA and SA channels has shown that the SA channel has a conductance of 71 pS for K^+, 50 pS for Na^+, and 23 pS for Ca^{2+}. Similar values for the RSA channel were 44 pS for Na^+ and 23 pS for Ca^{2+}. The reversal potential based on these values was 0 mV. Since the densities of the channels in the dendritic tips are unknown, some uncertainty still remains regarding the reversal potential. The SA channel was found to be voltage independent whereas the RSA showed a strong inward rectification. On the other hand, the stretch-induced current in the sensory neuron seems to be voltage-dependent (Fig. 8a,b) as shown by Brown et al. (1978), indicating an outward

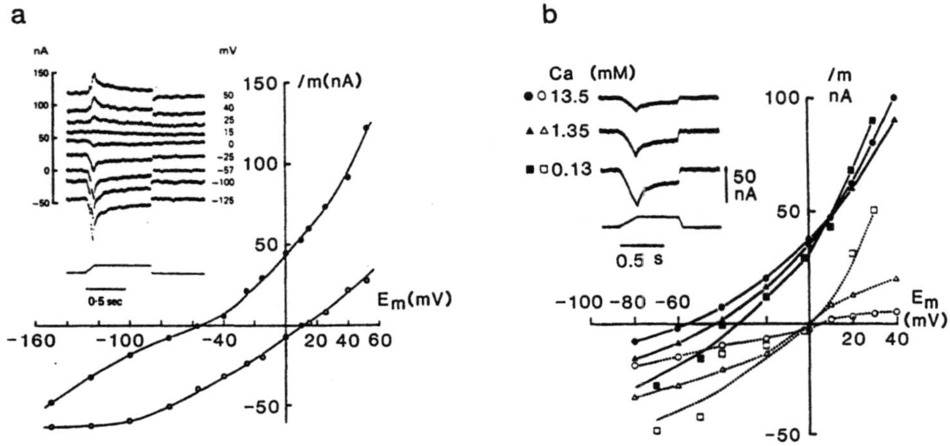

Fig. 8a. Current-voltage relationships for steady state current due to changes in membrane potential (*upper curve*) and the net peak stretch-induced current (*lower curve*). Normal saline. *Inset* Voltage-clamp records of stretch-induced currents at different holding potentials indicated to the *right*. *b* Current-potential relationships for three different Ca^{2+} concentrations. (Brown et al. 1978)

rectification. Also, when Ca^{2+} was reduced in the medium the receptor current increased (Fig. 8b). This is either due to a general effect on the receptor membrane or a direct effect on the SA channels. The SA channels or receptor current in crayfish or crab are not affected by 4-aminopyridine (4-AP) or tetraethylamonium (TEA) (Bush 1981; Erxleben 1989) when applied extracellularly. However, Ottoson and Swerup (1985b), while injecting TEA into the crayfish receptor neuron, found that both the peak and static phase of receptor current increased, indicating that TEA might be more effective from the inside.

4.3.2 Stimulus Response Relations

Contrary to what has been reported for the receptor potential or firing frequency (Wendler and Burkhardt 1961; Terzuolo and Washizu 1962; Obara 1968) the relation between extension and receptor current is nonlinear (Fig. 7d). If plotted in a log-log diagram, the slope of the relation between the stimulus and response is close to 1 when small (< 5%) extensions of the receptor muscle is used (Johansson and Rydqvist 1983). When extensions are within the physiological range (i.e. between 5 and 20%, Alexandrowicz 1967; Njå and Walloe 1973) the slope is 3. At large amplitude extensions, the receptor current saturates (Fig. 7c,d), resulting in a lower slope. This macroscopic behavior is consistent with the proposed properties of the SA channels, if the relation between tension and extension (Fig. 6) is taken into account. Since the relation between open probability of SA channels and the pressure can be described by a Boltzmann relation (Erxleben 1989, Eq. 1) in the crayfish MRO, care should be employed to use strict power laws for the macroscopic behavior. Njå and Walloe (1973), without measuring current, arrived at the same power coefficient (n = 3) between length and current response, but they concluded that this relation was due to a power of 3 between length and tension and a linear relationship between tension and current.

4.3.3 Generation of Receptor Potential

As was seen in Fig. 2c, there is an important difference between the time course of tension and current response to a ramp extension. In the crayfish this can easily be reconciled if the tension response is transformed by an exponential relation between (pressure)^2 and probability of openness as illustrated in 9a (cf. Chap. 4). The similarity between the transformed tension response and the recorded current response is striking. From Fig. 2 it is also evident that the time course of the receptor potential differs from the current response. If no other voltage-activated permeability systems are present in the membrane, the receptor potential would be a simple function of the receptor current and passive membrane properties. However, the presence of voltage-activated channels (other than Na^+) contribute to the potential response (cf. Swerup 1983). A distinct example of this is illustrated in Fig. 9b. The receptor potential reaches a peak which quickly decrease whereas the receptor current, which saturates at this extension, increases continuously (cf. the TCMRO, Bush 1976). This behavior of the receptor potential can be explained by the activation of a K^+ current and it has been shown that the kinetics of the delayed rectifier in the receptor neuron are sufficiently fast to account for this behavior (Rydqvist and Zhou 1989). Further support for the involvement of a K^+ current was obtained by injecting TEA which slowed down the time course of repolarization

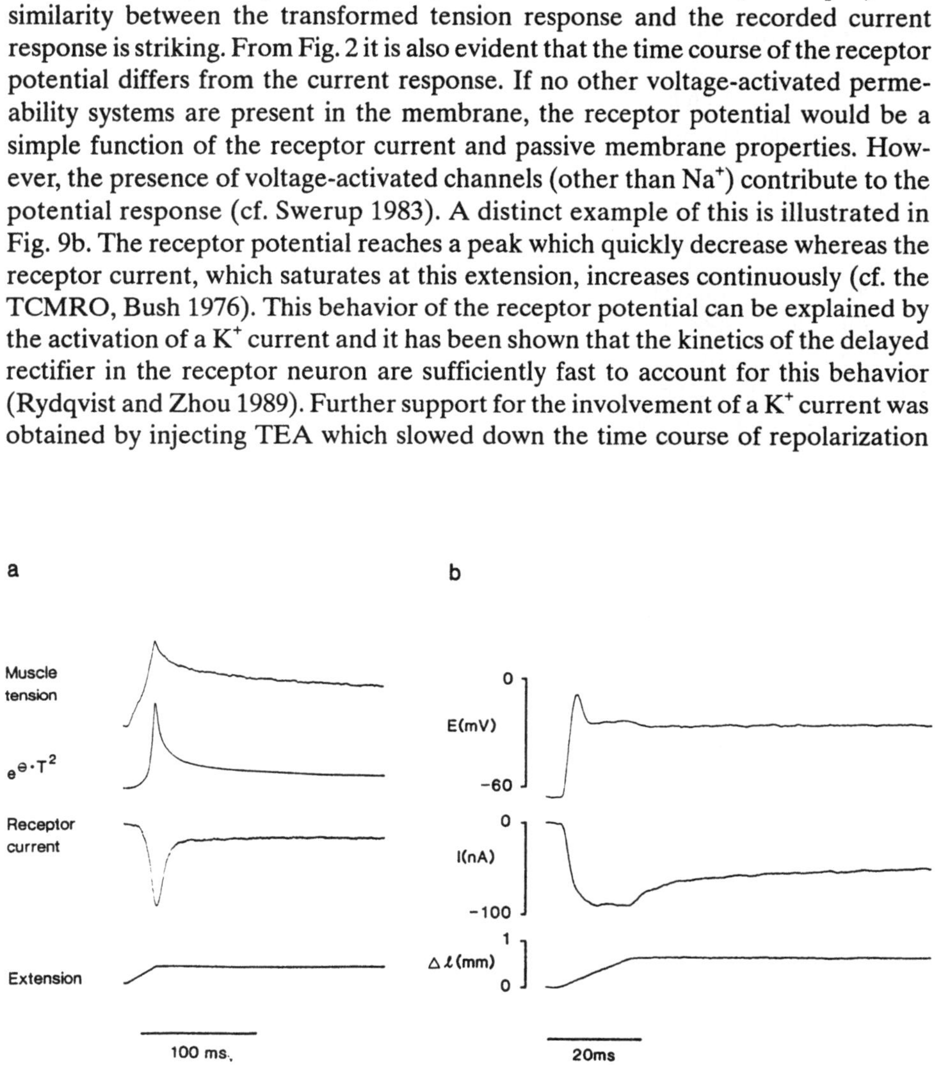

Fig. 9a. Top Typical tension response to ramp and hold extension. Enlarged from Fig. 6c. *Center* Tension multipled by exponential factor using a squared tension model. *T,* tension. *Below* Recorded receptor current response to identical ramp and hold extension. *b* Comparison of receptor potential (*above*) and receptor current (*center*) to a 25% ramp and hold extension. The receptor current saturates. (Rydqvist et al. 1990)

(Swerup 1983; Ottoson and Swerup 1985b). The possible involvement of a Ca^{2+} activated K^+ current (Ottoson and Swerup 1982) has also been suggested.

Similar mechanisms for the generation of the receptor current and receptor potential have been found in the TCMRO (Bush and Roberts 1971; Roberts and Bush 1971; Bush 1976, 1981; Bush and Laverack 1982).

4.4 General Electrical Properties of Receptor Neurons

4.4.1 Geometrical Considerations

In most systems, recordings are made at some location 50–100 µm away from the dendritic region which might affect the shape of receptor signals. However, in many organs the length constant seems to be fairly large. In TCMROs the length constant for S and T fibers in the posterior leg segment of *Carcinus* was around 10–16 mm (Bush 1976; Cannone and Bush 1980), but can be as long as 6 cm (Mirolli 1979a), indicating a specific membrane resistance (R_m) of about 1 $M\Omega cm^2$ which is very high compared to many other excitable membranes (cf. frog myelinated nerve 100 000 Ωcm^2). Similar studies of the leech stretch receptor gave a length constant of 2.4 mm and an specific membrane resistance (R_m) of 13 $k\Omega cm^2$ (Blackshaw and Thompson 1988). Recordings from the crayfish stretch receptor (Rydqvist and Swerup, unpubl.) suggest that the length constant is large, although no definite value can be determined due to the noncylindrical geometry (Fig. 10). Calculations according to Rall (1977) indicate that at least 95% of the receptor current, as determined by voltage-clamp technique, is seen in the soma (Johansson and Rydqvist 1983, see also Turner and Calvin 1981). On the other hand, care must by employed when recording voltage-activated currents in the soma. It is impossible to clamp membrane parts more than a few hundred microns away, and it is essential to keep the axon as short as possible to avoid regenerative contamination of clamp records (see Rydqvist and Zhou 1989).

Second, the exact distribution of different ionic channels is still an unsolved problem but it is obvious that, for example, Na^+ channels and delayed rectifier K^+ channels are present in the soma, as are SA channels (Erxleben 1989; Rydqvist and Zhou 1989). The generator current spreads electrotonically to a "trigger zone" rich in Na^+ channels, some 200–500 µm away in the axon, where the action potentials are set up (Edwards and Ottoson 1958; Grampp 1966; Ringham 1971).

4.4.2 Ionic Permeability Systems: Channels

The resting membrane potential for most MROs seems to be dependent on a K^+ permeability. Both in the TCMRO and the crayfish abdominal MRO, a significant Na^+ permeability is also present, since replacement of sodium leads to an initial hyperpolarization and a decreased leak conductance (Edwards et al. 1963; Bush 1981). Electrogenic pumps are also involved (see below). The leakage conductances in invertebrate sensory neurons differ considerably and only some examples will be given here. In the crayfish and lobster, the leak conductance typically ranges between 0.25 to 1 µS, whereas in the TCMRO S and T fibers it is in the order of 1 µS. The presence of voltage-dependent permeability systems or ionic channels have

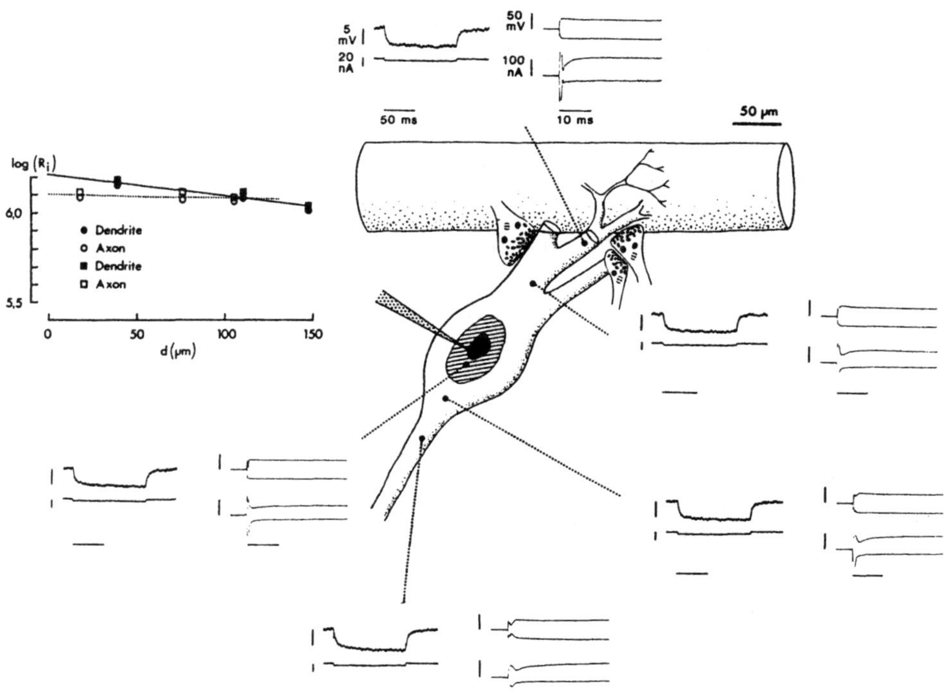

Fig. 10. Estimate of electrotonic properties of the crayfish (*Astacus astacus*) stretch receptor neuron. A current electrode was inserted in the soma and consecutive penetrations with a second membrane potential electrode was made. Stimulation using rectangular current or potential pulses were made in current clamp and voltage clamp, respectively. (Rydqvist and Swerup, unpubl.)

been studied mainly in the crayfish and lobster abdominal MRO. In the crayfish, a K^+ delayed rectifier system has been demonstrated in the slowly adapting MRO neuron (Rydqvist and Zhou 1989), probably correlated to the K^+ channels found by Erxleben (1989). A fast Na^+ permeability system is also present. A Ca^{2+} activated potassium permeability has been suggested by Swerup (1983), and Ottoson and Swerup (1985a). Contrary to that found in many invertebrate neurons, no transient fast outward potassium current (I_A) was found in the slowly adapting stretch activated neuron (Rydqvist and Zhou 1989). In the lobster (*Homarus gammarus*), Grampp and coworkers have made a detailed analysis of the different voltage activated currents in both the slowly and rapidly adapting neuron (Gestrelius et al. 1981; Gestrelius and Grampp 1983; Edman et al. 1986, 1987a) and recently an inwardly rectifying current (I_Q) was described in this preparation (Edman et al. 1987b) important for posttetanic restoration of excitability.

In the TCMRO, despite the basic electrotonic spread of the receptor potential to the thoracic ganglion, there seem to be voltage-activated Na^+ channels at least in the T fibers which give rise to small graded spikes at the onset and offset of stretch. These can be blocked by TTX but are either so sparsely distributed, alternatively permanently inactivated, or shunted through some potassium channels, that only abortive spikes are seen (See Mirolli 1979a; Bush 1981). The potassium channel blockers TEA or 3- or 4-AP, have no effect on the T-fiber responses to depolarizing current pulses in *Callinectes* (Blight and Llinas 1980) when applied extracellularly, so the presence of significant voltage-dependent potassium conductances that might prevent generation of all or none impulses remains unclear.

4.4.3 Ionic Pump Activities

Ionic pumping activities have been investigated in some detail in the abdominal MRO neuron of *Homarus, Procambarus, Orconectes* and *Astacus* (Nakajima and Takahashi 1966; Sokolove and Cook 1971; Edman et al. 1986; Gestrelius et al. 1981) and in the TCMRO of the crab *Scylla* (Mirolli 1979b). In the crustacean MRO, several lines of evidence indicate that the pumping activity for Na^+ and K^+ is electrogenic and contributes to firing adaptation and the development of post-tetanic hyperpolarization (Nakajima and Takahashi 1966; Nakajima and Onodera 1969a; Sokolove and Cook 1971; Fischer 1978). Possibly two components of the electrogenic pump are present in the crayfish, a fast one being Na^+-dependent and a slow one being voltage-dependent (Fohlmeister et al. 1977; Holloway and Poppele 1984). It has been shown that the influx of Na^+ during stretch activation is sufficient to affect the pump mechanism (Kaila et al. 1987). In the lobster, ouabain (5 mM) led to a rapid depolarization (5–10 mV) of the sensory neurons in both MRO1 and MRO2 (Gestrelius et al. 1981) and resting pumping current density was found to be 1.5 $\mu A/cm^2$ with a stoichiometry 3:2 (Na:K). The extrusion capacity for Na^+ was larger in the slowly than in the rapidly adapting neuron (Edman et al. 1986). The presence of a Na^+-H^+ pump is also established from studies on proton regulation in the slowly adapting MRO (Moser 1985).

In the crab TCMRO, cooling or ouabain (1 mM) depolarized the S fiber about 20 mV, indicating that in this preparation a substantial fraction of the resting potential stems from an electrogenic Na^+-K^+ pump (Mirolli 1979b; Mirolli in Bush 1981).

4.5 System Analysis of Invertebrate Muscle Receptors

One of the first system analysis using linear transfer models was made by Chapman and Smith (1963) on a cockroach mechanoreceptor. They found that both adaptation of the response (impulse discharge) to step changes in length and the transfer function as determined by sinusoidal stimulations could be described by a mathematical function which in the time domain predicts a power relation for adaptation. The slope for low frequencies in the Bode gain plot was accurately predicted, whereas the phase shift was more difficult to take into account. System analysis of the abdominal MRO in the crayfish and lobster (Borsellino et al. 1965; Brown and Stein 1966; Terzuolo and Knox 1971; Johansson and Rydqvist 1983, see also Buno et al. 1984) and in the moth (Weevers 1966) have shown that the transfer function for the impulse and receptor potential responses (Bode plots) had a slope of 1 to 3 dB/octave in the low frequency range and a slope of about 4–5 dB/octave in the high frequency range, with the peak occurring at about 10 Hz. Using the voltage-clamp technique, it was possible to determine the transfer functions for the

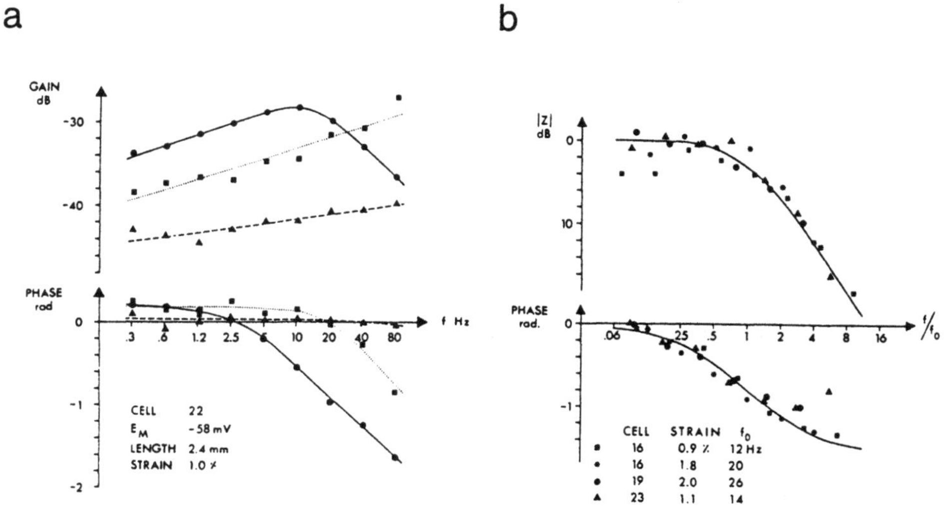

Fig. 11. Transfer function of crayfish stretch receptor organ as calculated from responses due to sinusoidal stimulation. *a* Bode plots of receptor current (*circles*), receptor potential (*squares*) and force (*triangles*) from the same preparation. *b* Bode plot of impedance (*Z*) computed from potential and current responses. Data from four cells normalized with respect to low impedance response and corner frequency (f_o). The *solid line* represent a first-order integrative network. (Johansson and Rydqvist 1983)

receptor current and receptor potential (Johansson and Rydqvist 1983), and the impedance of the transducer could be calculated from the current and potential responses (Fig. 11). The impedance transfer function was found to be a simple first order integrative network where the corner frequency could be related to the stretch-activated conductance and the capacitance of the receptor membrane. The corner frequency was linearly related to the stimulus amplitude (Fig. 11b), suggesting that with decreasing stimulus strength, the band width of the system decreases, increasing the signal to noise ratio.

4.6 Adaptation

As in most sensory systems, the mechanoreceptor response to a steady stimulus decays with time: it adapts. In the crustacean abdominal MRO, the overall difference in adaptive properties between the slowly (MRO1) and rapidly adapting receptor (MRO2) is striking (Nakajima and Onodera 1969b). Whereas the MRO1 can maintain firing for long periods of time, the MRO2 gives a brief impulse discharge when mechanically stimulated. The same adaptive behavior was seen when the receptor was stimulated electrically (Nakajima and Onodera 1969a,b) and it was concluded that mechanical factors were of minor importance. Today most investigators agree that adaptation is probably due to both mechanical and permeability factors, as was originally proposed by Wiersma et al. (1953). As was pointed out above (see Fig. 2), there are fundamental differences in adaptation between tension, receptor current and receptor potential of the crayfish MRO. The tension response does adapt (Fig. 6), but little difference between the MRO1 and MRO2 in the crayfish appears to be present (Nakajima and Onodera 1969b). If the properties of the SA channels (see above) of the crayfish stretch receptor are taken into consideration, the transformation of tension by the exponential relation between channel open probability and the square of membrane pressure (see above) results in a current response that is very close to what is actually recorded (Fig. 9a). Assuming similar SA channels in slowly and rapidly adapting receptors, this transformation cannot explain the overall difference in adaptive behavior. The transformation of current to potential response in the absence of voltage-activated permeabilities would be a simple function of the passive membrane properties. This does not appear to be the case, as indicated in Fig. 9a,b. Adaptation of the receptor potential seems to include time courses which are voltage-dependent (Swerup et al. 1983). The main candidate for this effect in TTX-treated preparations is K^+ currents (see Sect. 4.2). This is also supported by experiment where TEA was injected into the cell: the effect was a complete elimination of the early adaptation fall (Ottoson and Swerup 1985b).

A possible factor affecting adaptation is the influx of Ca^{2+} during mechanical stimulation. An increase in intracellular Ca^{2+} results in a decreased permeability in mechanoreceptor system (cf. Fig. 8b) and to a negative feedback system on the SA channels. This system could be different in slowly or rapidly adapting receptors.

In summary, several lines of evidence strongly suggest that the basic adaptive properties are a combination of (1) the viscoelastic properties of the accessory structures; (2) the properties of the SA channels in the dendrites of the sensory neuron; and (3) the set of voltage-activated ion channels present in the dendrites, soma and axon, including the ion channels generating the impulses which would constitute the "encoder". The difference between slowly or rapidly adapting receptors seems to be dependent mainly on the particular set of voltage-activated ion channels. In the lobster Edman et al. (1986) found that the difference in adaptive behavior is due to different kinetics of a slow Na^+ permeability system in the two receptors.

4.7 Efferent Control of Receptor Organs

Efferent control of the receptor neuron is common, but far from general. The arthropod MCO and TCMRO and insect MROs lack inhibitory control of the sensory neuron (Finlayson 1968). In the crayfish and the lobster MRO, stimulation of the inhibitory nerves (Kuffler and Eyzaguirre 1955) leads to a decreased rate of firing. It can be shown that stimulation of the inhibitory fibers sets up IPSPs in the sensory neuron, which are mostly hyperpolarizing at resting membrane potential (Kuffler and Eyzaguirre 1955; Jansen et al. 1971; Ozawa and Tsuda 1973) and consistent with the equilibrium potential of Cl^- (Deisz and Lux 1982). If recording electrodes low in Cl^- are used spontaneous miniature IPSPs can be observed (Iwasaki and Florey 1969). It is now well established that GABA activates Cl^- channels in the sensory neuron, leading to an IPSP when the inhibitory nerve is stimulated (see Deisz and Dose 1983). However, it has also been shown that bicarbonate ions (HCO_3^-) can pass through the GABA-activated channel which makes the inhibition sensitive to pH changes (Voipio et al. 1991).

Uptake of GABA is well established in the crayfish MRO (Deisz and Dose 1983) and probably contributes to the response to exogenous application of GABA, since depolarizing potential response or a biphasic depolarizing/hyperpolarizing have been observed. A considerable uptake of GABA in the sensory neuron and the glial cell helps to maintain a low GABA concentration in the synaptic cleft (Krause et al. 1981; Deiz et al. 1984). In the crustacean MRO Crealius (1976), using focal application of GABA, found the largest response close to the dendrites.

5 Summary

The development of proprioceptors in invertebrates, and MROs in particular, parallels the development of the motor apparatus which culminates in the arthropodan phyla (Crustacea, Chelicerata and Uniramia). The complex skeleton and muscle machinery of the arthropods and the demand for well-coordinated movements have made necessary an efficient set of muscle receptors and other proprioceptors.

The sensory neurons associated with the muscle receptor organs (MROs) are mostly of the multiterminal, nonciliary type in close contact with either skeletal muscles or associated with specialized accessory muscles, as in the case of the abdominal MRO of the crustaceans and some insects. In a few cases, e.g. TCMRO, the sensory neuron is located within the CNS.

The primary mechanotransduction takes place by stimulation of stretch-activated (SA) channels located in the fine dendritic tips in close contact with "ordinary" or accessory muscle. The resulting receptor current causes a local depolarization of the cell membrane which is electrotonically spread in the neuron. This is the normal way of signaling in the afferent axon of the TCMRO. However, the depolarization can also activate voltage-dependent channels, and thus modify the potential response and set up action potentials which are conducted through the axon to the CNS, as in the crustacean abdominal MRO. Furthermore, ionic pumping mechanisms can influence impulse firing. An inhibitory efferent innervations makes it possible for the CNS to control the sensory neurons.

Slowly and rapidly adapting muscle receptors can convey information about long- and short-term changes, respectively, leading to different kinds of reflex behavior and positional adjustments. These reflexes engage muscles both in direct relation to the MRO and in other appendages, and participate in the adjustment of muscle tone which contributes to feedback control of movements. For recent reviews see Bush and Clarac (1985) and Barnes and Gladden (1985).

It is thus clear that a thorough understanding of the proprioceptive contribution to motor control is dependent on a detailed knowledge of the primary transduction processes in these receptor organs.

Acknowledgments. The author wants to thank Drs. Barbara Canlon, Jan Lännergren and Christer Swerup for their valuable comments and criticism of the manuscript. Many thanks go to Monica Thunberg-Ericsson for her skillful work on the illustrations. This work was supported by grants from the Swedish Medical Research Foundation (Proj. 6838), Karolinska Institutet and Magnus Bergwall Foundation.

References

Alexandrowicz JS (1951) Muscle receptor organs in the abdomen of *Homarus vulgaris* and *Palinurus vulgaris*. Q J Microsc Sci 92:163–200

Alexandrowicz JS (1952a) Receptor elements in the thoracic muscles of *Homarus vulgaris* and *Palinurus vulgaris*. Q J Microsc Sci 93:315–346

Alexandrowicz JS (1952b) Muscle receptor organs in the Paguridae. J Mar Biol Assoc UK 31:277–286

Alexandrowicz JS (1954) Notes on the nervous system in the Stomatopods IV. Muscle receptor organs. Publ Stazione Zool Napoli 25:94–111

Alexandrowicz JS (1956) Receptor elements in the muscles of *Leander serratus*. J Mar Biol Assoc UK 35:129–144

Alexandrowicz JS (1960) A muscle receptor organ in *Eledone cirrohsa*. J Mar Biol Assoc UK 39:419–431

Alexandrowicz JS (1967) Receptor organs in thoracic and abdominal muscles of crustacea. Biol Rev 42:288–326

Alexandrowicz JS (1972) The comparative anatomy of leg proprioceptors in some decapod Crustacea. J Mar Biol Assoc UK 52:605–634

Alexandrowicz JS, Whitear M (1957) Receptor elements in the coxal region of Decapoda Crustacea. J Mar Biol Ass UK 36:603–628

Barber SB, Hayes WF (1964) A tendon receptor organ in *Limulus*. Comp Biochem Physiol 11:193–198

Barnes RSK (1984) A synoptic classification of living organisms. Blackwell, Oxford

Barnes WJP, Gladden MH (eds) (1985) Feedback and motor control in invertebrates and vertebrates. Croom Helm, London

Barth G (1934) Untersuchungen über Myochordotonalorgane bei dekapoden Crustaceen. Z Wiss Zool 145:576–624

Bazemore A, Elliott KAC, Florey E (1957) Isolation of factor I. J Neurochem 1:334–339

Berger CS, Bush BMH (1979) A non-linear mechanical model of a non-spiking muscle receptor. J Exp Biol 83:339–343

Blackshaw SE, Thompson SWN (1988) Hyperpolarizing responses to stretch in sensory neurones innervating leech body wall muscle. J Physiol 396:121–137

Blight AR, Llinas R (1980) The non-impulse stretch receptor complex of the crab – a study of depolarization-release coupling at a tonic sensorimotor synapse. Philos Trans R Soc Lond Ser B 290:219–276

Bodian D, Bergman RA (1962) Muscle receptor organs of crayfish functional anatomical correlations. Bull Johns Hopkins Hosp 110:78–106

Borsellino A, Poppele RE, Terzuolo CA (1965) Transfer functions of the slowly adapting stretch receptor organs of crustacea. Cold Spring Harbor Symp Quant Biol 30:581–586

Bowerman RF (1972) A muscle receptor organ in the scorpion postabdomen. I. The sensory system. J Comp Physiol 81:133–146

Brown MC (1967) Some effects of receptor muscle contraction on the responses of slowly adapting abdominal stretch receptors of the crayfish. J Exp Biol 46:445–458

Brown MC, Stein RB (1966) Quantitative studies on the slowly adapting stretch receptor of the crayfish. Kybernetik 3:175–185

Brown HM, Ottoson D, Rydqvist B (1978) Crayfish stretch receptor: an investigation with voltage-clamp and ion-sensitive electrodes. J Physiol 284:155–179

Bullock TH, Horridge GA (1965) Structure and function in the nervous systems of invertebrates. Freeman, San Francisco

Buno W Jr, Bustamante J, Fuentes J (1984) White noise analysis of pace-maker-response interactions and non-linearities in slowly adapting crayfish stretch receptor. J Physiol 350:55–80

Bush BMH (1976) Non-impulse thoracic-coxal receptors in crustaceans. In: Mill PJ (ed) Structure and function of proprioceptors in the invertebrates. Chapman and Hall, London, pp 115–151

Bush BMH (1981) Non-impulsive stretch receptors in crustaceans. In: Roberts A, Bush BMH (eds) Neurones without impulses: their significance for vertebrate and invertebrate nervous systems. Cambridge Univ Press, Cambridge, pp 147–176

Bush BMH, Clarac F (eds) (1985) Coordination of motor behaviour. Cambridge Univ Press, Cambridge

Bush BMH, Godden DH (1974) Tension changes underlying receptor potentials in non-impulsive crab muscle receptors. J Physiol 242:80–82P

Bush BMH, Laverack MS (1982) Mechanoreception. In: Atwood HL, Sandeman DC (eds) The biology of crustacea, vol 3. Neurobiology: structure and function. Academic Press, New York, pp 399–468

Bush BMH, Roberts A (1968) Resistance reflexes from a crab muscle receptor without impulses. Nature (Lond) 218:1171–1173

Bush BMH, Roberts A (1971) Coxal muscle receptors in the crab: the receptor potentials of S and T fibers in response to ramp stretches. J Exp Biol 55:813–832

Cannone AJ (1987) Crustacean elastic strand receptors that function without impulses. I. The depressor receptor in the shore crab, *Carcinus maenas*. J Comp Physiol A 160:599–611

Cannone AJ, Bush BMH (1980) Reflexes mediated by non-impulsive afferent neurones of thoracic-coxal muscle receptor organs in the crab, *Carcinus maenas*. J Exp Biol 86:275–303

Cannone AJ, Nijland MJM (1989) Curstacean elastic strand receptors that function without impulses. II. The levator receptor in the shore crab, *Carcinus maenas*. J Comp Physiol A 165:801–809

Chaplain RA (1975) Evidence for Ca^{2+} control of the transducer mechanism in crayfish stretch receptor. J Membr Biol 21:335–351

Chapman KM, Smith RS (1963) A linear transfer function underlying impulse frequency modulation in a cockroach mechanoreceptor. Nature (Lond) 197:699–700

Cohen MJ (1965) The dual role of sensory systems: detection and setting central excitability. Cold Spring Harbor Symp Quant Biol 30:587–599

Craelius W (1976) The distribution of GABA sensitivity on crayfish muscle receptor organs. J Physiol (Lond) 263:405–415

Deisz RA, Dose M (1983) Comparison of GABA analogues at the crayfish stretch receptor neurone. Brain Res Bull 11:283–288

Deisz RA, Lux HD (1982) The role of intracellular chloride in hyperpolarizing post-synaptic inhibition of crayfish stretch receptor neurones. J Physiol 326:123–138

Deisz RA, Dose M, Lux HD (1984) The time course of GABA action on the crayfish stretch receptor: evidence for a saturable GABA uptake. Neurosci Lett 47:245–250

Dorsett DA (1976) The structure and function of proprioceptors in soft-bodied invertebrates. In: Mill PJ (ed) Structure and function of proprioceptors in the invertebrates. Chapman and Hall, London, pp 443–479

Edman A, Gestrelius S, Grampp W (1986) Transmembrane ion balance in slowly and rapidly adapting lobster stretch receptor neurones. J Physiol 377:171–191

Edman A, Gestrelius S, Grampp W (1987a) Analysis of gated membrane currents and mechanisms of firing control in the rapidly adapting stretch receptor neurone. J Physiol 384:649–669

Edman A, Gestrelius S, Grampp W (1987b) Current activation by membrane hyperpolarization in the slowly adapting lobster stretch receptor neuron. J Physiol 384:671–690

Edwards C, Ottoson D (1958) The site of impulse initiation in a nerve cell of a crustacean stretch receptor. J Physiol 143:138–148

Edwards C, Kuffler SW (1959) The blocking effect of y-aminobutyric acid (GABA) and the action of related compounds on single nerve cells. J Neurochem 4:19–30

Edwards C, Terzuolo CA, Washizu Y (1963) The effect of changes of the ionic environment upon an isolated crustacean sensory neuron. J Neurophysiol 26:948–957

Edwards C, Ottoson D, Rydqvist B, Swerup C (1981) The permeability of the transducer membrane of the crayfish stretch receptor to calcium and other divalent cations. Neuroscience 7:1455–1460

Elekes K, Florey E (1987a) New types of synaptic connections in crayfish stretch receptor organs: an electron microscopic study. J Neurocytol 16:613–26

Elekes K, Florey E (1987b) Immunocytochemical evidence for the GABAergic innervation of the stretch receptor neurons in crayfish. Neuroscience 22:1111–22

Erxleben C (1989) Stretch-activated current through single ion channels in the abdominal stretch receptor organ of the crayfish. J Gen Physiol 94:1071–1083

Euteneuer U, Winter C (1979) The abdominal muscle receptor organ in *Astacus leptodactylus* (Crustaceae). Cell Tissue Res 202:41–61

Eyzaguirre C, Kuffler SW (1955a) Processes of excitation in the dendrites and in the soma of single isolated sensory nerve cells of the lobster and crayfish. J Gen Physiol 39:87–119

Eyzaguirre C, Kuffler SW (1955b) Further studyh of soma, dendrite, and axon excitation in single neurons. J Gen Physiol 39:121–153

Fields HL (1976) Crustacean abdominal and thoracic muscle receptor organs. In: Mill PJ (ed) Structure and function of proprioceptors in the invertebrates. Chapman and Hall, London, pp 65–114

Finlayson LH (1968) Proprioceptors in the invertebrates. Symp Zool Soc Lond 23:217–249

Finlayson LH (1976) Abdominal and thoracic receptors in insects, centipedes and scorpions. In: Mill PJ (ed) Structure and function of proprioceptors in the invertebrates. Chapman and Hall, London, pp 65–114

Finlayson LH, Loewenstein (1958) The structure and function of abdominal stretch receptors in insects. Proc R Soc B Lond 148:433–449

Fischer W (1978) Na-K-Ionenpumpe und Spike-Adaptation im langsam und schnell adaptierende Dehnungsrezeptorneuron des Flusskrebses *Orconectes limosus* (RAF). Acta Biol Med Germ 37:1681–1691

Florey E, Florey E (1955) Microanatomy of the abdominal stretch receptors of the crayfish (*Astacus Fluviatilis* L). J Gen Physiol 39:69–85

Fohlmeister JF, Poppele RE, Purple RL (1977) Repetitive firing: a quantitative study of feedback in model encoders. J Gen Physiol 69:815–848

Gestrelius S, Grampp W (1983) Kinetics of the TEA and 4-AP sensitive K^+ current in the slowly adapting lobster stretch receptor neurone. Acta Physiol Scand 118:125–134

Gestrelius S, Grampp W, Sjölin L (1981) Subthreshold and near-threshold membrane currents in lobster stretch receptor neurones. J Physiol 310:191–203

Grampp W (1966) The impulse activity in different parts of the slowly adapting stretch receptor neuron of the lobster. Acta Physiol Scand 66:3–35

Hirosawa K, Tao-Cheng J-H, Nakajima Y, Tisdale AD (1981) Thin-section and freeze-fracture studies of crayfish stretch receptor synapses including the reciprocal inhibitory synapse. J Comp Neurol 200:39–53

Holloway SF, Poppele RE (1984) Post-Tetanic hyperpolarization evoked by depolarizing pulses in crayfish stretch receptor neurones in tetrodotoxin. J Physiol 350:343–360

Iwasaki S, Florey E (1969) Inhibitory miniature potentials in the stretch receptor neurons of crayfish. J Gen Physiol 53:666–682

Jansen JRS, Njå A, Ormstad K, Walloe L (1971) On the innervation of the slowly adapting stretch receptor of the crayfish abdomen. An electrophysiological approach. Acta Physiol Scand 81:273–285

Johnasson B, Rydqvist B (1983) Electrical and mechanical properties of the crustacean stretch receptor during sinusoidal length changes. Acta Physiol Scand 117:183–194

Kaila K, Rydqvist B, Swerup C, Voipio J (1987) Stimulation-induced changes in the intracellular sodium activity of the crayfish stretch receptor. Neurosci Lett 74:53–57

Klie W, Wellhöner HH (1973) Voltage clamp studies on the stretch response in the neuron of the slowly adapting crayfish stretch receptor. Pflügers Arch Eur J Physiol 342:93–104

Komuro T (1981) Fine structural study of the abdominal muscle receptor organs of the crayfish (*Procambarus clarkii*). Fast and slow receptor muscles. Tissue Cell 13:79–92

Kosaka K (1969) Electrophysiological and electron microscopic studies on the neuromuscular junction of the crayfish stretch receptors. Jpn J Physiol 19:160–175

Krauhs JM, Mirolli M (1975) Morphological changes associated with stretch in a mechano-receptor. J Neurocytol 4:231–246
Krause DN, Ikeda K, Roberts E (1981) Dose-conductance relationships for GABA agonists and the effect of uptake inhibitors in crayfish stretch receptor neurons. Brain Res 225:319–332
Krnjevic K, van Gelder NM (1961) Tension changes in crayfish stretch receptors. J Physiol 159:310–325
Kuffler SW (1954) Mechanisms of activation and motor control of stretch receptors in lobster and crayfish. J Neurophysiol 17:558–574
Kuffler SW, Edwards C (1958) Mechanism of gamma aminobutyric acid (GABA) action and its relation to synaptic inhibition. J Neurophysiol 21:589–610
Kuffler SW, Eyzaguirre C (1955) Synaptic inhibition in an isolated nerve cell. J Gen Physiol 39:155–184
Markl H (1965) Ein neuer Propriorezeptor am Coxa-Trochanter-Gelenk der Honigbiene. Naturwissenschaften 52:460
Mill PJ (ed) (1976) Structure and function of proprioceptors in the invertebrates. Chapman and Hall, London, pp 115–151
Mirolli M (1979a) The electrical properties of a crustacean sensory dendrite. J Exp Biol 78:1–27
Mirolli M (1979b) Electrogenic Na^+ transport in a crustacean coxal receptor. J Exp Biol 78:29–45
Moser H (1976) Musclereceptororgans (MRO) in Isopoda (Crustacea) – histological observations. Mikroskopie 31:350–62
Moser H (1985) Intracellular pH regulation in the sensory neurone of the stretch receptor of the crayfish (*Astacus fluviatilis*). J Physiol 362:23–38
Moulins M (1976) Ultrastructure of chordotonal organs. In: Mill PJ (ed) Structure and function of proprioceptors in the invertebrates. Chapman and Hall, London, pp 115–151
Nadol JB Jr, DeLorenzo AJD (1968) Observations on the abdominal stretch receptor and the fine structure of associated axo-dendritic synapses and neuromuscular junctions in *Homarus*. J Comp Neurol 132:419–444
Nakajima S, Onodera K (1969a) Membrane properties of the stretch receptor neurones of crayfish with particular reference to mechanisms of sensory adaptation. J Physiol 200:161–185
Nakajima S, Onodera K (1969b) Adaptation of the generator potential in the crayfish stretch receptors under constant length and constant tension. J Physiol 200:187–204
Nakajima Y, Reese T (1983) Inhibitory and excitatory synapses in crayfish stretch receptor organs studied with direct rapid-freezing and freeze-substitution. J Comp Neurol 213:66–73
Nakajima S, Takahashi K (1966) Post-tetanic hyperpolarization and electrogenic Na pump in stretch receptor neurone of crayfish. J Physiol 187:105–127
Nakajima S, Tisdale AD, Henkart MP (1973) Presynaptic inhibition at inhibitory nerve terminals. A new synapse in the crayfish stretch receptor. Proc Natl Acad Sci USA 70:2462–2466
Njå A, Walloe L (1973) In situ performance of slowly adapting stretch receptor of the crayfish. J Neurophysiol 36:1062–1070
Obara S (1968) Effects of some organic cations on generator potential of crayfish stretch receptor. J Gen Physiol 52:363–386
Osborne MP (1970) Structure and function of neuromuscular junction and stretch receptors. In: Neville AC (ed) Insect ultrastructure. Blackwell, Oxford, pp 77–100
Osborne MP, Finlayson LH (1962) The structure and topography of stretch receptors in representatives of seven orders of insects. Q J Microsc Sci 103:227–242
Ottoson D, Swerup C (1982) Studies on the role of calcium in adaptation of the crustacean stretch receptor. Effects of intracellular injection of calcium, EGTA and TEA. Brain Res 244:337–341
Ottoson D, Swerup C (1985a) Ionic dependence of early adaptation in the crustacean stretch receptor. Brain Res 336:1–8
Ottoson D, Swerup C (1985b) Effects of intracellular TEA injections on early adaptation of crustacean stretch receptor. Brain Res 336:9–17
Ozawa S, Tsuda K (1973) Membrane permeability changes during inhibitory transmitter action in crayfish stretch receptor cell. J Neurophysiol 36:805–816
Peterson RP, Pepe FA (1961) The fine structure of inhibitory synapses in the crayfish. J Biophys Biochem Cytol 11:157–169
Pilgrim RLC (1960) Muscle receptor organs in some decapod crustacea. Comp Biochem Physiol 1:248–257

Pringle JWS (1961) Proprioception in arthropods. In: Ramsay JA, Wigglesworth VB (eds) The cell and the organism. Cambridge Univ Press, Cambridge, pp 256–282
Rall W (1977) Core conductor theory and cable properties of neurons. In: Kandel ER (ed) Handbook of physiology, Sect 1. The nervous system, vol 1. Cellular biology of neurons, vol 1, Part 1. Am Physiol Soc, Bethesda, pp 39–97
Ringham GL (1971) Origin of nerve impulse in slowly adapting stretch receptor of crayfish. J Neurophysiol 34:773–784
Ripley SH, Bush BMH, Roberts A (1968) Crab muscle receptor which responds without impulses. Nature (Lond) 218:1170–1171
Roberts A, Bush BMH (1971) Coxal muscle receptors in the crab: the receptor current and some properties of the receptor nerve fibres. J Exp Biol 54:515–524
Rydqvist B, Zhou J-Y (1989) Potential dependent outward potassium currents in the slowly adapting crayfish stretch receptor neuron. Acta Physiol Scand 137:409–419
Rydqvist B, Swerup C, Lännergren J (1990) Viscoelastic properties of the slowly adapting stretch receptor muscle of the crayfish. Acta Physiol Scand 139:519–527
Sokolove PG, Cooke IM (1971) Inhibition of impulse activity in a sensory neuron by an electrogenic pump. J Gen Physiol 57:125–163
Swerup C (1983) On the ionic mechanisms of adaptation in an isolated mechanoreceptor – an electrophysiological study. Acta Physiol Scand Suppl 520:1–43
Swerup C, Rydqvist B, Ottoson D (1983) Time characteristics and potential dependence of early and late adaptation in the crustacean stretch receptor. Acta Physiol Scand 119:91–99
Tao-Cheng J-H, Hirosawa K, Nakajima Y (1981a) Ultrastructure of the crayfish stretch receptor in relation to its function. J Comp Neurol 200:1–21
Tao-Cheng J-H, Hirosawa K, Nakajima Y, Peng HB (1981b) Freeze-fracture study of the crayfish stretch receptor. J Comp Neurol 200:23–38
Terzuolo CA, Knox CK (1971) Static and dynamic behavior of the stretch receptor organ of crustacea. Handbook of sensory physiology I. Springer, Berlin Heidelberg New York, pp 500–522
Terzuolo CA, Washizu Y (1962) Relation between stimulus strength, generator potential and impulse frequency in stretch receptor of crustaceae. J Neurophysiol 25:56–66
Turner DA, Calvin WH (1981) Dendritic analysis of lobster stretch receptor neurons: electrotonic properties with single and distributed inputs. Cell Molec Neurobiol 1:189–207
Uchizono K (1967) Inhibitory synapses on the stretch receptor neurone of the crayfish. Nature (Lond) 214:833–834
Voipio J, Pasternack M, Rydqvist B, Kaila K (1991) Effect of γ-aminobutyric acid on intracellular pH in the crayfish stretch-receptor neurone. J Exp Biol 156:349–361
Wales W (1976) Receptors of the mouthpart and gut of arthropods. In: Mill PJ (ed) Structure and function of proprioceptors in the invertebrates. Chapman and Hall, New York, pp 65–114
Weevers RdeG (1966) The physiology of a lepidopteran muscle receptor. I. The response to stretching. J Exp Biol 44:177–194
Wells MJ (1978) Octopus. Physiology and behaviour of an advanced invertebrate. Chapman and Hall, London
Wendler L, Burkhardt D (1961) Zeitlich abklingende Vorgänge in der Wirkungskette zwischen Reiz und Erregung (Versuche an abdominalen Streckreceptoren dekapoder Krebse). Z Naturforsch 166:464–469
Whitear M (1965) The fine structure of crustacean proprioceptors. II. The thoracico-coxal organs in *Carcinus, Pagurus*, and *Astacus*. Philos Trans R Soc Ser B 248:437–456
Wiersma CAG, Furshpan E, Florey E (1953) Physiological and pharmacological observations on muscle receptor organs of the crayfish, Cambarus Clarkii Girard. J Exp Biol 30:136–150
Wiersma CAG, Pilgrim RLC (1961) Thoracic stretch receptors in crayfish and rocklobster. Comp Biochem Physiol 2:51–64

Chapter 12

Muscle Mechanoreceptors in Nonmammalian Vertebrates

F. Ito, M. Sokabe, and *N. Fujitsuka*[1]

Contents

1 Introduction

Barker (1974) gives a comprehensive review of the anatomy of muscle receptors of mammalian and nonmammalian vertebrates. In the vertebrates, all skeletal, smooth and heart muscles are innervated by free nerve endings. They are supplied by nonmyelinated axons (0.2–1 μm in diameter; classified into group C in vertebrate or group IV in mammals), and some thin myelinated axons (2–6 μm in diameter; group A_δ of vertebrates in general or group III or II in mammals). The endings consist of numerous parallel varicose threads either on the surface of the extrafusal muscle fiber and tendon, or lying in the connective tissue in muscle and joint capsule as well as in the spindle sensory ending of frogs (cf. Barker 1974). The number and length of the nonmyelinated terminal branches varies widely between species and even between individual preparations.

Most parts of the lungs and airways in mammals, except the nose, are distensible and contain proprioceptors in their walls. Free nerve endings of the pulmonary stretch receptor are located in airway smooth muscles, and innervated by C fibers (Widdicombe 1986). About 80–90% of the axons in the abdominal vagi are afferent and unmyelinated in non-ruminant animals. They innervate muscle mechanorecep-

[1]Department of Physiology, Nagoya University School of Medicine, Nagoya 466, Japan

Advances in Comparative and Environmental Physiology, Vol. 10

tors in esophagus, stomach and intestine (Andrews 1986). Those axons and terminals are known to contain many kinds of peptides (Dockray and Sharkey 1986). There is no evidence that these mechanoreceptors of free nerve endings in mammals are different from those in nonmammalian vertebrates.

Sensory endings that are almost as widespread in vertebrates possess axon branches that form plaques of terminals on the surface of skeletal muscle fibers or tendon fascicles. The endings are termed "simple-branching sensory endings" in fish, "leaflike endings" (*terminaisons en buissons*) in amphibia (Kulchitsky 1924; Ito et al. 1964), and "non-encapsulated plaque endings" in reptiles (cf. Barker 1974) and birds (Huber and Dewitt 1897). The leaflike endings of the frog are found mostly in superficial muscles, for example, sartorius; and they are innervated by myelinated sensory axons of 5–11 μm in diameter (Ito et al. 1964). It is not clear whether or not these receptors are homologous with Ruffini endings or Dogiel endings in mammals. The diameter range of the supplying myelinated afferents is 5–10 μm, i.e. an average of 6.3 μm (cf. Barker 1974). The number and length of the nonmyelinated terminal branches were considerably less than those in free nerve endings. The smooth muscle of airways or gastrointestinal tract is also innervated by myelinated vagal afferent axons, with diameters ranging from 1–14 μm in mammals (Andrews 1986). The presence of visceral reflexes indicates that similar sensory endings may be present in the visceral tracts in nonmammalian vertebrates (Nieuwenhuys and Opdam 1976).

Muscle spindles are found in most skeletal muscles in vertebrates, but not in smooth and heart muscles. Simple (monofibral) spindles have been reported in the jaw-closing muscle of the Japanese salmon (Maeda et al. 1983), but thereafter it has been unable to confirm them. The structure and function of the frog muscle spindle have been established by Ottoson (1976) and Ottoson and Shepherd (1971). The frog has many kinds of muscle spindles, from tri-fibral spindles within fast muscles like sartorius to complex tandem spindles or spindle systems in slow muscles, like the red part of semitendinosus (Diwan and Ito 1989). They are innervated by myelinated sensory axons of 7–18 μm in diameter. Although the number and length of nonmyelinated sensory terminals in the spindle vary greatly in relation to the structural complexity of the spindle, they are in general considerably larger than in the leaflike receptors. This implies that the density and/or absolute number of the ionic channels are larger than those in leaflike receptors. Two kinds of monofibral spindles (short- and long-capsular spindles) occur in snake skeletal muscle (Fukami and Hunt 1970). Avian muscle spindles are also classified into two types: thin and thick capsular spindles (Adal and Cheng 1980), which are innervated by thin and thick afferent axons, respectively. There is no evidence of a correlation between the thick axon and Ia afferent in mammalian spindles. This review explores linkages between mechanosensitive channels and functional architecture in the processing of sensory information in the frog muscle spindle.

2 Spindle Preparation and Parameters of Sensory Axon Terminal

Muscle spindles were isolated from hindlimb muscles of frogs (*Rana catesbeiana* or *R. nigromaculata*) for electrical recordings of membrane potential from the sensory axon, using an air-gap method described in detail by Ito et al. (1982), or from the terminal node of the undivided branch of the sensory axon, using microelectrodes (Ito et al. 1974). Mean resting membrane potential of the terminal node was about –65 mV. Mean amplitudes of spontaneous action potentials were about 5 mV at the air gap and 90 mV intracellularly. The attenuation may be due to leakage of action current across the air gap, since a similar attenuation was measured by application of current across the gap. The attenuation was neither a function of the distance between the gap and the terminal node, nor of the transgap resistance (Ito and Fujitsuka 1983). Possibly it was related to differences in the branching patterns, including branch lengths and diameters along the subdivided branch.

Depolarization induced at the terminal node in response to imposed transgap current was on average 18 times less than the transgap depolarization. Assuming proportionality of current and potential, the threshold current needed to evoke an overshooting spike at the terminal node was calculated to be 2 nA (Fig. 1A). This is five times larger than for frog sciatic node (Tasaki 1968), and may be due to the ramification at the terminal and second nodes, where nonmyelinated filaments and myelinated branches arise, respectively. Assuming the attenuation to be applicable to orthodromic spread from the terminal node to the gap, the calculated current at the terminal node during static stretch was 0.9 nA, based on the transgap current of 0.05 nA, as estimated from the mean value of the spindle potential and the transgap resistance (Fig. 1C). This is less than half the spike threshold at the terminal node. Therefore, it is insufficient to provoke afferent impulses at the node. It is concluded that the spindle potential cannot be an actual generator potential, because afferent impulses during the static component cannot be triggered by the spindle potential alone.

3 Distinction Between Spindle Potential and the Actual Receptor Potential

Katz (1950) recorded a depolarization from an isolated sensory axon at a point close to the terminal during stretch of the frog's fourth toe extensor, calling it the spindle potential. He estimated the length constant of the nonmyelinated terminal to be about 70 μm, from an approximate equivalent electric circuit based on parameters derived from his ultrastructural study (Katz 1961). As the lengths of the nonmyelinated axons range up to 300–400 μm and the diameter ranges from 0.1–0.2 μm, the receptor potential evoked in distal regions may contribute little to the initiation of impulses unless active mechanisms are involved. However, Ottoson and Shepherd

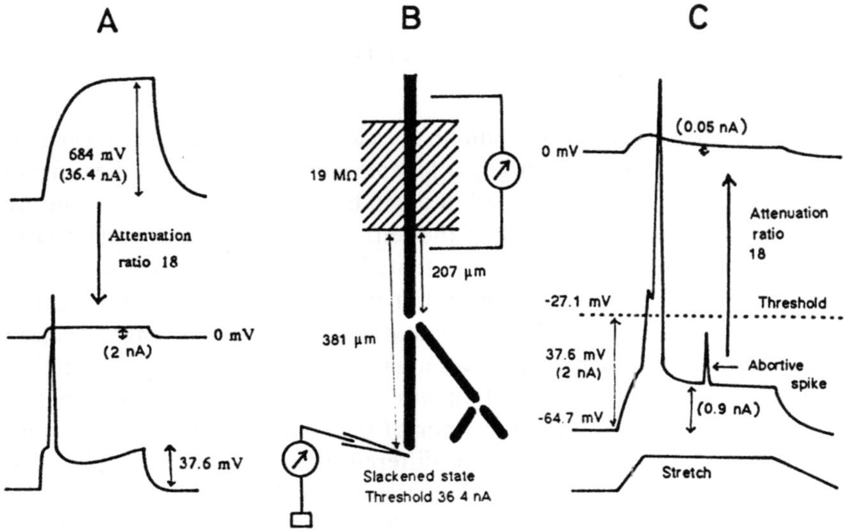

Fig. 1. Schematic diagrams showing inferred and observed current and potential changes associated with the *B* preterminal branches of the afferent axon of the frog muscle spindle in response to transgap application of *A* current or to *C* stretch. *A upper trace* transgap potential recorded during application of a 36.4 nA (spike threshold) transgap current; *middle trace* intracellular zero potential level at the first node with calculated attenuated current superimposed; *lower trace* resting potential and an action potential triggered by the threshold current. *B* Position of recording systems (air gap is shaded) in relation to the afferent preterminal branches. *C Upper trace* transgap current estimated from spindle potential and transgap resistance with its baseline positioned at zero membrane potential level; *middle trace* resting membrane potential and an overshooting action potential superimposed on spindle potential; calculated as though recorded intracellularly during stretch (*lower trace*)

(1971) showed spindle potential of up to 4 mV or more in amplitude from isolated frog spindle and regarded this as a receptor/generator potential.

In addition to the possible discrepancy between spindle and receptor potentials described in the preceding section, experiments on single-type spindles in the frog sartorius provided the following evidence against regarding the spindle potential as a receptor potential: (1) In semi-isolated spindles, afferent spikes were often superimposed on a hyperpolarized spindle potential induced by various procedures (Ito 1969, 1970a,b,c). (2) The amplitude of the spindle potential decreased with increase in temperature (Oyama 1970), whereas the discharge rate increased (Ito and Yokoyama 1978). (3) In isolated spindles, potential changes along the nonmyelinated terminals were virtually unaffected by antidromic polarizing current, as were Ca^{2+} – and abortive spikes (Ito et al. 1980a); whereas full-sized spikes generated at the terminal or branching nodes were modifiable by such currents. (4) Ito et al. (1990b) have recently demonstrated distinctions between the effects of ions or drugs on the spindle-potential amplitude and the afferent-discharge rate. If the discharge rate is linearly dependent on the amplitude of the receptor potential as is generally believed (Loewenstein 1971), the spindle and receptor potentials must be distinguished.

Application of 10 mM TEA induced a slight depolarization in the transgap resting potential and increased the transgap impedance and the amplitudes of spontaneous afferent discharges and of spindle potential, as compared with those in normal Ringer's solution (Ito et al. 1985a). There was a distinct increase in the amplitude of the static component of the spindle potential during maintained stretch, and a relatively small increase in the dynamic component. In Ringer's solution containing both TTX and TEA, afferent discharges were abolished, but the enhanced spindle potential remained (Ito et al. 1985a).

Ca^{2+}-channel blockers (nifedipine, verapamil, $CdCl_2$, $CoCl_2$, or $MnCl_2$) strongly depressed the static component of the spindle potential, although discharges survived for 5–10 min. During the next 5 min a slow hyperpolarization occurred, following a small depolarizing deflection on which a few afferent impulses were superimposed (Fig. 2B). As the hyperpolarization was abolished by addition of K^+-channel blocker (for example, 10 mM TEA as shown in Fig. 2C), the potential seems to be due to the outflow of K^+ from the sensory nerve terminal. The dynamic component of the spindle potential decreased in amplitude and the afferent discharges disappeared 20–30 min later. If the Ca^{2+}-channel blockers were removed within 15 min, afferent discharges recovered in the absence of the spindle potential (Fig. 2D). The small depolarizing deflection during dynamic stretch that remained after treatment with Ca^{2+}-channel blockers may have been induced by influx of cations other than calcium ion. Thus in normal Ringer's solution, the increase in K^+ conductance following the inflow of cations (mainly Na^+ and Ca^{2+}) may attenuate the depolarization potential induced by the inward currents.

Our recent experiments with monovalent cations (Ito et al. 1990b) have demonstrated that N-methyl-D-glucosamine (NMDG) was impermeant in frog spindle sensory terminal activated by stretch. An intact muscle spindle stretched in isotonic NMDG solution produced a hyperpolarizing spindle potential. Consequently, the depolarizing component in normal Ringer's solution may be due to calcium-ac-

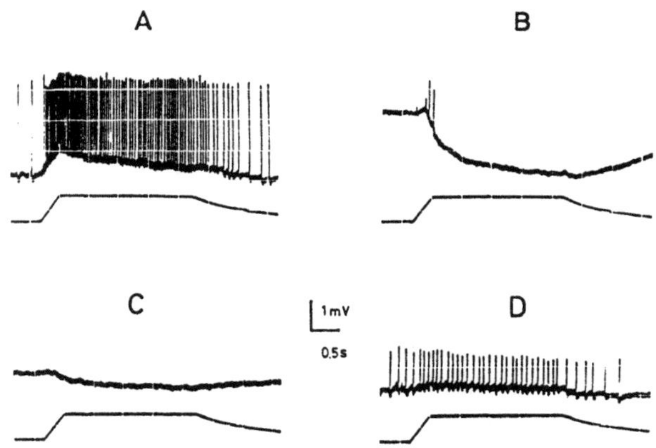

Fig. 2. Effects of Ca^{2+} and K^+ channel blockers on the responses (*upper traces*) of a spindle sensory terminal during a ramp-and-hold stretch (*lower traces*) *A* In normal Ringer's solution. *B* In a Ringer's solution containing 5 mM $CoCl_2$. *C* In a Ringer's solution containing both 5 mM $CoCl_2$ and 10 mM TEA. *D* After returning to normal Ringer's solution

tivated nonspecific channels (I_{NS}^{Ca}), because voltage-gated Ca^{2+} channels have been shown to be present in the distal regions of the axon (Ito and Komatsu 1979). The mechanosensitive channels may also be permeant to calcium ions (Ottoson 1976). Similar selective impermeability to NMDG has been observed in I_{NS}^{Ca} in *Aplysia* pacemaker neuron which helps to boost the burst by generating depolarizing after-potentials that trigger successive spikes (Thompson and Smith 1976; Smith and Thompson 1987; Partridge and Swandulla 1988). However, in frog spindle the current does not trigger successive spikes superimposed on the depolarizing deflection of spindle potential, because the amplitude of the current is insufficient to attain the threshold at the site of origin of the spindle potential (see Fig. 1).

4 The Site of Origin and Supposed Role of the Spindle Potential

The greatest peak depolarization of the spindle potential in intact spindles was 3.4 ± 1.8 mV, when the transgap resistance was 39 meg ohm, which was about twice the mean. The amplitude of superimposed action potentials was reduced by up to 50% during stretch (Figs. 2A, 3A), so that the membrane impedance should decrease with increase in the amplitude of the spindle potential. The lowest impedance, as measured by attenuation of the amplitude of repetitive square pulses applied across the gap, occurred at the peak spindle potential (at the end of the ramp stretch) and was on average 88% of that associated with the resting level (Ito et al. 1985a). The impedance may have changed less than the attenuation of action potentials because of a leakage current across the air gap.

When the distance from the spindle capsule to the gap was increased, the amplitude of the spindle potential decreased after an initial slight increase at 1 mm distance, as shown in Fig. 3D (filled circles). A similar relationship between the spindle-potential amplitude and distance was also observed after elimination of Na^+ spikes with TTX (Fig. 3C and open circles in D), although the amplitude was larger than normal. This increase in amplitude may be due to removal of a shunt circuit by activated channels during high-frequency repetition of spikes. The slight increase in amplitude at 1 mm distance from the capsule implies that the source of the spindle current was located there, where the first node outside the capsule usually occurs.

Microapplication of procaine or crush to a node immediately outside the capsule abolished afferent spikes and spindle potential, whereas abortive spikes remained (Fig. 3B). The amplitude of the abortive spikes decreased exponentially with an increase in the capsule-gap distance (Fig. 3D, crosses). These results support the hypothesis that the spindle potential may be generated at or close to the first node outside the capsule, and also that the abortive spikes may be initiated at a more distal region.

The question naturally arises as to the role played by the spindle potential. Therefore, a study was made to test a working hypothesis that a mass action current due to extrafusal muscle excitation, invading antidromically into the sensory af-

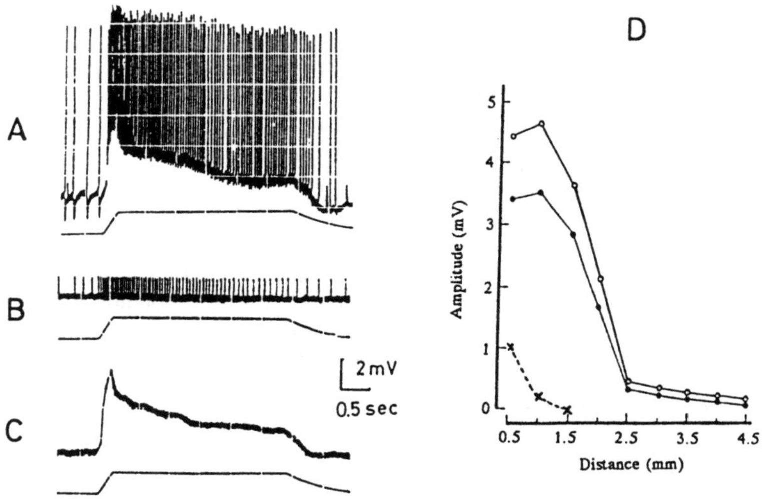

Fig. 3. Spindle potential and abortive spikes. *A* Responses (*upper trace*) of an intact muscle spindle during stretch (*lower trace*). *B* Abortive spikes of muscle spindle with procain applied to a node immediately outside the spindle capsule. *C* Spindle potential of the same preparation as *A* after perfusion with TTX. *D* The peak amplitudes (ordinate) of spindle potential with (*filled circles*) or without (*open circles*) full-sized afferent discharges and of abortive spikes (*crosses*) in relation to distance between the air gap and spindle capsule (*abscissa*)

ferent terminal, may be shortcircuited by the spindle potential so as to prevent misfiring of the muscle spindle. A dorsal-root filament was isolated containing a functionally single muscle-spindle afferent from the semitendinosus muscle of bull frog. A ventral-root bundle was stimulated, while isometric twitch tension, surface EMG, and afferent discharges were recorded simultaneously. In normal Ringer's solution, afferent discharges did not occur during the twitch contraction, apart from the falling phase, when the spindle was stretched by the relaxation of extrafusal muscle (Fig. 4A). However, in Ringer's solution containing TEA, early afferent discharges occurred simultaneously with EMG activity (Fig. 4B). Usually (without TEA treatment) no spikes appeared in the falling phase, probably due to postspike depression (Ito 1968) and also to long-lasting activation of K^+ channels after the mass current of the EMG (Ito and Fujitsuka 1983). After TEA treatment has blocked, certain K^+ channels in the extracapsular nodal membrane where the spindle potential may be initiated, the mass action current of the extrafusal muscle fibers may be able to invade the axon terminal in the absence of a short circuit through the K^+ channels, and may modulate the excitability of the encoding site in the terminal. As the K^+ channels may act as a significant short circuit even in the resting state in normal Ringer's solution, a considerably larger effect to short out the exogenous current might be expected during the spindle potential. When K^+ channels in the extracapsular nodal membrane are activated orthodromically by activation of Ca^{2+} gated nonselective channels, the amplitudes of spindle potential and afferent spikes are reduced considerably, and misfiring due to an exogenous electrotonic current is prevented.

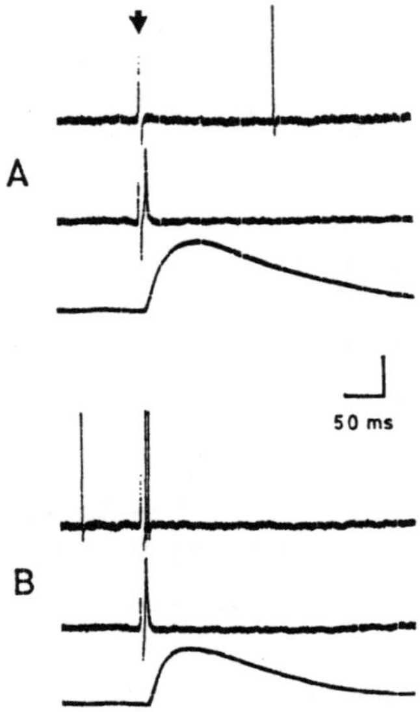

Fig. 4. Misfiring of *B* afferent spikes 45 min after injection of 10 mM TEA Ringer's solution into a semitendinosus muscle, compared with *A* control after injection of normal Ringer's solution. A single shock (*arrows*) was applied to a bundle of the ventral root. Simultaneous records consist of afferent impulses (*upper traces*) recorded from a functionally isolated filament of dorsal root (calibration: 1 mV), EMG (*middle traces*) recorded with bipolar silver electrode on the surface of the muscle (calibration: 10 mV), and isometric tension (*lower traces*) recorded with a strain gauge attached to the pelvic end of the muscle (calibration: 5 g)

5 A model of Primary Processes in Sensory Encoding

Since the ratio of amplitude between abortive and full-sized spikes was similar whether recorded intracellularly from the first node or extracellularly across a paraffin gap on the axon (Ito et al. 1974), the abortive spike was identified as a nonovershooting spike generated at or near the recording site. This assumption is supported by an experiment in which the amplitude of full-sized spikes increased during hyperpolarizing current applied antidromically across the gap and decreased during depolarization, while that of abortive spikes remained unchanged. Current applied across the air gap, bridged by a myelinated segment, may flow mostly through the nodes just outside the gap, and the current density may be attenuated according to distance from the nodes and terminal branching complexity. The density of the applied transmembrane currents at the first node of the nonsubdivided myelinated branch, at which the full-sized spike seems to be initiated, might be much larger than that along individual nonmyelinated filaments. The abortive spikes have been shown to be due to activation of voltage-dependent Na^+ channels, since they were abolished by application of TTX (Ito and Ito 1976). Consequently, they seem to be sodium spikes that are not overshooting due to a low density of Na^+ channels. This also suggests that the density of voltage-dependent Na^- channels declines gradually towards the distal portions of the nonmyelinated filaments (Fig. 5A).

In contrast to Na^+ channels, increase in the density of Ca^{2+} channels can be presumed to occur in the nonmyelinated terminals, since calcium spikes were elicited along the sensory terminal during stretch in Ringer's solution containing TTX (Ito and Komatsu 1979). The amplitude of these spikes depended upon the external Ca^{2+} concentration, and they were blocked by some Ca^{2+} channel blockers ($CoCl^2$ or $MnCl_2$). The calcium spike could also be elicited by electrical pulses, antidromically applied across the air-gap, after the sodium spike had been blocked

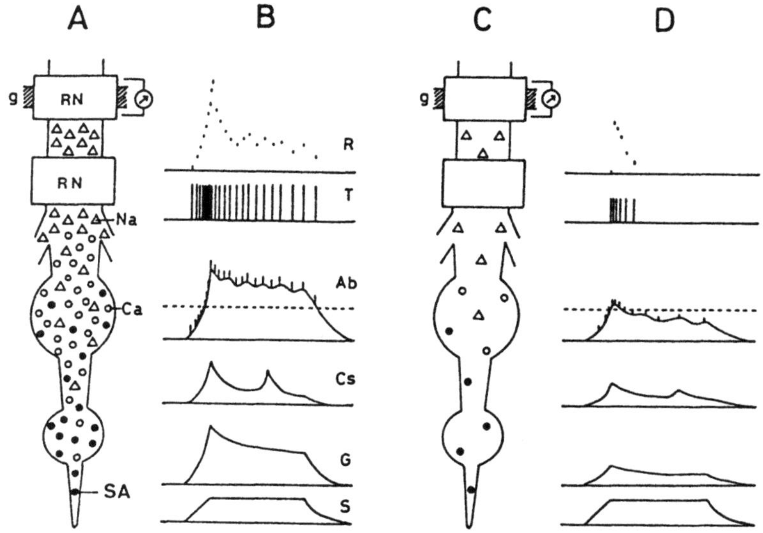

Fig. 5. Model for distribution of *A* ionic channels and *B* supposed partial responses along the sensory terminal of the intact frog muscle spindle, and for those (*C* and *D*) of denervated spindle. When the spindle is stretched (*S*), activation of stretch-activated (*SA*) channels provokes receptor potential (*G*), by which voltage-dependent Ca^{2+} channels (*Ca*) are activated leading to calcium spikes (*Cs*). The calcium spikes and attenuated receptor potential spread electrotonically and converge on the branching point of nonmyelinated branches, then abortive spikes (*Ab*) are elicited by activation of low density Na^+ channels (*Na*). *RN* myelinated segment. *g* Air gap. When the sum of those depolarizing currents exceeds the threshold for high density Na^+ channels, afferent impulses are initiated (*T*). The density of all kinds of ionic channels decreases rapidly after denervation or treatment with protein-synthesis blockers (*C*). The threshold, which may be increased because of the decay in the density of the Na^+ channels, is too high to be exceeded by the lower depolarizing currents, except the peak at the completion of stretch. Consequently, a phasic response occurs during dynamic stretch (*D*)

by TTX (Ito et al. 1981). The threshold current was higher than that for the sodium spikes, suggesting that the voltage-gated calcium channels were distributed distally to the site of afferent impulse initiation, or that the density of Ca^{2+} channels at the encoding site was less than that along more distal portions. Since the amplitude of the calcium spikes was not modifiable by applying polarization currents, they may be generated along the nonmyelinated terminal filaments, as was concluded for the abortive spikes on similar evidence (Ito et al. 1980a).

These results led us to a working hypothesis that ionic channels are distributed asymmetrically along the sensory nerve terminal, as illustrated in Fig. 5A. Na^+ channels occupy the terminal and more proximal nodes, but their density declines gradually towards the tip of the nonmyelinated axon. The density of voltage-dependent Ca^{2+} channels increases from the terminal node to the tip. During stretch of the muscle spindle, a receptor potential occurs by activation of the displacement-dependent ionic channels (cf. Boyd and Rosenberg 1985) and spreads electrotonically towards the proximal protion along the nonmyelinated axon. Voltage-dependent Ca^{2+} channels would then be activated by the receptor potential, producing calcium spikes that are conducted along the numerous branches of the nonmyelinated axon to converge on a branch point, inducing the dynamic component of spindle potential, which drives the membrane potential to threshold for activation of Na^+ channels, producing abortive spikes (see Fig. 5B). When the sum of the depolarizing currents converging on the myelinated axon from its nonmyelinated branches exceeds the threshold for Na^+ channels, whose density increases from the first node to more proximal nodes, full-sized afferent spikes occur.

Several varieties of K^+ channel are also present in the nonmyelinated axon, although they are not illustrated in Fig. 5A. In addition to the description of Section 2 in this chapter, a delayed rectification was observed following a large shift in potential during application of a depolarizing current in Ringer's solution containing TTX (Ito et al. 1980b). Calcium activated potassium conductance was found to provoke a rhythmic hyperpolarization in the terminal during imposed depolarizing current (Ito et al. 1982). The rhythmic hyperpolarization is associated with the static component of the potential, and may be the immediate cause of it.

6 Transduction and Mechanosensitive Channel: Ultrastructural Aspects of Frog Muscle Spindle

There is no evidence regarding the transduction in the frog muscle spindle, as the actual receptor potential has not yet been identified. We therefore are attempting to identify mechanosensitive (MS) channels in intramembrane particles (IMP).

In freeze-fractured muscle spindle, the nonmyelinated axons were identified as cylindrical processes embedded in the inner capsular space, as shown in Fig. 6A. They were divided into two regions based on diameter; trunks (> 320 nm in diameter) and branches (< 320 nm) which included chains of expanded bulbs connected by fine links 80–200 nm in diameter (cf. Katz 1961; Karlsson et al. 1966; Fujitsuka and Ito 1983). The bulbs can be readily identified in the replicas as ball-like enlargements connected by fine tubes. The majority of the IMPs along the nonmyelinated axons were associated with the protoplasmic (P) face, and were usually evenly distributed (Fig. 6B,C).

Average IMP density in the P face was 919 μm^{-2}. In the fine tubes and bulbs the density was less, while in the middle and large branches it was more (Ito et al. 1985b). Neither significant clumping nor linear groups (cf. Flower 1971) of IMPs could be detected. The mean density in the nodal P face of the parent axon was 1308 μm^{-2} (cf. Fujitsuka et al. 1990).

The IMP density declined rapidly after crush of the sciatic nerve to less than 10% of normal after 7 days. Spontaneous discharges were absent and the static response to stretch was reduced in all spindles 3 days after denervation. A phasic response was still present after 4 days, as shown schematically in Fig. 5D (cf. Fujitsuka et al. 1987). Similar reductions of IMP density and of static response were found within a few days after injection of protein-synthesis blockers (Ito et al. 1989), suggesting that the IMPs are turned over quite rapidly. In a state of low density of IMPs after denervation or treatment with protein-synthesis blockers, the low amplitudes of abortive and calcium currents are presumed to be insufficient to exceed

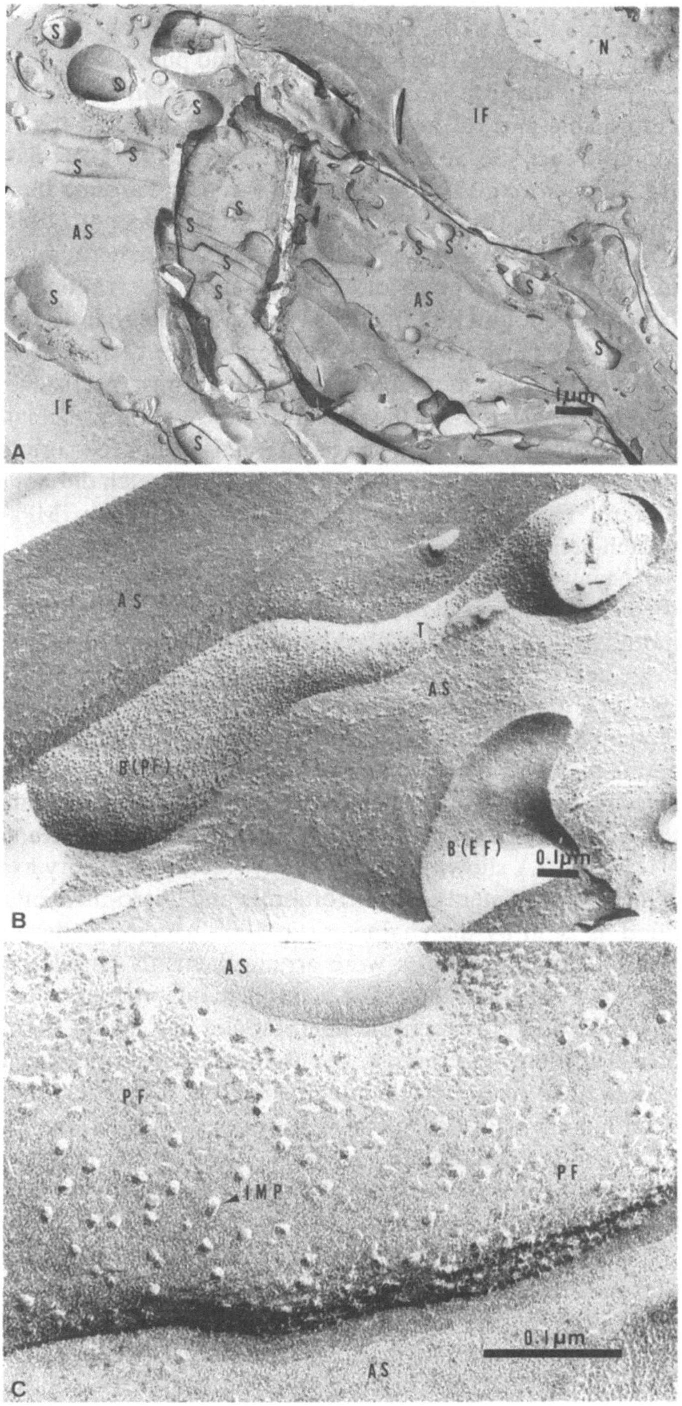

Fig. 6. Freeze-fracture replica of sensory nerve terminal along the reticular zone of intact frog muscle spindles. *A* Low-power picture; *S* sensory terminal; *AS* axial space; *IF* intrafusal muscle fibers; *N* nucleus of the intrafusal fiber. *B* High magnification of a bulb region; *B(PF)* protoplasmic face of bulb region; *B(EF)* external face of bulb region; *T* fine tube; *C* p-face (*PF*) membrane of nonmyelinated axon showing dense IMPs (*IMP*)

normal spike threshold, except during the dynamic phase of stretch. It is concluded that some IMPs in the spindle terminal are MS channels, although many may represent several kinds of ionic channels and pumps.

Using electron microscopy, the bulb region has been shown to be the most deformable part of the nonmyelinated terminal and the volume of the bulb was found to increase during stretch (Kim et al. 1985). A relatively higher density of IMPs larger than 12 nm in diameter was found around the bulb region than in the large branches (Fujitsuka et al. 1989). Fujitsuka et al. (1987) have identified IMPs larger than 12 nm in diameter as the MS channel, because such large IMPs survived until the phasic response disappeared after denervation. Recently, Fujitsuka et al. (1989) have found that the density of IMPs less than 12 nm in diameter increased by 100% or more 1 week after tenotomy, while there were no changes in density of IMPs of 12 nm or more in diameter. By contrast, persistent lengthening for 8 days resulted in a decrease in the small IMP density to approximately 60–70% of control, without any changes in the large IMP density. For the periods of the tenotomy or of the persistent lengthening, the responses to stretch did not differ from the normal pattern. This indicates that the density of the large IMPs is associated with the stretch-activated function.

Many recent reports have suggested that neurofilaments and microtubules of the cytoskeleton could play an important role in neuronal activities, for example, motility of growth cone, plasticity, excitation and axonal transport, and also that those activities are influenced both extracellularly and intracellularly through the cytoskeleton (Mills and Lubin 1986; Forscher 1989; Hoffman and Simonsen 1989; Hollenbeck 1989). Inspired by the suggestion that the arrangement of actin filaments may be closely related to synaptic transmission (Hirokawa et al. 1989), we investigated the arrangement of the cytoskeleton in the sensory axon terminals of the frog muscle spindle by quick-freeze, deep-etch electron microscopy.

Figure 7A shows a longitudinally fractured sensory axon terminal. The main cytoskeletal elements are microtubules and micro-filaments. The latter were identified as actin by the characteristic 5-nm banding pattern (Heuser and Kirshner 1980). The microtubules were arranged mainly parallel to the axis of the nonmyelinated sensory axon. The actin filaments were arranged transversely, forming loose networks between the microtubules and axonal membrane. Extracellularly fine bridgelike structures linked the sensory-axon membrane with the intrafusal muscle-fiber membrane and the filamentous network in the periaxial space via the basal laminae (Fig. 7). We have no evidence yet on the connection between the IMPs and the cytoskeletal elements. However, the meshwork consisting of extra- and intra-axonal filaments may transmit externally applied stretch leading to deformation of the sensory terminal.

The basal lamina or extracellular matrix has also been considered as a factor in transduction. Fujitsuka and Ito (1983) have shown the presence of the basal lamina outside the sensory axon terminal with electron-microscopy of ultra-thin sections, after incubation with 0.5 mM ruthenium red, which is known to selectively stain the extracellular matrix. A layer of small electron-dense deposits outside the terminal membrane was found. Similar deposits could also be demonstrated inside and outside of intrafusal muscle fibers and spindle capsules.

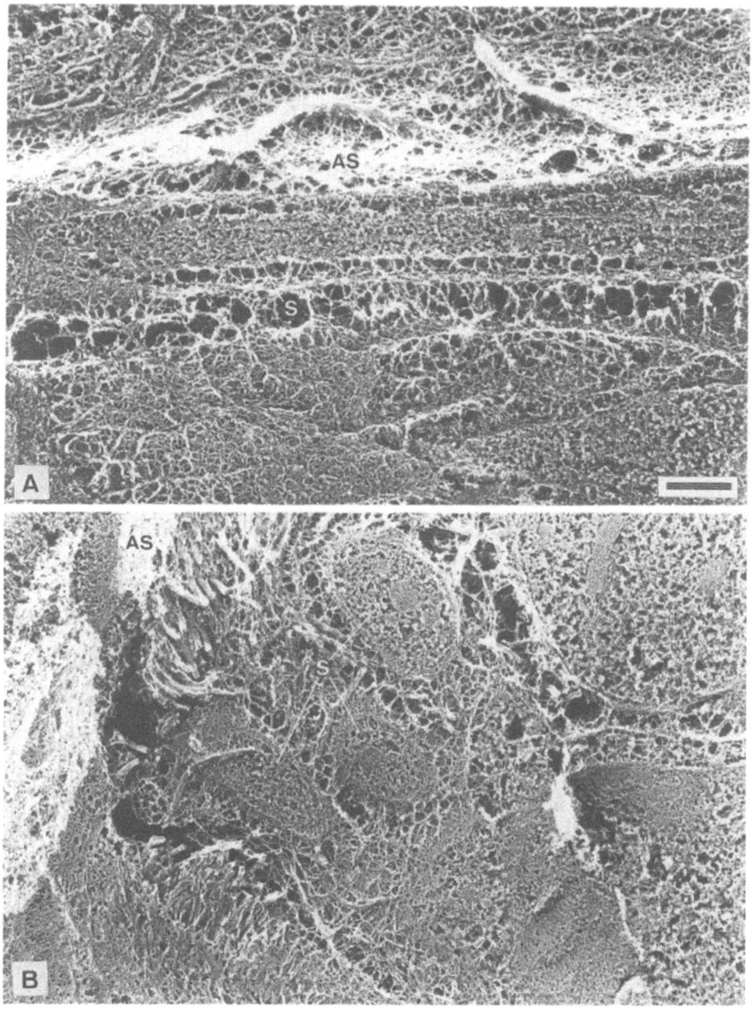

Fig. 7. Electron micrographs of deep-etched sensory terminals of intact frog muscle spindles. *A* Longitudinal fractured tip of the nonmyelinated terminal. *B* Transverse fractured aspect of the terminal. *AS* Axial space; *S* sensory axon terminal. Calibration: 200 nm

7 Second Messengers in Transduction and Encoding Processes

Cytosolic calcium and cyclic nucleotides are regulators for a wide range of cellular functions (Rasmussen and Tenenhouse 1968; Rasmussen 1970; Rasmussen and Goodman 1977). In sensory systems, both cyclic nucleotides and Ca^{2+} have been proposed as intracellular messengers for coupling light activation of rhodopsin to channel activity (Lipton et al. 1977; Miller and Nicol 1979; George and Hagins 1983; Fesenko et al. 1985), and also for coupling odorant or flavor receptors and olfactory or taste transduction (Pace et al. 1985; Tonosaki and Funakoshi 1988; Roper 1989).

In decapsulated frog muscle spindles (Fujitsuka et al. 1988), application of dibutyryl cyclic AMP and forskolin increased the rate of spontaneous discharges and decreased the responsiveness to stretch (Fig. 8), as did the Ca^{2+} ionophore A23187 (Ito et al. 1988; Sokabe et al. 1988). These effects were prevented by addition of $CaCl_2$ or $BaCl_2$ to the drugs. Similar changes in the afferent discharges followed treatment with isobutylmethyl-xanthine, which enhances cAMP activity by inhibiting cAMP phosphodiesterase (Beavo et al. 1970; Montague and Cook 1971). Carbonyl cyanide chlorophenyl-hydrazone, which elevates intracellular calcium levels by uncoupling mitochondrial transport (Glagoleva et al. 1970; Bygrave 1978), increased the rate of spontaneous discharges, but both the discharge rate and spindle-potential amplitude decayed gradually during stretch (Sokabe et al. 1988). Quercetin, a calcium pump blocker, produced a prolonged increase in the spontaneous discharge rate (Sokabe et al. 1988). Theophilline, the phosphodiesterase inhibitor, increased the amplitudes of afferent spikes and the spindle potential as did K^+ channel blockers by enhancing the release of Ca^{2+} from a storage site, perhaps intra-axonal or in the axial space (Ito et al. 1984). The rate of spontaneous discharges increased after exposure to prostaglandin Bx, which acts as a potent Ca^{2+} ionophore by complexing with Ca^{2+} (Ohnishi and Devlin 1979). These results suggest that cAMP is involved in the regulation of the mechanoelectric encoding processes of the frog muscle spindle through a modification of intracellular Ca^{2+}.

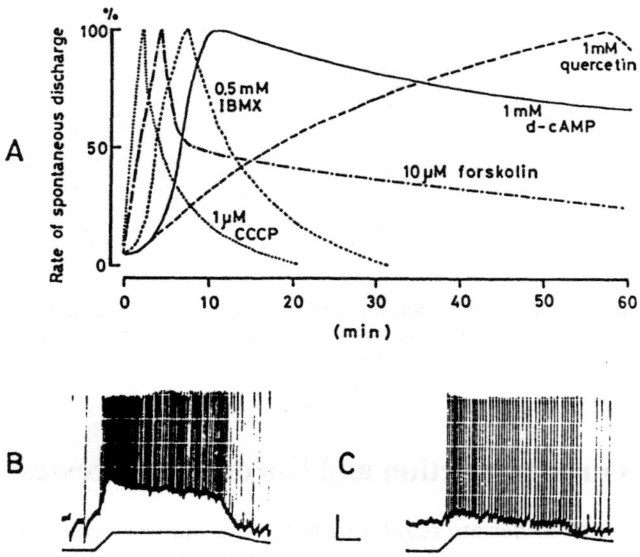

Fig. 8. Effects of drugs modulating intracellular Ca^{2+} concentration, including cAMP. *A* The rate of spontaneous discharges from slackened muscle spindles were plotted with time after application of the drugs. The rates are normalized to the maximum (100%). *B* and *C* Responses (*upper traces*) of a muscle spindle during a ramp-and-hold stretch (*lower traces*) in *B* normal Ringer's solution and *C* in a Ringer's solution containing 2 mM dibutyryl cAMP. Calibration: 1 mV, 0.5 s

The fact that increasing the external calcium concentration results in a decrease of the rate of spontaneous discharges suggests that the rate may be determined by a balance between external and internal Ca^{2+} activities across the sensory terminal membrane (Ito et al. 1988). A possible explanation is that Ca^{2+} may affect the rate of discharges by altering the membrane potential of sensory terminals through screening, binding membrane surface charges or by activating some kinds of channels via second messengers; for example, inositol phosphates and protein kinase (Berridge and Irvin 1989).

Protein phosphorylation is an important mechanism in the transduction of extracellular signals that modulate diverse cellular processes such as metabolism, excitability, neurotransmitter release, and synaptic plasticity (Zigmond et al. 1989). Two major classes of protein kinase, Ca^{2+}- and cyclic nucleotide-dependent, have been described in the nervous system (Nairn et al. 1985). Influx of Ca^{2+} into synaptosomes leads to a Ca^{2+}-dependent phosphorylation of many endogenous proteins (Krueger et al. 1977; Wu et al. 1982; Robinson and Dunkley 1983; Wang et al. 1988) and the activation of Ca^{2+}/calmodulin-, Ca^{2+}/phospholipid-dependent protein kinases and protein phosphatases (Dunkley et al. 1986; Robinson et al. 1987). Ca^{2+} influx also activates adenyl cyclase (Cheung et al. 1975), leading to an increase in intracellular cAMP which could activate cAMP-dependent protein kinase (cf. Yip and Kelly 1989).

We have recently reported that the rate of spontaneous discharge, the amplitude of spindle potential, and responses during stretch were reduced by application of H-7 or retinoic acid, inhibitors of protein-kinase C. The enhanced rate of spontaneous discharge by cAMP was also minimized by addition of H-7 (Ito et al. 1990a). Phorbolester, an activator of protein kinase C, caused a delayed increase in the rate of spontaneous discharges (Table 1). These results suggest that the decreased channel activity caused by cAMP may be attributed to a phosphorylation-related reaction of a channel protein itself, a membrane regulatory protein closely associated with the channel or a protein component of the intracellular cytoskeletal network closely adherent to the channel protein molecule.

It is known that cAMP opens Ca^{2+}-activated K^+ channels by protein phosphorylation in the neurons of the *Helix* oesophageal ganglion (Ewald et al. 1985), and also that a cAMP-dependent protein kinase closes the serotonin-sensitive K^+ channel by a phosphorylation-related mechanism (Shuster et al. 1985). This suggests that channels normally gating inward cation current and several different types of outward K^+ channels can be affected differently by cAMP-induced phosphorylation of the channel protein, or a membrane-bound regulation protein.

Although increased levels of cAMP have been documented in spindle sensory terminals following spontaneous discharge activation by the enhancement of intracellular Ca^{2+}, the role of cAMP in the regulation of the mechano-electric transduction has mainly been studied by its action at the genomic level to increase protein synthesis. As to whether cAMP displaces Ca^{2+} from the Ca^{2+} binding site of the channel protein, it is possible to speculate that the change in the channel protein conformation by cAMP, may conceal the Ca^{2+}-binding site which would therefore reduce the stimulatory action of Ca^{2+} on the channel, as has been supposed in bovine pituitary gonadotrophs by Sikdar et al. (1989). They have

Table 1. Effects of second messengers on responses of frog muscle spindle

		Concent. (mM)	Spont.d.r.[a]	Stretch r.[b]	Latency[c] (min)
C kinase					
Activators[d]	Phorbol 12-myristate 13 acetate	1	↑	↑	30
	Phorbol 12,13-didecanoate	1	↑	↑	35
Inhibitors	retinoic acid	1	↓	↓	10
	H–7	0.1	↓	↓	10
	Acetylsalicylic acid	2	↓	↓	0.5
A kinase					
Activators	Dibutyryl cyclic AMP	0.1	↑	↓	20
	Forskolin	0.1	↑	↓	2
	IBMX	0.5	↑	↓	2
	Dibutyryl cyclic GMP	1	–[e]	–	
	Pertussis toxin	0.01	–	–	
$(Ca^{2+})_1$					
Enhance	A23187	0.1	↑	↓	2
	Prostaglandin Bx	0.002	↑	↓	2
	Carbonyl cyanide m-chlorophenyl-hydrazone (CCCP)	0.001	↑	↓	0.5
	Quercetin	0.1	↑	↓	1
	W–7	0.1	–	–	
	$[Ca^{2+}]_0$	2	↓	↑	0.5

[a]Rate of spontaneous discharge.
[b]Rate of discharge during stretch.
[c]Latent time for occurrence of the effect.
[d]Upward arrows, increasing the rate of discharge. Downward arrows, decreasing the rate of discharge.
[e]No effect on the rate of discharge.

demonstrated that Ca^{2+} promoted an increase in the open probability (P_o) of Ca^{2+}-activated K^+ channel, possibly by affecting gating kinetics of the channel (Magleby and Pallotta 1983), while cAMP decreased P_o. The inhibitory influence of cAMP on the channel was retained in the presence of heightened channel activity promoted by Ca^{2+}, in contrast to the results from the spindle. This suggests two different modes of channel modulation, acting differently on the same channel in the gonadotrophs, whereas in the spindle only one mode acts on the same channel.

References

Adal MN, Cheng BC (1980) The sensory ending of duck muscle spindles. J Anat 131:657–668
Andrews PLR (1986) Vagal afferent innervation of the gastrointestinal tract. Prog Brain Res 67:65–86
Barker D (1974) The morphology of muscle receptors. In: Hunt CC (ed) Handbook of sensory physiology, III/2. Springer, Berlin Heidelberg New York, pp 1–190

Beavo JA, Rogers NL, Crofford OB, Hardman JG, Sutherland EW, Newman EV (1970) Effects of xanthine derivatives on lipolysis and on adenosine 3',5'-monophosphate phosphodiesterase activity. Mol Pharmacol 6:597–603

Berridge MJ, Irvine RF (1989) Inositol phosphates and cell signalling. Nature (Lond) 341:197–205

Boyd IA, Rosenberg J (1985) Review. In: Boyd IA, Gladden MH (eds) The muscle spindle. Macmillan, London, pp 327–352

Bygrave FL (1978) Mitochondria and the control of intracellular calcium. Biol Rev 53:43–74

Cheung WY, Bradham LS, Lynch TJ, Lin YM, Tallan EA (1975) Peptide activation of cyclic 3':5'-nucleotide phosphodiesterase of bovine or rat brain also activates its adenylate cyclase. Biochem Biophys Res Commun 66:1055–1062

Diwan FH, Ito F (1989) Intrafusal muscle fibre types in frog spindles. J Anat 163:191–200

Dockray GJ, Sharkey KA (1986) Neurochemistry of visceral afferent neurones. Prog Brain Res 67:133–148

Dunkley PR, Baker CM, Robinson PJ (1986) Depolarization-dependent protein phosphorylation in rat cortical synaptosomes: characterization of active protein kinases by phosphopeptide analysis of substrates. J Neurochem 46:1692–1703

Ewald DA, Williams A, Levitan IB (1985) Modulation of single Ca^{2+}-dependent K^{+}-channel activity by protein phosphorylation. Nature (Lond) 315:503–506

Fesenko EE, Kolensnikov SS, Lyubarsky AL (1985) Induction by cyclic GMP of cationic conductance in plasma membrane of retinal rod outer segment. Nature (Lond) 313:310–313

Flower NE (1971) Particles within membranes – a freeze-etch view. J Cell Sci 9:435–441

Forscher P (1989) Calcium and polyphosphoinositide control of cytoskeletal dynamics. Trend Neurosci 12:468–474

Fujitsuka N, Ito F (1983) Uptake of ruthenium red by the sensory nerve terminal of the frog muscle spindle. Okajimas Folia Anat Jpn 60:1–16

Fujitsuka N, Hama K, Ito F, Sokabe M (1987) Intramembrane particles and terminal responses following denervation of frog muscle spindles. J Neurocytol 16:185–194

Fujitsuka N, Kori AA, Sokabe M, Ito F (1988) Modification of sensory-terminal responses and membrane structure of frog muscle spindle by collagenase. Brain Res 443:243–253

Fujitsuka N, Yoshimura A, Sokabe M, Ito F (1989) Intramembrane particles associated with stretch-activation function in the sensory axon terminal of frog muscle spindle. Neurosci Res Suppl 9:165

Fujitsuka C, Fujitsuka N, Diwan FH, Hama K, Sokabe M, Ito F (1990) Intramembrane particles and responses of sensory axon terminals during reinnervation of the frog muscle spindle. J Neurocytol 19:176–186

Fukami Y, Hunt CC (1970) Structure of snake muscle spindles. J Neurophysiol 33:9–27

George JS, Hagins WA (1983) Control of Ca^{2+} in rod outer segment disks by light and cyclic GMP. Nature (Lond) 303:344–348

Glagoleva IM, Liberman EA, Khashayev ZKh (1970) Effect of uncoupling agents of oxidative phosphorylation on the release of acetylcholine from nerve endings. Biofizika 15:76–82

Heuser JE, Kirshner MW (1980) Filament organization resolved in platinum replica of freeze-dried cytoskeleton. J Cell Biol 86:212–234

Hirokawa N (1989) The arrangement of actin filaments in the postsynaptic cytoplasm of the cerebellar cortex revealed by quick-freeze deep-etch electron microscopy. Neurosci Res 6:269–275

Hoffmann EK, Simonsen LO (1989) Membrane mechanisms in volume and pH regulation in vertebrate cells. Physiol Rev 67:315–382

Hollenbeck PJ (1989) The transport and assembly of the axonal cytoskeleton. J Cell Biol 108:223–227

Huber GC, DeWitt LM (1897) A contribution on the motor nerve endings and the nerve endings in muscle spindle. J Comp Neuol 7:169–230

Ito F (1968) Recovery curves of thresholds of muscle spindle, leaflike and tendon receptors in the frog sartorius muscle after an antidromic discharge. Jpn J Physiol 18:731–745

Ito F (1969) The effect of ions on the steady and generator potential of frog muscle spindles. Proc Jpn Acad 45:409–412

Ito F (1970a) Spindle potential of the frog muscle spindle depending upon the exclusion of extrafusal muscle fiber. J Physiol Soc Jpn 32:249–250

Ito F (1970b) Effects of polarizing currents on long-lasting discharges in the frog muscle spindle. Jpn J Physiol 20:697–710
Ito F (1970c) The behavior of frog muscle spindle in hyper- and hypotonic solutions. Jpn J Physiol 20:394–407
Ito F, Fujitsuka N (1983) Electrical threshold of the sensory nerve terminal of the frog muscle spindle: a role of spindle potential for generating afferent impulses. Neurosci Lett 37:233–237
Ito F, Ito Y (1976) Selective blockage of the sensory terminal activities by micro-application of tetrodotoxin in the frog muscle spindle. Brain Res Bull 1:363–366
Ito F, Komatsu Y (1979) Calcium-dependent regenerative responses in the afferent nerve terminal of the frog muscle spindle. Brain Res 175:160–164
Ito F, Yokoyama N (1978) Potential deflections at the terminal of the frog muscle spindle during stretch. Nagoya J Med Sci 40:13–23
Ito F, Toyama K, Ito R (1964) A comparative study on structure and function between the extrafusal receptor and the spindle receptor in the frog. Jpn J Physiol 14:12–33
Ito F, Kanamori N, Kuroda H (1974) Structural and functional asymmetries of myelinated branches in the frog muscle spindle. J Physiol 241:389–405
Ito F, Komatsu Y, Kaneko N (1980a) Site of origin of calcium spike in frog muscle spindle. Brain Res 202:459–463
Ito F, Komatsu Y, Kaneko N (1980b) Delayed rectification in the afferent nerve terminal of frog muscle spindle. Integrated Control Function of Brain 3:65–66
Ito F, Komatsu Y, Fujitsuka N (1981) Calcium spike induced by electrical stimulation to the sensory nerve terminal of the frog muscle spindle. Neurosci Lett 27:135–137
Ito F, Komatsu Y, Fujitsuka N (1982) $G_{K(Ca)}$-dependent cyclic potential changes in the sensory nerve terminal of frog muscle spindle. Brain Res 252:39–50
Ito F, Fujitsuka N, Hanaichi T (1984) Effects of dantrolene and methylxanthines on the sensory nerve terminal of the frog muscle spindle. Brain Res 294:269–280
Ito F, Fujitsuka N, Fan XL (1985a) Reversal of the static component of spindle potential by imposed depolarizing current in the frog muscle spindle. Brain Res 326:107–116
Ito F, Fujitsuka N, Funahashi A, Hama K (1985b) Ionic channels in the sensory terminal of the frog muscle spindle. In: Boyd IA, Gladden MH (ed) The muscle spindle. Macmillan, London, pp 353–358
Ito F, Sokabe M, Diwan FH, Fujitsuka N, Yoshimura A (1988) Effects of intracellular Ca^{2+} on the frog muscle spindle in relation to cyclic AMP action. In: Hnik P, Soukup T, Vejsada R, Zelena J (eds) Mechanoreceptors. Plenum Press, New York, pp 195–199
Ito F, Fujitsuka N, Sokabe N, Yoshimura A (1989) Effects of protein-synthesis blocker on intramembrane particles in sensory axon terminal of frog muscle spindle. Neurosci Res (Suppl) 9:167
Ito F, Sokabe M, Fujitsuka N, Yoshimura A (1990a) Activities of sensory nerve terminal of frog muscle spindle and C kinase. Neurosci Res 11:S130
Ito F, Sokabe M, Nomura K, Naruse K, Fujitsuka N, Yoshimura A (1990b) Effects of ions and drugs on the responses of sensory axon terminals of decapsulated frog muscle spindle. Neurosci Res 12:S15–26
Karlsson U, Andersson-Cedergren E, Ottoson D (1966) Cellular organization of the frog muscle spindles as revealed by serial sections for electron microscopy. J Ultrastruct 14:1–35
Katz B (1950) Depolarization of sensory terminal and the initiation of impulses in the muscle spindle. J Physiol 111:261–282
Katz B (1961) The terminations of the afferent nerve fibre in the muscle spindle of the frog. Philos Trans R Soc Lond B 243:221–240
Kim N, Fujitsuka N, Ito F (1985) Ultrastructural changes of the sensory nerve terminals in frog muscle spindle during dynamic stretch. J Neurocytol 14:105–112
Krueger BK, Forn J, Greengard P (1977) Depolarization-induced phosphorylation of specific proteins, mediated by calcium ion influx, in rat synaptosomes. J Biol Chem 252:2764–2773
Kulchitsky N (1924) Nerve endings in muscles of the frog. J Anat 59:1–43
Lipton SA, Rasmussen H, Dowling JE (1977) Electrical and adaptive properties of rod photoreceptors in *Bufo marinus*. II. Effects of cyclic nucleotides and prostaglandins. J Gen Physiol 70:771–791

Loewenstein WR (1971) Mechano-electric transduction in the Pacinian corpuscle. Initiation of sensory impulses in mechanoreceptors. In: Loewenstein WR (ed) The handbook of sensory physiology, vol 1. Springer, Berlin Heidelberg New York, pp 269–290

Maeda N, Miyoshi S, Toh H (1983) First observation of a muscle spindle in fish. Nature (Lond) 302:61–62

Magleby KL, Pallotta BS (1983) Calcium dependence of open and shut interval distributions from calcium-activated potassium channels in cultured rat muscle. J Physiol 344:585–604

Miller WH, Nicol GD (1979) Evidence that cyclic GMP regulates membrane potential in rod photoreceptors. Nature (Lond) 280:64–66

Mills FW, Lubin M (1986) Effect of adenosine 3′,5′-cyclic monophosphate on volume and cytoskeleton of MDCK cells. Am J Physiol 250:C319–C324

Montague W, Cook JR (1971) The role of adenosine 3′,5′-cyclic monophosphate in the regulation of insulin release by isolated rat islets of Langerhans. Biochem J 122:115–120

Nairn AC, Hemmings Jr NC, Greengard P (1985) Protein kinases in the brain. Annu Rev Biochem 54:931–976

Nieuwenhuys R, Opdam P (1976) Structure of the brain stem. In: Llinas R, Precht W (eds) Frog neurobiology. Springer, Berlin Heidelberg New York, pp 811–855

Ohnishi ST, Devlin TM (1979) Calcium ionophore activity of a prostaglandin B_1 derivative (PGBx). Biochem Biophys Res Commun 89:240–245

Ottoson D (1976) Morphology and physiology of muscle spindle. In: Llinas R, Precht W (eds) Frog neurobiology. Springer, Berlin Heidelberg New York, pp 643–675

Ottoson D, Shepherd GM (1971) Transducer properties and integrative mechanisms in the frog's muscle spindle. In: Loewenstein WR (ed) Handbook of sensory physiology. Springer, Berlin Heidelberg New York, pp 442–499

Oyama O (1970) The effects of temperature on the dynamic and static sensitivities of the frog muscle spindle. Nagoya J Med Sci 33:13–25

Pace U, Hanski E, Salomon Y, Lancet D (1985) Odorant-sensitive adenylate cyclase may modify olfactory reception. Nature (Lond) 316:255–258

Partridge LD, Swandulla D (1988) Calcium-activated non-specific cation channels. Trends Neurosci 11:69–72

Rasmussen H (1970) Cell communication, calcium ion, and cyclic adenosine monophosphate. Science 170:404–412

Rasmussen H, Goodman DBP (1977) Relationship between calcium and cyclic nucleotides in cell activation. Physiol Rev 57:421–509

Rasmussen H, Tenenhouse A (1968) Cyclic-adenosine monophosphate, calcium and membranes. Proc Natl Acad Sci USA 59:1364–1370

Robinson PJ, Dunkley PR (1983) Depolarization-dependent protein phosphorylation in rat cortical synaptosomes: factors determining the magnitude of the response. J Neurochem 41:909–918

Robinson PJ, Hauptschein R, Lovenberg W, Dunkley P (1987) Dephosphorylation of synaptosomal proteins P96 and P139 is regulated by both depolarization and calcium, but not by a rise in cytosolic calcium alone. J Neurochem 48:187–195

Roper SD (1989) The cell biology of vertebrate taste receptors. Annu Rev Neurosci 12:329–353

Shuster MJ, Camardo JS, Siegelbaum SA, Kandel ER (1985) Cyclic AMP-dependent protein kinase closes the serotonin-sensitive K^+ channels in *Aplysia* sensory neurones in cell-free membrane patches. Nature (Lond) 313:392–395

Sikdar SK, McIntosh RP, Mason WT (1989) Differential modulation of Ca^{2+}-activated K^+ channels in ovin pituitary gonadotrophs by GnRH, Ca^{2+} and cyclic AMP. Brain Res 496:113–123

Smith SJ, Thompson SH (1987) Slow membrane current in bursting pace-maker neurones of *Tritonia*. J Physiol 382:425–448

Sokabe M, Fujitsuka N, Kori AA, Ito F (1988) Effects of cyclic nucleotides and calcium on transduction and encoding processes in frog muscle spindle. Brain Res 443:254–260

Tasaki I (1968) Nerve Excitation. Charles Thomas, Illinois

Thompson SH, Smith SJ (1976) Depolarizing afterpotentials and burst production in molluscan pacemaker neurons. J Neurophysiol 39:153–161

Tonosaki K, Funakoshi M (1988) Cyclic nucleotides may mediate taste transduction. Nature (Lond) 331:354–356

Wang JKT, Walaas SI, Greengard P (1988) Protein phosphorylation in nerve terminals: comparison of calcium/calmodulin-dependent and calcium/diacylglycerol-dependent systems. J Neurosci 8:281–288
Widdicomb JG (1986) Sensory innervation of the lungs and airways. Prog Brain Res 67:49–64
Wu WC-S, Walaas SI, Greengard P (1982) Calcium/phospholipid regulates phosphorylation of a M_r "87K" substrate protein in brain synaptosomes. Proc Natl Acad Sci USA 79:5249–5253
Yip RK, Kelly PT (1989) In situ protein phosphorylation in hippocampal tissue slices. J Neurosci 9:3618–3630
Zigmond RE, Schwarzschield MA, Rittenhous AR (1989) Acute regulation of tyrosine hydroxylase by nerve activity and by neurotransmitters via phosphorylation. Annu Rev Neurosci 12:415–461

Chapter 13
Muscle Receptors in Mammals

M.H. Gladden[1]

Contents

1 Introduction

Hunt (1990) has recently written a review on mammalian muscle spindles. His review would have fitted well in this Volume, as it includes discussion of Hunt's own work on receptor potentials in muscle spindles and also comparative information. It may be asked what another review could add, coming so shortly afterwards. Rather than offering an alternative version, this chapter focusses on the work of one man, I.A. Boyd, who had a powerful influence on spindle research during his

[1]Institute of Physiology, Glasgow University, Glasgow, G12 8QQ, Scotland

Advances in Comparative and Environmental Physiology, Vol. 10

lifetime, an influence still felt at symposia which have taken place since his death (on the fusimotor system in Helsinki, July 1989, and on proprioceptive feedback in motor control in Glasgow, June 1990). Although Boyd died in September, 1987, it is still possible to reproduce the views that he held at that time from work published just prior to his death, and from unpublished work. He had the reputation for stating his views clearly, sometimes too clearly, so that the situation was oversimplified. But in retrospect this had the beneficial effect of offering a challenge to others, and as a result exacting and difficult work was done which might never have been embarked upon had Boyd not provided the stimulus. It is perhaps an opportune time to examine his work now, before the subject has moved on too far, to identify pointers which may help in the development of the subject.

An individual's concept of muscle spindles depends of course on past experience, and is influenced considerably by the methods which that individual employed. Like Hunt, Boyd worked not only with spinal root preparations, recording from spindle afferents in dorsal root filaments while stimulating fusimotor efferents in ventral root filaments, a technique which he in fact learned from Hunt as a young man, but also on dissected, or "isolated" spindles, and in this technique Boyd was himself a pioneer. However, Boyd's approach was much influenced by his early work on the structure of cat muscle spindles, during which he discovered that spindle muscle fibres fell into two groups, nuclear bag fibres and nuclear chain fibres. For the rest of his life his research was dominated by a kind of mission to unravel what physiological significance the various intrafusal muscle fibre types, with their differing mechanical properties, had for the mammalian CNS.

We now know that there are two types of bag fibres, bag_1 and bag_2, and that all mammals possess the three basic types of intrafusal fibre. The core of Boyd's work was concerned with the linkage between the mechanical properties of the fibres and the form of the afferent output. He worked exclusively upon cat spindles, but it is reasonable to suppose that this work is generally true for other mammals, although there have been only two studies in which spindles of other species have been directly visualized. Smith (1966) isolated rat spindles long before the two types of bag fibre were recognized, and got comparable results to Boyd, who at that time was just beginning to work on dissected cat spindles. No further work was done on isolated rat spindles, possibly because of their smaller size; the size of cat spindles is about the maximum attained in mammals. Gladden (1985) isolated human spindles and activated their intrafusal fibres by stimulating motor axons in intramuscular nerve branches, showing that cat spindles are good models for human spindles. In the opinion of the writer there is no particular problem about accepting Boyd's work on intrafusal fibre mechanical properties and their action on the sensory endings (Sect. 4).

The controversial aspect of Boyd's work concerns how the CNS can bring the different fibre types into play. There would be no problem if each fibre type was independently innervated, but they are not, and it is in assessing the incidence and significance of this shared, or "non-selective" innervation that problems do arise (Sect. 5).

2 Types of Intrafusal Muscle Fibre

It was established in the late 1950s by light microscopy that mammalian muscle spindles contain distinct "nuclear bag" and "nuclear chain" fibres (Boyd 1956, 1962; Cooper and Daniel 1956, 1963). The names were introduced by Boyd because of the arrangement of centrally-sited nuclei in the region where each muscle fibre was overlain by the primary sensory ending. The discovery that tonic stimulation of γ-axons could exert two effects, dynamic and static (see later), on the discharge of the Group Ia axons from primary sensory endings (Matthews 1962) was followed by a period of uncertainty about the provenance of these effects. One idea was that bag fibre contraction caused dynamic effects and chain fibre contraction caused static effects (Crowe and Matthews 1964). Although well founded (see Matthews 1972), the proposal turned out to be an oversimplification. If it were true, bags and chains would need to be independently innervated, and although Boyd (1962) had stated that this was the case, he later admitted privately that he had encountered instances where there was some doubt. However, eventually Barker et al. (1973) did establish unequivocably that γ-axons exerting static effects could innervate bag fibres as well as chain fibres (see Laporte and Emonet-Dénand 1973).

Resolution of the problem began when Ovalle and Smith (1972) subdivided the bag fibres into "bag_1" and "bag_2"; they found that two histochemically distinct staining patterns were present along the length of these fibres. The significance of this was not appreciated until Bessou and Pagès (1975) showed that contraction of single bag fibres could have a static effect on the primary afferent discharge. However, these particular bag fibres were never innervated by γ-axons which generated dynamic effects. Boyd et al. (1975, 1977) confirmed this, and proposed that in a "typical" cat muscle spindle containing two bag fibres, one bag fibre was responsible for dynamic effects, and was innervated by dynamic γ-axons; this they named the "dynamic bag_1 fibre". The second bag fibre, the "static bag_2 fibre", was involved in static effects either alone or together with chain fibres; static γ-axons innervated these fibre types. The bag fibre morphology was matched with the physiological response by investigating the structural characteristics of bag fibres which had been marked during tests for their physiological identity (Gladden 1976). This basic plan of muscle spindle motor innervation survived, requiring only additions in the succeeding years (Fig. 1).

2.1 Nomenclature

The dynamic bag_1/static bag_2 nomenclature has not been universally adopted (see reviews by Laporte 1978; Matthews 1981; Barker and Banks 1986; Hulliger 1984; Hunt 1990). The simpler bag_1/bag_2 terms will be used in this review because of the existence of "paradoxical" static and dynamic effects. In special conditions dynamic effects have been recorded from spindle afferents during stimulation of static γ-axons (Emonet-Dénand et al. 1972; Baumann et al. 1982; Hulliger et al. 1985; see Section 5), and static effects during stimulation of dynamic γ-axons (Goodwin et al. 1975; Hulliger et al. 1977). A reasonable explanation is that the same bag fibre(s)

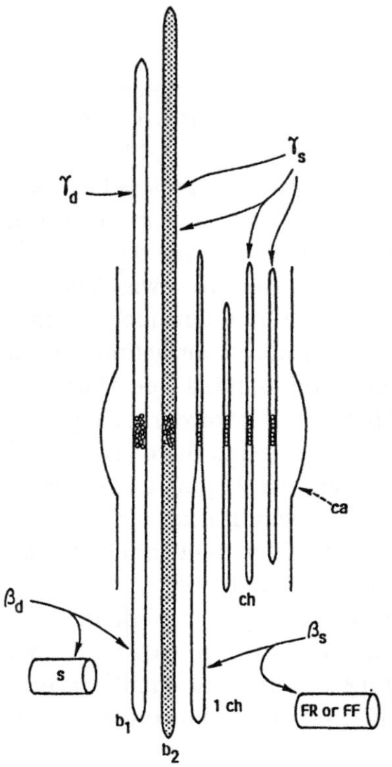

Fig. 1. Diagram illustrating the structure and innervation of a cat muscle spindle containing two nuclear bag fibres, a bag$_1$ (b_1) and a bag$_2$ (b_2), a long chain fibre (*1ch*) and three typical chain fibres (*ch*). Capsule *ca*. Motor innervation: fusimotor, γ_s and γ_d; skeleto-fusimotor, β_s and β_d, supplying extrafusal muscle of the fast, fatigue resistant (*FR*) or fast fatiguable (*FF*) and slow (*S*) types respectively, see Section 3 for description. Sensory innervation (not shown): primary sensory ending (PSE), supplied by a Group Ia axon, would overlie nucleated central portion ≈ 300 μm long. Up to 5 secondary sensory endings (SSE), ~ 400 μm long, supplied by Group II axons, lie on either side of the PSE if present. The PSE surrounds all intrafusal fibres whereas the SSE is sited mainly, or totally on the chain fibres. Total length of spindle ≈ 8 mm, transverse scale about 10 times vertical scale

can be induced to produce static effects under some conditions, and dynamic in others, although by classical testing the response would be clearly either static or dynamic, depending on the type of bag fibre. So the label "static", for example, would not be appropriate as it would imply that the fibre was capable of static effects only. The disadvantage in using only the bag$_1$/bag$_2$ terms is that the physiologist is put in the uncomfortable position of using histochemical labels for fibres which he recognizes only by their physiological properties. The physiologist may identify a bag fibre whose contraction has a dynamic effect, in response to classical tests, but may be unable in the same experiment to confirm that it is truly, histochemically, a bag$_1$ fibre. The dual nomenclature has the advantage that the assumptions are plain: in a solely physiological investigation the "dynamic bag$_1$ fibre" has a dynamic effect and is assumed to be bag$_1$. However, as paradoxical effects could be physiologically important, the dual nomenclature should be dropped to save future confusion.

2.2 Bag$_1$ Type of Nuclear Bag Fibre

In this chapter "bag$_1$" implies a fibre which has a histo- and immunohistochemical profile similar to an extrafusal slow tonic fibre (Rowlerson et al. 1985), the mechanical properties of Boyd's (1976) "slow bag fibres", and nonpropagated evoked potentials can be recorded from them (Barker et al. 1978; Gladden 1981). The innervation is described in Sections 3 and 5. The biophysical properties of these fibres have not yet been fully worked out. Emonet-Dénand, Hunt and Wilkinson's (1983) study was confined to the central portion of the fibres, and the structural inhomogeneity along the length of the fibres (Adal 1985) may be reflected in the membrane properties also changing along the length. The extracapsular portion of some of these fibres behaves oddly during stimulation of dynamic γ-axons innervating them. Close to the end of the fibre the sarcomeres converge as if at an endplate, but no endplate is actually present (Banks et al. 1978; Arbuthnott et al. 1982). The real motor ending is in the sleeve region where the spindle capsule closely envelopes the intrafusal fibres.

2.3 Bag_2 Type of Nuclear Bag Fibre

In histochemical and immunological profile, bag_2 fibres have similarities with slow twitch extrafusal fibres (Rowlerson et al. 1985). Their mechanical properties were described by Boyd (1976), but he called them "fast bag fibres" at that time. As with bag_1 fibres, nonpropagated potentials have usually been recorded, though in the study by Barker et al. (1978) propagated potentials were recorded from one of eight histologically identified bag_2 fibres.

2.4 Nuclear Chain Fibres

Chain fibres contain myosin which is immunologically identical to neonatal fast myosin (Rowlerson et al. 1985) and their mechanical characteristics are therefore correspondingly faster than those of bag_1 or bag_2 (Boyd 1976). Some chain fibres project more than a millimetre past the end of the capsule, and are termed long chain fibres. Usually only one end, or pole, is long. The diameter of these long chain fibres may be comparable with that of bag fibres for all of the fibre length, or just the extracapsular portion may be thick. Barker et al. (1978) recorded action potentials from four chain fibres, and junctional potentials from two others. But one fibre with junctional potentials was a long chain, and was innervated by a dynamic γ-axon, unlike the others which were innervated by static γ-axons. The other fibre was a typical chain but was said to be damaged, so it is not certain whether all typical chains have action potentials under physiological conditions. However, some chain fibres do not twitch if stimulated indirectly at low frequency as in classical descriptions, so it is possible that potentials do not propagate in these typical chain fibres.

2.5 Development

Bag_2 fibres form first during development (Milburn 1984), but they are not the progenitors of bag_1 and chain fibres because, by delaying afferent innervation of developing muscle, spindles with bag_1 and chain, but no bag_2 fibres can be formed (Kucera and Walro 1989). The primary myotubes from which the bag_2 fibres develop are not thought to be genetically committed to be intrafusal until contacted by sensory axons (Zelena 1957, 1976; Kozeka and Ontell 1981; Kucera and Walro 1990; Antoniou et al. 1990). Bag_1 and chain fibres develop from secondary myotubes, which also apparently are not genetically predetermined (Kucera and Walro 1989, 1990).

3 Fusimotor Innervation: The Basic Pattern

The following general description applies to the muscle spindles about which most is known, namely, those of the tenuissimus muscle of the cat. Departures from this pattern of motor innervation in other muscles of the cat, and in other species, are explained in Section 4.

Bag_1 fibres are selectively innervated by dynamic γ- and/or β-axons in the vast majority of spindles (Fig. 1). γ-axons are purely fusimotor, whereas β-axons are skeleto-fusimotor. When dynamic γ- or β-axons are stimulated singly, the response is quite stereotyped, since the same fibre type is the effector. Bag_2 fibres are innervated by static γ-axons, either selectively, or in common with chain fibres, which may also be selectively innervated by static γ-axons. In many spindle poles (or half spindles) there is one non-selective static γ-axon, while about 25% of poles receive two non-selective axons (Boyd and Gladden 1985). Thus, although static γ-axons are grouped together, the effect elicited by a single static γ-axon depends on the fibre type which it innervates, as will be shown in Section 4. Static β-axons, however, innervate a single fibre type, chain fibres, usually long chain fibres; up to three long chain fibres may be innervated in the same spindle (Harker et al. 1977). It should be noted that all long chain fibres are not innervated by static β-axons; thus the incidence of long chain fibres in a muscle may not indicate the density of β-innervation. Dynamic β-axons co-innervate type S extrafusal fibres (Barker et al. 1977), while static β-axons co-innervate type FR or type FF extrafusal fibres (Jami et al. 1979; Kucera and Hughes 1983; Kucera 1984a; Kucera et al. 1984).

4 Effects of Contraction of Intrafusal Fibres on Sensory Endings

This section summarizes Boyd's work on the effect of the contraction of each type of fibre, bag_1, bag_2 and chains, on both primary and secondary sensory endings. It was based on experiments in which afferent responses to stimulation of single static γ-axons, and dynamic γ- and β-axons, were recorded from spindles which were subsequently isolated to permit identification of the types of intrafusal fibre contracting. In some cases, these spindles were serially sectioned to confirm the motor innervation pattern deduced from observation. This histological follow-up is particularly useful to distinguish between weak contractions and passive movements set up by other actively contracting fibres. The tenuissimus muscle of the cat was used in all the experiments.

4.1 The Bag$_1$ Fibre

The typical response of a primary sensory ending to stimulation of a dynamic γ-axon innervating a bag$_1$ fibre at constant length is illustrated in Fig. 2a,d. Either the stimulation frequency was increased from 0 to 150 impulses per second (ips) and then held at 150 ips (Fig. 2a, ramp stimulation) or different constant frequencies varying from 10–250 ips were tested (Fig. 2d, tonic stimulation).

The maximal biassing effect of the bag$_1$ fibre is relatively small, and "driving" is absent. In "driving" the Ia afferent frequency locks onto the stimulus frequency, or a subharmonic of this. There may be no effect on the secondary ending, as with the example shown in Fig. 2g, though a very small biassing action is quite common (Fig. 2j).

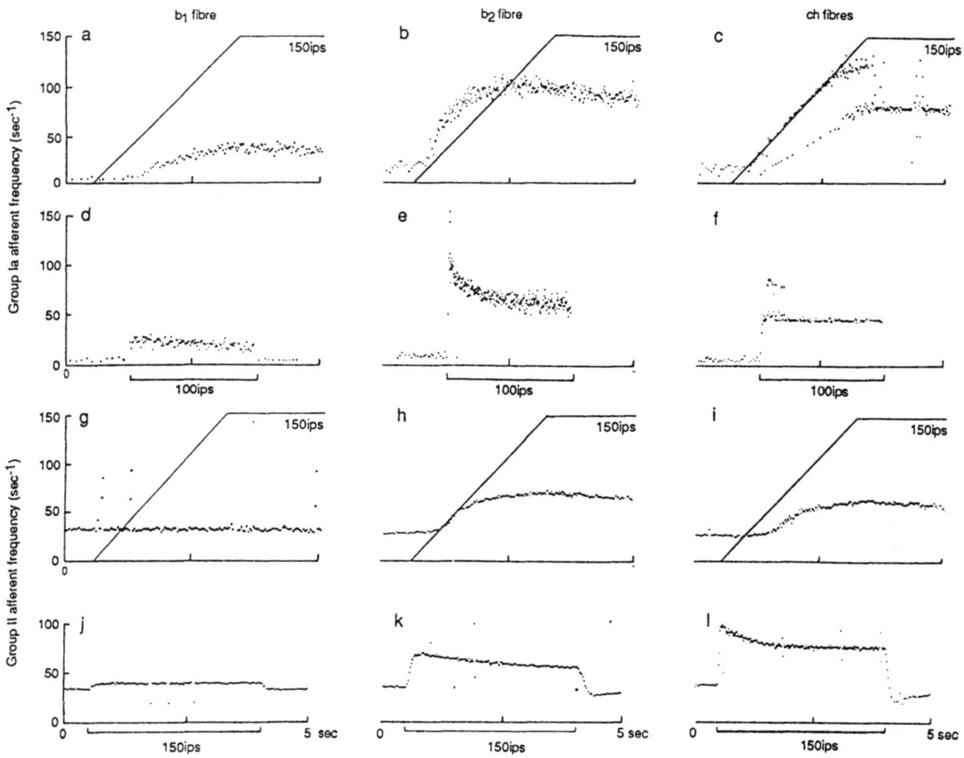

Fig. 2. Typical effects of contraction of bag$_1$ fibres (b_1), bag$_2$ fibres (b_2) and nuclear chain fibres (ch) on the responses of primary and secondary endings in tenuissimus muscle spindles. The effect of ramp frequency stimulation, 0–150 ips *a-c*, and tonic stimulation at 100 ips (*d-f*), and 150 ips (*j-l*), of fusimotor fibres selectively innervating the three intrafusal fibre types at constant length. Records *a-i* from the same spindle; *j-l* from a different spindle

The typical Ia afferent response to stretch during activation of the bag_1 fibre is shown in Fig. 3b. There is a marked increase in dynamic index (cf. Fig. 3a). The dynamic index is the difference between the peak frequency at the end of the stretching phase and the frequency 0.5-s later with the spindle held at the extended length (Matthews 1962). This is due primarily to an increase in the length sensitivity of the primary ending during the dynamic phase of stretch ("dynamic length sensitivity") rather than an increase in the amplitude of the fast rise and fast fall phases which are relatively little affected by bag_1 fibre activity, but are greatly affected by the velocity of stretch (Boyd et al. 1985b). The pronounced phase of slow adaptation apparently reflects the mechanical "creep", or back-slippage in the bag_1 fibre. This back-slippage is seen when isolated spindles are held at constant length immediately after stretching them. During stretching, the bag_1 sarcomeres move rapidly away from the primary sensory ending, and then slide back slowly towards it when the hold phase begins (Boyd et al. 1988; Dickson et al. 1989). Dynamic β-axons have a similar effect on the primary ending to dynamic γ-axons (see Laporte and Emonet-Dénand 1976).

Activation of the bag_1 fibre by either dynamic γ- or β-axons may increase the dynamic length sensitivity of secondary endings which have an appreciable number of terminals on this fibre whether or not there is a detectable biassing. The effect tends to be small, however, and depends upon the initial length of the spindle (Boyd

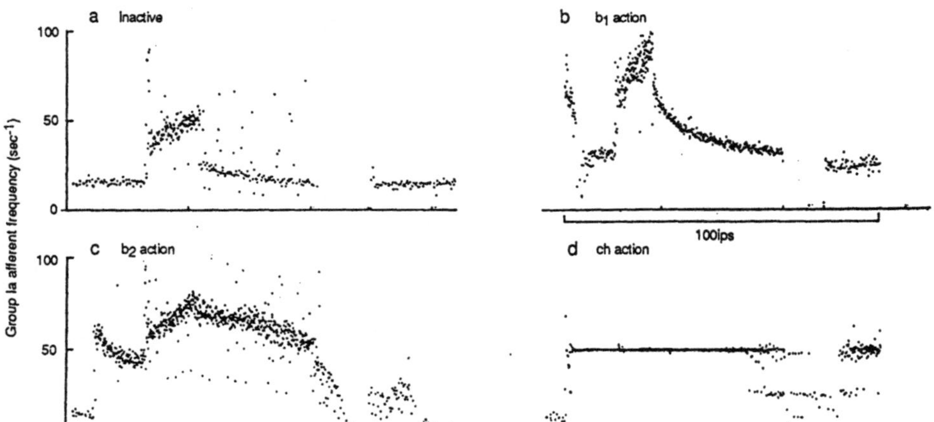

Fig. 3. Typical effects of the three types of intrafusal fibre on the primary sensory ending of the same spindle. Ramp stretch applied during tonic stimulation of fusimotor axons selectively innervating each type of fibre. The peak response at the commencement of dynamic γ-axon stimulation in *b* is unusually large. The dynamic length sensitivity is increased by bag_1 fibre contraction (*a,b*), unchanged by bag_2 fibre contraction (*a,c*), and the length sensitivity is abolished by "driving" due to chain fibre contraction (*a,d*)

et al. 1985c). Two out of three secondary endings have a small number of terminals on bag_1 fibres.

4.2 The Bag_2 Fibre

4.2.1 At Constant Length

The typical response of a primary sensory ending to ramp and to tonic stimulation of a static γ-axon supplying solely the bag_2 fibre at constant length is illustrated in Fig. 2b,e. A powerful biassing action, most of which occurs over the stimulus frequency range of 30–75 ips, is evident, and there is a pronounced phase of adaptation following the commencement of tonic stimulation. The extent of the bias varies with the strength of contraction in the bag_2 fibre produced by a particular static γ-axon. The strength of contraction may be markedly different in the two poles of these fibres even when the same static γ-axon innervates both poles. Clearly, indifferent physiological conditions cannot be blamed for the weak contraction in these cases, yet there are no obvious structural differences in the motor endings. A maximal strong contraction at both poles results in a Ia bias of 100 ips or more (Fig. 2e). There is no driving of the Ia discharge at any stimulus frequency (Fig. 2b). It is not possible to distinguish between contraction of the bag_1 fibre and a weak contraction of the bag_2 fibre without applying stretch to the muscle.

The extent of the biassing effect of bag_2 contraction on secondary sensory endings depends on the proportion of secondary terminals on the particular fibre. About one in four secondary endings have no such terminals. The maximum biassing effect rarely exceeds 40 ips, most of it occurring over a stimulus frequency range of 30–75 ips, and there is no driving (Fig. 2h,k). There is a much less pronounced peak at the onset of tonic stimulation than in the case of the primary ending (cf. Fig. 2k,e).

4.2.2 During Muscle Lengthening or Shortening

The response of the primary ending during a ramp and hold stretch when the bag_2 fibre alone is active is illustrated in Fig. 3c. Comparison with the corresponding response from the inactive spindle (Fig. 3a) shows that the dynamic index is reduced primarily because of a reduction in the amplitude of the fast rise and fall phases of the Ia response (Boyd et al. 1985b). The dynamic length sensitivity is unchanged (Fig. 3c), or, most commonly, slightly reduced (Boyd 1981a). Under static conditions the length sensitivity (position sensitivity) is usually decreased, though it can occasionally be increased as in Fig. 3c.

Secondary endings with terminals on the bag_2 fibre are similarly affected by its activity, but to a much smaller degree. A slight reduction in both dynamic and static length sensitivities is common (Boyd 1985; Boyd et al. 1985c). Confirmation that the bag_2 fibre does not increase the length sensitivity of secondary endings is provided by experiments in which the intrafusal fibres supplied by single static γ-axons were depleted of glycogen by repetitive stimulation to identify them in a histological follow-up (Jami et al. 1980).

4.3 The Action of Chain Fibres on the Primary Sensory Ending

4.3.1 At Constant Length

The typical response of a primary sensory ending to ramp, and to tonic stimulation of a static γ-axon supplying solely chain fibres at constant length is illustrated in Fig. 2c,f. The Ia afferent response is driven 1:1 by stimulus frequencies up to about 50 ips, sometimes to 120 ips, and 1:2 by higher frequencies. At a constant stimulus frequency of 150 ips the discharge varies around a mean of 75 ips. Driving of primary endings occurs at all muscle lengths in many tenuissimus spindles, though it may be length-dependent in the case of some primary endings in other muscles (Boyd et al. 1985a).

The nature of driving depends upon the number of chain fibres innervated, whether one or both poles are active- and possibly on the nature of the electrical response, whether propagated or not (see Sect. 2.4). It is rare for all the chain fibres to be activated by a single static γ-axon, even at one pole, and one to three fibres are usually involved (Boyd and Gladden 1985, Table 2). Even one chain fibre can maintain 1:1 driving up to 50 ips. Two or three chain fibres can maintain driving up to about 75 ips and 1:2 thereafter. If most chain fibres are active, the effect is usually 1:1 driving to about 120 ips, or 1:1 driving alternating with 1:2 driving, the discharge settling down to 1:2 driving at 150 ips. 1:2 driving at low stimulus frequencies can often be converted to 1:1 driving by increasing the length of the muscle. Two static γ-axons supplying chain fibres of the same spindle may produce 1:1 driving when stimulated together, when either alone produces 1:2 driving (Boyd et al. 1985a).

Static β-axons can also produce powerful driving of the primary sensory ending (Jami et al. 1980). Since these axons are known to innervate long chain fibres at one pole, it follows that this fibre can alone drive the primary ending at 1:1. However, it is unlikely that under physiological conditions the discharge in static β-axons would reach frequencies in excess of 30 ips.

If the bag_2 fibre is innervated together with several chain fibres by a non-selective axon, chain fibre driving dominates at low stimulus frequencies, while the interaction between the two types of fibre at higher frequencies leads to a highly irregular discharge with a mean which usually exceeds the stimulus frequency. Sometimes, however, the response during stimulation of a non-selective axon at frequencies in excess of 150 ips resembles that of an axon supplying the bag_2 fibre only. Thus it is necessary to test the effect of stimulation at a range of tonic frequencies to distinguish between the two situations, or preferably, to apply ramp stimulation which reveals the effect of all frequencies of stimulation in a single test.

4.3.2 During Muscle Lengthening or Shortening

Chain fibre driving at 1:1 tends to suppress the response of primary sensory endings to a ramp and hold stretch either almost completely (Fig. 3d), or partially. As might be expected the effect depends upon the amplitude of the dynamic response to stretch in the passive spindle, peak responses greater than the driving frequency tending to 'break through' the driving. This is not always so, however. For example, driving at 35 ips can completely suppress a response to stretch reaching 50 ips in the passive spindle. Also, 1:2 driving by one or two chain fibres is less effective than 1:1 driving unless the peak stretch response is less than half the stimulation frequency.

What may be more significant in voluntary movement is that chain fibres can keep the Group Ia input to the CNS constant during muscle shortening. Single static γ-axons innervating several chain fibres can achieve this (Boyd et al. 1985a), as can single static β-axons (Jami et al. 1980), but it is more usual for the discharge to fall during muscle shortening, often to 1:2 driving. However, activation of two single static γ-axons, each producing driving of the same primary ending, can maintain the Group Ia input at the stimulation frequency, even during rapid muscle shortening (Boyd et al. 1985a).

4.4 The Action of Chain Fibres on the Secondary Sensory Ending

4.4.1 At Constant Length

Chain fibres have a powerful biassing action on secondary sensory endings, but rarely produce driving in cat spindles. In the rat, driving of secondary sensory endings may be much more common (Leslie and Part 1990). If the length of the spindle is such that the Group II afferent frequency during stimulation of a static γ-axon supplying chain fibres is close to the stimulation frequency, then the response may lock onto this frequency, giving 1:1 driving. This is true for some secondary endings in the cat, and for most secondary endings in the baboon (Cheney and Preston 1976).

It should be noted that the pull that chain fibres exert on secondary sensory endings is less well transmitted if there is a primary sensory ending between the contracting fibres and the secondary. The typical Group II response to tonic stimulation of all the chain fibres at constant length at the pole containing the secondary ending is shown in Fig. 2l. The fast rise to a peak, followed by adaptation, is not unlike the response of the primary ending to the onset of contraction in the bag_2 fibre (Fig. 2e). The ramp stimulation test shows that most of the effect occurs over a frequency range from 50–150 ips (Fig. 2i), considerably higher than the range for the bag_2 fibre in the same spindle (e.g. Boyd 1986, Fig. 9), and the chain fibres tend to produce some irregularity in the Group II discharge. If only some of the chain fibres are active (Fig. 2i) then the maximum bias may not be greater than that produced by the bag_2 fibre on a secondary ending, which likely has a considerable number of terminals on this fibre (cf. Fig. 2h,i). These two situations may be difficult to distinguish unless the recordings are from the same secondary ending, since the effective range of stimulus frequency depends upon the temperature of the spindle, and tends to vary somewhat between one spindle and another.

If the bag_2 fibre is active in addition to some chain fibres at the pole containing the secondary ending, and there is a marked contribution from the bag_2 fibre, then the response to ramp stimulation reaches a peak at about 100 ips, and the peak discharge may reach or exceed the stimulation frequency.

4.4.2 During Muscle Lengthening or Shortening

Marked increases in the length sensitivity of secondary endings, such as those shown in Fig. 4, are always due to activity in chain fibres, usually at the pole containing the secondary ending (Boyd 1981a, 1986; Boyd et al. 1985c). Chain fibres in the pole opposite to that containing a secondary ending usually bias the Group II discharge frequency to a very small degree, though sometimes the dynamic length sensitivity increases markedly at high frequencies of stimulation (Fig. 4c). The simultaneous stimulation of two axons to the chain fibres enhances both the biassing effect on the secondary ending and the increase in dynamic length sensitivity (Fig. 4d).

The length sensitivity of the secondary ending under static conditions (position sensitivity) is also increased to almost the same extent, particularly if both poles of the chain fibres are active (Fig. 4d). For this reason the dynamic index is largely unchanged and the axons are classified as static, despite the large increase in sensitivity under dynamic conditions (see Boyd 1985a).

Not all static γ-axons to chain fibres, however, produce an increase in the length sensitivity of secondary endings. The glycogen depletion experiments of Jami et al. (1980) suggest that the involvement of more than two chain fibres is necessary, since in three of their spindle poles depletion of one or two chain fibres was associated with no change in sensitivity. These chain fibres could, however have been sited in the pole not containing the secondary ending.

Boyd et al. (1985c) found that a marked increased in sensitivity could be produced by some static γ-axons which innervated only a few chain fibres, and both Jami et al. (1980) and Boyd et al. (1985c) noted that the effect on sensitivity did not correlate with the biassing effect of an axon on the Group II discharge. This is

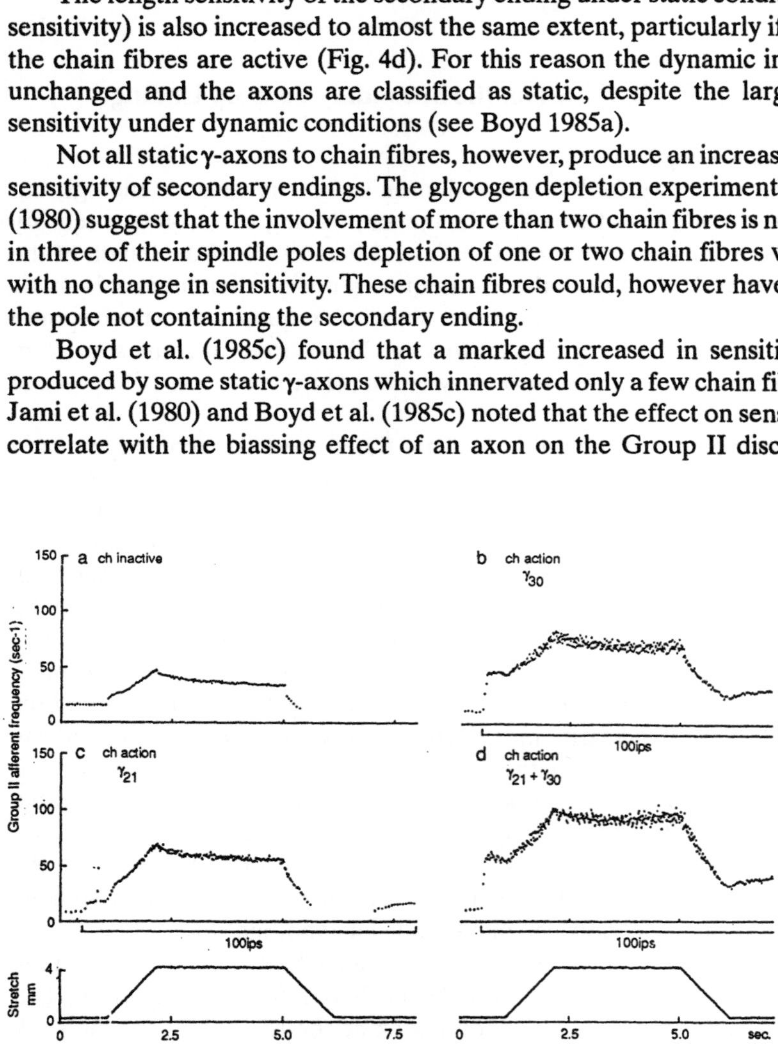

Fig. 4. The effect of stimulating two static γ-axons, γ_{30} and γ_{21}, separately (*b,c*) and together (*d*) on the response of a secondary sensory ending to a ramp and hold stretch. The passive response is shown in *a*. Both the dynamic and static length sensitivities of this ending were increased, and the effects were additive in *d*, when both poles of the chain fibres contracted together

surprising as the biassing action should be greatest when all the chain fibres are active. Possibly if several axons supplying chain fibres are activated at the same time in vivo, then the sensitivity of secondary endings under both dynamic and static conditions will be greatly enhanced, as in Fig. 4d.

5 Nonselectivity and the Subdivision of Static γ-Axons

γ-axons are usually distributed to more than one spindle in a muscle. In physiological experiments this distribution tends to be underestimated because all the Ia afferents may not be found in the dorsal root filaments, and some fusimotor effects might be missed if they are weak. However, it is not uncommon to find that a γ-axon can affect six Ia afferents. β-axons appear to have a more restricted distribution than γ-axons. Soon after Matthews (1962) showed that there were two effects, static or dynamic, on the Ia discharge when single γ-axons were stimulated, it became clear that a γ-axon had the same effect on all the Ia afferents tested (Crowe and Matthews 1964; Brown et al. 1965; Bessou et al. 1966). Consideration of Sections 4.1 to 4.3 suggests that γ-axons with a dynamic action ought to innervate bag_1 fibres only in every spindle, and that γ-static axons should innervate only the fibres mediating static effects, the bag_2 and chain fibres, as illustrated in Fig. 1. Unfortunately, as will be explained, this simple scheme does not hold entirely true, even for the cat, in which the γ-axon distribution seems to be the most straight-forward.

It is quite common for γ-dynamic axons in the cat to innervate one single chain fibre in addition to the bag_1 fibre (Kucera 1984b). Little seems to have been published about the form of the Ia afferent discharge when this non-selective innervation occurs, but both published and unpublished evidence suggests that the additional chain fibre contraction does not dominate the output. It may be relevant that, as mentioned in Section 2.4, junctional potentials were recorded from a chain fibre innervated by a dynamic γ-axon (Barker et al. 1978), rather than the action potentials more commonly recorded from chain fibres. Nevertheless, activation of these single chain fibres does set up visible movement (Boyd and Gladden, unpubl.). However, it does seem unlikely that this additional innervation will prove to be greatly significant physiologically.

It is rare in cat tenuissimus muscles for dynamic γ-axons to innervate bag_2 fibres. Barker et al. (1977) found one instance when they recorded intracellularly during dynamic γ-axon stimulation. A bag_2 fibre was depleted together with a bag_1 fibre in one of 17 spindles innervated by dynamic γ-axons in a glycogen depletion study by Barker et al. (1976); in all the other spindles, the bag_1 fibre was depleted, in two cases with a single chain fibre in addition.

The innervation of bag_1 fibres by static γ-axons may be more important because apparently it is more prevalent in other mammals compared with the cat. The use of the glycogen depletion method initially overestimated bag_1 innervation by static γ-axons in the cat (see Boyd 1981b). Decorte et al. (1984) showed that this was because fatigue, arising from repetitive stimulation of static γ-axons, depleted bag_1

fibres even though the axons did not innervate them. They still found one instance of bag_1 depletion in addition to bag_2 and chain fibre depletion in 13 spindles depleted by repetitively stimulating a total of 9 static γ-axons. However, the incidence of this spill-over of static γ-innervation onto bag_1 fibres must be low in the tenuissimus muscle. On tracing fusimotor axons through serially sectioned spindles, Banks et al. (1982), Arbuthnott et al. (1982) and Kucera (1985a) found no co-innervation of bag_1 and bag_2 in spindles from these muscles. This confirms conclusions reached from physiological studies. During stimulation of static γ-axons, no active movements of bag_1 fibres were observed (Bessou and Pagès 1975; Boyd et al. 1975, 1977; Arbuthnott et al. 1982), neither were any intracellular potentials recorded from bag_1 fibres (Barker et al. 1978; Gladden 1981). Some selection process must operate during development because bag_1 and bag_2 fibres are innervated together commonly in newborn kitten tenuissimus muscles (Gladden et al. 1989).

β-dynamic innervation is almost exclusive to bag_1 fibres. Barker et al. (1977) studied the innervation by β-axons of 18 spindles of the tenuissimus and peroneus brevis muscles by the glycogen depletion method; bag_1 fibres were depleted in every case, but in addition a bag_2 fibre was depleted in one spindle, and one chain fibre in another. β-static axons, however, specifically innervate chain fibres, though not necessarily only long chain fibres (Harker et al. 1977; Jami et al. 1978; Kucera et al. 1984).

In other cat hindlimb muscles, the overspill of γ-static axons onto bag_1 fibres, and of γ-dynamic axons onto bag_2 and chain fibres might be greater than in the tenuissimus muscle. Emonet-Dénand et al. (1977) classified the effects of stimulating γ-axons on primary sensory endings of the peroneus brevis muscle into six categories, ranging from purely dynamic (category I) to purely static (category VI). In comparing this work with his own on the tenuissimus muscle Boyd thought that the dynamic effects in categories I and II were due to contraction of bag_1 fibres alone, with the larger biassing effects in category II being due to the contraction of both poles of these fibres. Bag_2 fibre contraction would be responsible for most category VI, and many category V effects. Non-selective axons innervating bag_2 and one or two chain fibres would also cause some category VI effects. Chain fibre contraction would be implicated in those effects with 1:1 driving mostly included in category VI. This still, however, left intermediate effects, some of which, as Emonet-Dénand et al. (1977) showed, could be mimicked by simultaneously stimulating γ-axons in the purely dynamic and static categories. When Boyd analyzed his own results from tenuissimus in the same way, he found less than 3% of effects in categories III and IV, and very few in category V. But in peroneus tertius, 12% of effects were in categories III and IV, and 15% in category V. The proposal that the bag_1 fibre might be more frequently innervated by static γ-axons in muscles other than the tenuissimus is supported by an analysis of silver-stained spindles (Barker and Stacey 1981). Whereas only 8% of tenuissimus bag_1 fibres received trail (static) innervation, it was present in 17% of bag_1 fibres in peroneus brevis.

In other species, studies in which motor axons have been traced through serial sections of muscle spindles have indicated that a complete nonselectivity is more common than in the cat. 27% of motor axons in spindles of the lumbrical muscles of the monkey co-innervated bag_1 and bag_2 (Kucera 1985b). Two of 13 motor axons in a human biceps muscle spindle co-innervated bag_1 and bag_2 (Kucera 1986). On the other hand, Gladden et al. (1985) traced 10 motor axons in a spindle of a child and found the bag_1 fibre to be separately innervated, except for one axon which innervated a chain fibre in addition, a type of innervation which, as already mentioned, commonly occurs in the cat. In rat soleus muscles, Walro and Kucera (1985) found 18% of motor axons co-innervated bag_1 and bag_2, and Arbuthnott et al. (1989) also found such innervation, but with a lower incidence (8%).

Physiological investigations have not suggested a significant overlap between static and dynamic fusimotor effects. In baboon soleus and gastrocnemius muscles, Cheney and Preston (1976) could separate static and dynamic groups clearly. Gladden (1985) did not observe co-innervation of bag_1 and bag_2 fibres in eight human muscle spindles on recruiting their motor axons in intramuscular nerve branches. In the rat soleus muscle, Andrew et al. (1978) had difficulty in classifying only one of 59 responses to γ-axon stimulation.

The apparent discrepancy between histological and physiological results could be because the effects on the afferents resulting from any overspill innervation are weak. For example, Cheney and Preston (1976) commented that it was necessary to stimulate γ-axons in the baboon at 200 ips to separate dynamic from static effects, as stimulation at 50 ips did not differentiate clearly when the effects were weak. Possibly the high frequency obscured the contribution of some intrafusal fibres to the response in these cases. Chain fibres are especially prone to fatigue, and neuromuscular block can occur at intrafusal endplates. Alternatively, the effects of over-spill innervation may be of a nature which does not occlude the main effect. This can be illustrated by an experiment in which three γ-axons innervating a human spindle were stimulated in an intramuscular nerve, one supplying chain fibres, a second supplying a bag_2 fibre and a third supplying a bag_1 fibre (Fig. 5). The dynamic effect of the bag_1 contraction was not occluded (Gladden 1985). It is also possible that axons may have overspill innervation in some spindles, and be selective in innervating the fibres mediating dynamic or static effects in others. This is suggested by the classification of effects exerted by axons supplying more than one spindle (Emonet-Dénand et al. 1977, Fig. 14). Thus, of 19 axons with effects on Ia afferents intermediate between purely dynamic and purely static only four had effects restricted to intermediate categories; all the rest had a "pure" effect on at least one other spindle. The intermediate effects may have little impact for the CNS when all

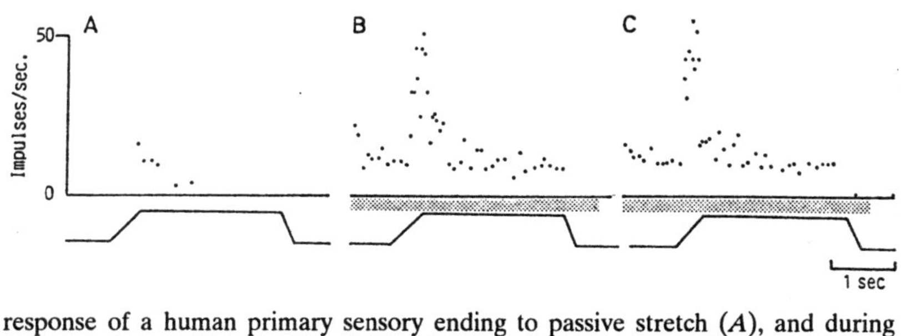

Fig. 5. The response of a human primary sensory ending to passive stretch (*A*), and during stimulation of three fusimotor axons at 20 ips (*B*) and 30 ips (*C*). (Gladden 1985)

the spindles supplied by a γ-axon are activated together, so that the aggregate action is either static or dynamic.

After taking into account these problems in establishing whether γ-axons adhere to innervating either bag_1 fibres, which are responsible for dynamic effects, or bag_2 and/or chain fibres, which are responsible for static effects, it is difficult to accept that a further level of sophistication in fibre type specificity exists. However, Boyd (1986) became convinced that while bag_2 and chain fibres clearly did not have innervation approaching the specificity to fibre type of bag_1 fibres, nevertheless static γ-axons in the main supplied either bag_2 fibres or chain fibres. His evidence is summarized in Table 1. Of 16 γ-static axons innervating three or more spindles he found that 9 innervated chains and 6 innervated bag_2 fibres in every spindle, although there was some overlap. The innervation was deduced from the afferent responses using the criteria described in Section 4. These "diagnoses" were checked by isolating a proportion of the spindles, and for some of these spindles the observations were later confirmed by tracing the motor axons through serial sections of the spindles. It should be noted that Boyd's (1986) results were for spindles in the tenuissimus muscle, in fact, only the central portion of this muscle, in the region where the nerve enters the muscle.

Criticism of the two static γ hypothesis came from two sources. Banks (1988) proposed that fast-conducting γ-static axons tend to innervate bag_2 fibres, while the slowest conducting supply chain fibres, there being a continuum between the two extremes. His methods differed from Boyd's in that he did not use the ramp frequency test, and followed up his physiological experiments by silver-staining the spindles without isolating them to observe the intrafusal contractions directly during stimulation of single static γ-axons as Boyd had done. The second objection arose from an analysis by Gladden and Sutherland (1989) of the distribution of γ-static axons and the form of their endings determined by tracing them through

Table 1. Type of intrafusal fibre activated by 16 static γ-axons in up to 6 spindles in the same cat tenuissimus muscle, diagnosed from modification of the response of primary and/or secondary sensory endings (After Boyd 1986)

	Static chain γ-axons[a]										Static bag γ-axons[a]					
Spindle	$γ_1$	$γ_2$	$γ_3$	$γ_4$	$γ_5$	$γ_6$	$γ_7$	$γ_8$	$γ_9$	$γ_{10}$	$γ_{11}$	$γ_{12}$	$γ_{13}$	$γ_{14}$	$γ_{15}$	$γ_{16}$
1	ch[b]	ch	ch(b_2)[b]	ch+b_2	ch	ch	ch+b_2	ch	ch	b_2	b_2[b]	b_2	b_2	b_2[b]	b_2	b_2
2	ch	ch(b_2)	ch(b_2)	ch	ch	ch+b_2[b]	ch	ch	ch	b_2	b_2	b_2	b_2	b_2	b_2	b_2
3	ch	ch	ch	ch	ch	ch	ch+b_2	?ch	?ch	b_2+ch	b_2	b_2	b_2	b_2	b_2(ch)	b_2+ch
4	?ch	ch	ch	ch						ch	b_2	b_2	b_2			
5	ch+b_2	ch+b_2	ch								b_2	b_2+ch				
6	ch+b_2										b_2	b_2+ch				

[a]Boxes, diagnosis confirmed by observation in isolated spindle; weak actions in parentheses.
[b]Diagnosis confirmed by serial electron microscopy of the spindle.

serial sections of spindles. Some of these spindles had been studied physiologically by Boyd. Boyd (1986) had suggested that his "static chain" γ-axons had complex endings on chain fibres, called "m_c" by Arbuthnott et al. (1982), and that the "static bag" γ-axons had simpler "m_a" plates. But the analysis showed that it was not possible to tie up ending structure with type of γ-axon in this way. There were, for example, too many axons with type m_a endings distributed solely to chain fibres.

Recently, however, evidence that the CNS could control bag_2 and chain fibres separately has come from experiments in which exteriorized tenuissimus spindles were observed during central stimulation in anaesthetized cats. In earlier experiments, (Gladden and McWilliam 1977; Gladden 1981) γ-motoneurones with axons exclusively innervating bag_2 or chain fibres in the observed spindle had always been recruited together or sequentially. But localized areas have now been found both in the mesencephalon and on the cortical surface where stimulation recruited γ-motoneurones innervating bag_2 fibres, but simultaneously inhibited spontaneously active γ-motoneurones innervating chain fibres (Dickson and Gladden 1990; Asgari-Khozankalaei and Gladden 1990). Interestingly, Wand and Schwarz (1985) found that in cat hindlimb flexor muscle spindles, removal of inhibition in the reticular part of the substantia nigra by picrotoxin injection reduced a static bias in the discharge from primary, but not secondary sensory endings. Their interpretation was that spontaneous activity of γ-static motoneurones innervating bag_2 fibres was inhibited by the picrotoxin injection, leaving activity in γ-static motoneurones innervating chain fibres unaffected. The presence of spontaneous activity in the γ-motoneurones innervating chain fibres was indicated by a reduction in the bias when the motor supply to the muscles was cut.

These central studies clearly favour Boyd's view that static γ-motoneurones are subdivided in a fashion which allows the expression of the individual properties of the bag_2 and chain fibre types. Could this be achieved without selective innervation? So long as γ-static axons innervate bag_2 and chain fibres specifically in at least some spindles, even though both types are innervated together in others, might this not suffice, because a proportion of the spindle afferent output would thus be influenced by "pure" bag_2 and chain effects? According to Boyd, when an axon innervates both bag_2 and chain fibres together, the chain fibre activity dominates the Ia afferent response at low stimulus frequencies, causing driving (Sect. 4.3.1). Since the physiological range of frequencies is low, then bag_2 effects would tend to be swamped by chain fibre driving. Even in tenuissimus muscles, the incidence of shared bag_2-chain innervation is quite high, so that when several static γ-motoneurones are active together, the proportion of Ia afferents reflecting the properties of active bag_2 fibres ought to be very low indeed. This raises questions for which there are as yet no firm answers. Do the chain fibres always win at physiological frequencies of static γ-motoneurone output, and if they do does it matter? Alternatively, is there a specific innervation of bag_2 as Boyd advocated, together with an organization of the static γ-motoneurone pool to turn off the chain contraction in appropriate circumstances?

It is possible that the passive properties of bag_2 fibres might be more important physiologically than the active properties. In fact, new evidence is emerging which suggests that bag_2 fibres might indeed be responsible for much of the passive response of primary sensory endings. Price and Dutia (1989) and Gioux et al. (1991) showed that the dynamic responses to stretch of primary sensory endings were indistinguishable whether there were terminations only on bag_2 and chain fibres or on all three fibre types. These authors recorded from tandem spindles in cat neck muscles, tenuissimus and peroneus tertius. The bag_1 fibre is frequently absent from one encapsulation of a tandem spindle. These are complexes in which two encapsulations, both supplied by a primary sensory ending, are linked by a bag_2 fibre, though each encapsulation has its own chain fibres (Banks et al. 1982; Richmond et al. 1986). It is unlikely that the dynamic response to stretch arises in Ia afferent terminals on chain fibres because chain fibres do not have the appropriate mechanical properties: they stretch easily, and crumple at short muscle lengths (Boyd 1976).

Active bag_2 fibres have a biassing effect on the Ia afferent frequency, and a variable effect on length sensitivity under both dynamic and static conditions (Sect. 4.2.1). These effects seem rather uninteresting, but it should be noted that the tests employed high stimulus frequencies; this state might not appertain physiologically unless several static γ axons were active simultaneously. Hunt (1990) has suggested that bag_2 fibres might be responsible for the paradoxical dynamic effects of static γ-axon stimulation (Emonet-Dénand et al. 1972; Hulliger et al. 1985), and Boyd (1985a) apparently thought so too. At low stimulus frequencies and short muscle lengths, there was an increase in the dynamic sensitivity of Ia afferents to small amplitude sinusoidal stretching. This phenomenon was present on about half of the occasions in which proven static γ-axons were tested against Ia afferents of the soleus muscle (Hulliger et al. 1985). Moreover there was a residual increase in dynamic sensitivity which survived as an after-effect following stimulation, and which was remarkably similar to an after-effect of dynamic γ-axon stimulation (Baumann et al. 1982). Active bag_2 fibres might be important in maintaining the dynamic sensitivity of Ia afferents to small perturbations at the shorter muscle lengths when there is no dynamic activity or any residual effect of dynamic activity, and in spindles with no bag_1 fibres. If these effects can be attributed to the bag_2 fibre, it would be interesting to know whether concomitant chain fibre contraction would obscure them.

6 Summary and Conclusions

I.A. Boyd's work on the action of each type of intrafusal fibre on the responses of primary and secondary endings at constant, or changing, length of the muscle is described more fully than previously. Particular emphasis has been placed on the differences between bag_2 and chain fibres. This work is particularly relevant for the understanding of intrafusal mechanisms. But it is hoped that the account will also be useful to those wishing to interpret spindle afferent responses in terms of the types of fibre active when it is not possible to uncover the actual spindles in order to observe individual intrafusal fibres in action. The account also exposes the gaps in our knowledge and understanding. We need to know more about the afferent responses to contraction of individual fibre types at low frequency, within the physiological range, and when different fibre types are contracting simultaneously in combinations which occur during normal movements.

References

Adal MN (1985) Variation in the T-system in different types of intrafusal muscle fibre of cat spindles. In: Boyd IA, Gladden MH (eds) The muscle spindle. Macmillan, London

Andrew BI, Leslie GC, Part NJ (1978) Some observations on efferent innervation of rat soleus muscle spindles. Exp Brain Res 31:433–443

Antoniou A, Shafi AI, Rowlerson A (1990) Soleus muscle fibre and spindle development in Swiss White and C3H Brown Mice. J Physiol 429:32P

Arbuthnott ER, Ballard KJ, Boyd IA, Gladden MH, Sutherland FI (1982) The ultrastructure of cat fusimotor endings and their relationship to foci of sarcomere convergence in intrafusal fibres. J Physiol 331:285–309

Arbuthnott ER, Gladden MH, Sutherland FI (1989) The selectivity of motor innervation in muscle spindles of the rat studied by light microscopy. J Anat 163:183–190

Asgari-Khozankalaei A, Gladden MH (1990) Recruitment of intrafusal muscle fibres from the sensorimotor cortex in tenuissimus muscles of cats under barbiturate anaesthesia. J Physiol 429:9P

Banks RW (1981) A histological study of the motor innervation of the cat's muscle spindle. J Anat 133:571–591

Banks RW (1988) Studies on the motor innervation of tenuissimus muscle spindles in the anaesthetised cat. J Physiol 406:70P

Banks RW, Barker D, Bessou P, Pagès B, Stacey MJ (1978) Histological analysis of cat muscle spindles following direct observation of the effects of stimulating dynamic and static motor axons. J Physiol 283:605–619

Banks RW, Barker D, Stacey MJ (1982) Form and distribution of sensory terminals in cat hindlimb muscle spindles. Philos Trans R Soc B 299:329–364

Barker D, Banks RW (1986) The muscle spindle. In: Engel AG, Banker BQ (eds) Myology. McGraw Hill, New York, pp 309–341

Barker D, Stacey MJ (1981) On the innervation of bag_1 fibres in cat muscle spindles by static γ-axons. J Physiol 320:93P

Barker D, Emonet-Dénand F, Laporte Y, Proske U, Stacey MJ (1973) Morphological identification and intrafusal distribution of the endings of static fusimotor axons in the cat. J Physiol 230:405–42

Barker D, Emonet-Dénand F, Harker DW, Jami L, Laporte Y (1976) Distribution of fusimotor axons to intrafusal muscle fibres in cat tenuissimus muscle spindles as determined by the glycogen depletion method. J Physiol 261:49–70

Barker D, Emonet-Dénand F, Harker DW, Jami L, Laporte Y (1977) Types of intra- and extrafusal muscle fibre innervated by dynamic skeletofusimotor axons in cat peroneus brevis and tenuissimus muscles as determined by the glycogen depletion method. J Physiol 266:713–726

Barker D, Bessou P, Jankowska E, Pagès B, Stacey MJ (1978) Identification of intrafusal muscle fibres activated by single fusimotor axons and injected with fluorescent dye in cat tenuissimus spindles. J Physiol 275:149–165

Baumann TK, Emonet-Dénand F, Hulliger M (1982) After-effects of fusimotor stimulation on spindle Ia afferents' dynamic sensitivity, revealed during slow movements. Brain Res 232:460–465

Bessou P, Pagès B (1975) Cinematographic analysis of contractile events produced in intrafusal muscle fibres by stimulation of static and dynamic fusimotor axons. J Physiol 252:397–427

Bessou P, Laporte Y, Pagès B (1966) Similitude des effets (statiques ou dynamiques) exercés par les fibres fusimotrices uniques sur les terminations primaires de plusieurs fuseaux chez le chat. J Physiol (Paris) 58:31–39

Boyd IA (1956) The tenuissimus muscle of the cat. J Physiol 133:35–36P
Boyd IA (1962) The structure and innervation of the nuclear bag muscle fibre system and the nuclear chain muscle fibre system in mammalian muscle spindles. Trans R Soc Lond B 245:81–136
Boyd IA (1976) The response of fast and slow nuclear bag fibres and nuclear chain fibres in isolated cat muscle spindles to fusimotor stimulation and the effect of intrafusal contraction on the sensory endings. Q J Exp Physiol 61:203–254
Boyd IA (1981a) The action of the three types of intrafusal fibre in isolated cat muscle spindles on the dynamic and length sensitivities of primary and secondary sensory endings. In: Taylor A, Prochazka A (eds) Muscles receptors and movement. Macmillan, London, pp 17–32
Boyd IA (1981b) The muscle spindle controversy. Sci Progr Oxford 67:319–335
Boyd IA (1985a) Part 2 review: internal working of muscle spindles. In: Boyd IA, Gladden MH (eds) The muscle spindle. Macmillan, London, pp 129–150
Boyd IA (1985b) Muscle spindles and stretch reflexes. In: Swash M, Kennard C (eds) Scientific basis of clinical neurology. Livingstone, London, pp 74–96
Boyd IA (1986) Two types of static gamma axon in cat muscle spindles. Q J Exp Physiol 71:307–327
Boyd IA, Gladden MH (1985) Part 1 review: morphology of mammalian muscle spindles. In: Boyd IA, Gladden MH (eds) The muscle spindle. Macmillan, London, pp 3–22
Boyd IA, Gladden MH, McWilliam PN, Ward J (1975) 'Static' and 'dynamic' nuclear bag fibres in isolated cat muscle spindles. J Physiol 250:11–12P
Boyd IA, Gladden MH, McWilliam PN, Ward J (1977) Control of dynamic and static nuclear bag fibres and nuclear chain fibres by gamma and beta axons in isolated cat muscle spindles. J Physiol 265:133–162
Boyd IA, Gladden MH, Halliday D, Dickson M (1988) Stroboscopic, cinematographic and videorecording of dynamic bag_1 fibres during rapid stretching of isolated cat muscle spindles. In: Hnik P, Soukup T, Vejseda R, Zelena J (eds) Mechanoreceptors; development, structure and function. Plenum Press, New York, pp 215–220
Boyd IA, Murphy P, Mann C (1985a) The effect of chain fibre 'driving' on the length sensitivity of primary sensory endings in the tenuissimus, peroneus tertius and soleus muscles. In: Boyd IA, Gladden MH (eds) The muscle spindle. Macmillan, London, pp 195–200
Boyd IA, Murphy P, Moss VA (1985b) Analysis of primary and secondary afferent responses to stretch during activation of the dynamic bag_1 fibre or the static bag_2 fibre in cat muscle spindles. In: Boyd IA, Gladden MH (eds) The muscle spindle. Macmillan, London, pp 153–158
Boyd IA, Sutherland FI, Ward J (1985c) The origin of the increase in length sensitivity of secondary endings produced by some fusimotor axons. In: Boyd IA, Gladden MH (eds) The muscle spindle. Macmillan, London, pp 207–214
Brown MC, Crowe A, Matthews PBC (1965) Observations on the fusimotor fibres of the tibialis posterior muscle of the cat. J Physiol 189:545–550
Cheney PD, Preston JB (1976) Classification of fusimotor fibres in the primate. J Neurophysiol 39:9–19
Cooper S, Daniel PM (1956) Human muscle spindles. J Physiol 133:1P
Cooper S, Daniel PM (1963) Muscle spindles in man: their morphology in the lumbricals and the deep muscles of the neck. Brain 86:563–586
Crowe A, Matthews PBC (1964) Further studies of static and dynamic fusimotor fibres. J Physiol 174:132–151
Decorte L, Emonet-Dénand F, Harker DW, Jami L, Laporte Y (1984) Glycogen depletion elicited in tenuissimus intrafusal muscle fibres by stimulation of static γ-axons in the cat. J Physiol 346:341–352
Dickson M, Gladden MH (1990) Dynamic and static gamma effects in tenuissimus muscle spindles during stimulation of areas in the mes- and diencephalon in anaesthetised cats. J Physiol 423:73P
Dickson D, Gladden MH, Halliday D, Ward J (1989) Fusimotor mechanisms determining the afferent output of muscle spindles. In: Allum JHJ, Hulliger M (eds) Afferent control of posture and locomotion. Elsevier, Amsterdam, pp 9–18
Emonet-Dénand F, Joffroy M, Laporte Y (1972) Fibres fusimotrices dont l'action sur la sensibilité phasique des terminations primaires dépend de leur fréquence de stimulation. CR Hebd Séannces Acad Sci (III) 275:89–91

Emonet-Dénand F, Laporte Y, Matthews PBC, Petit J (1977) On the subdivision of static and dynamic fusimotor actions on the primary ending of the cat muscle spindle. J Physiol 268:827–861

Emonet-Dénand F, Hunt CC, Wilkinson RS (1983) Current-voltage relations and contractile responses in identified nuclear bag$_1$ and nuclear bag$_2$ fibres in the cat. J Physiol 334:137P

Goodwin GM, Hulliger M, Matthews PBC (1975) The effects of fusimotor stimulation during small amplitude stretching on the frequency-response of the primary ending of the mammalian muscle spindle. J Physiol 253:175–206

Gioux M, Petit J, Proske U (1991) Responses of cat muscle spindles which lack a dynamic fusimotor supply. J Physiol 432:557–571

Gladden MH (1976) Structural features relative to the function of intrafusal muscle fibres in the cat. Prog Brain Res 44:51–59

Gladden MH (1981) The activity of intrafusal muscle fibres during central stimulation in the cat. In: Taylor A, Prochazka A (eds) Muscle receptors and movement. Macmillan, London, pp 109–122

Gladden MH (1985) Efferent control of human muscle spindles. In: Boyd IA, Gladden MH (eds) The muscle spindle. Macmillan, London, pp 161–166

Gladden MH, McWilliam PN (1977) The activity of intrafusal fibres during cortical stimulation in the cat. J Physiol 273:28–29P

Gladden MH, Sutherland F (1989) Do cats have three types of static gamma axon? J Physiol 414:19P

Gladden MH, Wallace W, Craigen ML (1985) Movement convergence and motor end plate location in a human muscle. In: Boyd IA, Gladden MH (eds) The muscle spindle. Macmillan London, pp 115–120

Gladden MH, Spike RC, Sutherland FI (1989) Immaturity of fusimotor innervation at birth in the cat. J Physiol 414:19P

Harker DW, Jami L, Laporte Y, Petit J (1977) Fast conducting skeleto-fusimotor axons supplying intrafusal chain fibres in the cat peroneus tertius muscle. J Neurophysiol 40:791–799

Hulliger M (1984) The mammalian muscle spindle and its central control. Rev Physiol Biochem Pharmacol 101:1–110

Hulliger M, Matthews PBC, Noth J (1977) Static and dynamic fusimotor action on the response of Ia fibres to low frequency sinusoidal stretching of widely ranging amplitude. J Physiol 267:811–838

Hulliger M, Emonet-Dénand F, Baumann TK (1985) Enhancement of stretch sensitivity of cat primary spindle afferents by low-rate static gamma-action. In: Boyd IA, Gladden MH (eds) The muscle spindle. Macmillan, London, pp 189–194

Hunt CC (1990) Mammalian muscle spindle: peripheral mechanisms. Physiol Rev 70:643–663

Jami L, Lan-Couton D, Malmgren K, Petit J (1978) "Fast" and "slow" skeleto-fusimotor innervation in cat tenuissimus spindles: a study with the glycogen depletion method. Acta Physiol Scand 103:284–298

Jami L, Lan-Couton D, Malmgren K, Petit J (1979) Histophysiological observations on fast skeleto-fusimotor axons. Brain Res 164:53–59

Jami L, Lan-Couton D, Petit J (1980) A study with the glycogen-depletion method of intrafusal distribution of γ-axons that increase sensitivity of spindle secondary endings. J Neurophysiol 43:16–26

Kozeka K, Ontell M (1981) The three-dimensional cytoarchitecture of developing murine muscle spindles. Dev Biol 87:133–147

Kucera J (1984a) Histological identification of (static) skeletofusimotor innervation to a cat muscle spindle. Brain Res 294:390–395

Kucera J (1984b) Non-selective motor innervation of cat nuclear bag$_1$ intrafusal fibres. Cell Tissue Res 236:383–391

Kucera J (1985a) Histological study of motor innervation to nuclear bag intrafusal muscle fibres in the cat. J Comp Neuro 232:331–346

Kucera J (1985b) Characteristics of motor innervation of muscle spindles in the monkey. Am J Anat 173:113–125

Kucera J (1986) Reconstruction of the nerve supply to a human muscle spindle. Neurosci Lett 63:180–184

Kucera J, Hughes R (1983) Histological study of motor innervation to long nuclear chain intrafusal fibres in the muscle spindle of the cat. Cell Tissue Res 228:535–547
Kucera J, Walro JM (1989) Alteration of intrafusal bundle composition by nerve crush in neonatal rats. Neurosci Lett 105:52–56
Kucera J, Walro JM (1990) Myosin heavy chain expression in developing rat intrafusal muscle fibres. Neurosci Lett 109:18–22
Kucera J, Hammar K, Meek B (1984) Ultrastructure of dynamic and static skeletofusimotor endings in a cat muscle spindle. Cell Tissue Res 238:151–158
Laporte Y (1978) The motor innervation of the mammalian muscle spindle. In: Porter R (ed) Studies in neurophysiology. Cambridge Univ Press, Cambridge, pp 45–59
Laporte Y, Emonet-Dénand F (1973) Evidence for common innervation of bag and chain muscle fibres in cat spindles. In: Stein RB, Pearson KB, Smith RS, Redford JB (eds) Control of posture and locomotion. Plenum Press, New York
Laporte Y, Emonet-Dénand F (1976) The skeletofusimotor or β-innervation of cat muscle spindle. Prog Brain Res 44:99–106
Leslie GC, Part NJ (1990) "Driving" of muscle spindle secondary endings by ramp frequency stimulation of fusimotor fibres to the soleus muscle of the anaesthetised rat. J Physiol 429:33 p
Matthews PBC (1962) The differentiation of two types of intrafusal fibre by their effects on the dynamic response of muscle spindle primary endings. Quart J Exp Physiol 47:324–333
Matthews PBC (1972) Mammalian muscle receptors and their central actions. Arnold, London
Matthews PBC (1981) Evolving views of the internal operation and functional role of the muscle spindle. J Physiol 320:1–30
Milburn A (1984) Stages in the development of cat muscle spindles. J Embryol Exp Morphol 82:177–216
Ovalle WK, Smith RS (1972) Histochemical identification of three types of intrafusal muscle fibres in the cat and monkey based on the myosin ATPase reaction. Can J Physiol Pharmacol 50:192–202
Price RF, Dutia MB (1989) Physiological properties of tandem muscle spindles in neck and hindlimb muscles. Prog Brain Res 80:47–56
Richmond FJR, Bakker GJ, Bakker DA, Stacey MJ (1986) The innervation of tandem muscle spindles in the cat neck. J Comp Neurol 245:483–497
Rowlerson A, Gorza L, Schiafino S (1985) Immunohistochemical identification of spindle fibre types in mammalian muscle using type-specific antibodies to isoforms of myosin. In: Boyd IA, Gladden MH (eds) The muscle spindle. Macmillan, London, pp 29–34
Smith RS (1966) Properties of intrafusal muscle fibres. In: Granit R (ed) Nobel Symp I. Muscular afferents and motor control. Almquist and Wiksell, Stockholm, pp 69–80
Walro JM, Kucera J (1985) Motor innervation of intrafusal fibres in rat muscle spindles: incomplete separation of static and dynamic systems. Am J Anat 173:55–68
Wand P, Schwarz M (1985) Two types of cat static fusimotor neurones under separate central control? Neurosci Lett 58:145–149
Zelena J (1957) Morphogenetic influence of innervation on the ontogenetic development of muscle spindles. J Embryol Exp Morphol 5:283–292
Zelena J (1976) The role of sensory innervation in the development of mechanoreceptors. Prog Brain Res 43:59–64

Subject Index

MIX
Papier aus verantwortungsvollen Quellen
Paper from responsible sources
FSC® C105338

FSC
www.fsc.org

If you have any concerns about our products,
you can contact us on
ProductSafety@springernature.com

In case Publisher is established outside the EU,
the EU authorized representative is:
Springer Nature Customer Service Center GmbH
Europaplatz 3, 69115 Heidelberg, Germany

Printed by Libri Plureos GmbH
in Hamburg, Germany